CONCEPTUAL STRUCTURES IN PRACTICE

Chapman & Hall/CRC
Studies in Informatics Series

SERIES EDITOR

G. Q. Zhang
Case Western Reserve University
Department of EECS
Cleveland, Ohio, U.S.A

PUBLISHED TITLES

Stochastic Relations: Foundations for Markov Transition Systems
Ernst-Erich Doberkat

Conceptual Structures in Practice
Pascal Hitzler and Henrik Schärfe

Context-Aware Computing and Self-Managing Systems
Waltenegus Dargie

Chapman & Hall/CRC
Studies in Informatics Series

CONCEPTUAL STRUCTURES IN PRACTICE

Edited by

Pascal Hitzler
Henrik Schärfe

CRC Press
Taylor & Francis Group
Boca Raton London New York

CRC Press is an imprint of the
Taylor & Francis Group, an **informa** business

A CHAPMAN & HALL BOOK

The following figures were provided by Springer -Verlag, Berlin:

Figure 13.4. (Schärfe, Henrik; Hitzler, Pascal; Ohrstrom, Peter [Eds.] Conceptual Structures: Inspiration and Application. 14th International Conference on Conceptual Structures, July 16-21, 2006, p. 292. Used with permission.)

Figure 13.5. (Schärfe, Henrik; Hitzler, Pascal; Ohrstrom, Peter [Eds.] Conceptual Structures: Inspiration and Application. 14th International Conference on Conceptual Structures, July 16-21, 2006, p. 298. Used with permission.)

Chapman & Hall/CRC
Taylor & Francis Group
6000 Broken Sound Parkway NW, Suite 300
Boca Raton, FL 33487-2742

First issued in paperback 2017

© 2009 by Taylor & Francis Group
Chapman & Hall/CRC is an imprint of Taylor & Francis Group, an Informa business

No claim to original U.S. Government works

ISBN 13: 978-1-138-11464-7 (pbk)
ISBN 13: 978-1-4200-6062-1 (hbk)

Library of Congress Cataloging-in-Publication Data

Conceptual structures in practice / editors, Pascal Hitzler and Henrik Schärfe. -- 1st ed.
     p. cm. -- (Chapman & Hall/CRC studies in informatics series ; 2)
   Includes bibliographical references and index.
   ISBN 978-1-4200-6062-1 (alk. paper)
   1. Conceptual structures (Information theory) 2. Knowledge representation
(Information theory) I. Hitzler, Pascal. II. Schärfe, Henrik.

Q387.2.C68 2009
003'.54--dc22   2008036234

Visit the Taylor & Francis Web site at
http://www.taylorandfrancis.com

and the CRC Press Web site at
http://www.crcpress.com

Contents

Preface

This book is about structures: Conceptual Structures. Following standard definitions, we may tentatively define a structure as consisting of elements of something, and relations connecting those elements. And so it becomes obvious that structures are ubiquitous; they occur everywhere in nature and in culture. In fact, it is hard to think of anything that is not – some way or another – arranged in structures. In this sense, if one is to study structure, the whole world may become his laboratory, and going through the chapters of this book, we easily see that the examples and illustrations given by the authors range considerably. Here, however, we are not just interested in structures. We seek out and investigate a special kind of structure, namely the conceptual ones. A concept may be considered as a basic element of thought, as opposed to an element of perception. Conceptual Structures, therefore, is the name of a field of research that investigates various aspects of thought by means of structures. The scientific community of Conceptual Structures goes back more than 20 years and has brought together a wide array of researchers from mathematics, computer science, linguistics, social sciences, and philosophy to ponder the nature and implications of Conceptual Structures. This diversity in approaches and legacies of thought has become a hallmark of the community, and the very reason that many researchers return to the annual conferences year after year. The workshop proceedings from the early years (1986–1992) and the conference proceedings published by Springer in the esteemed Lecture Notes on Artificial Intelligence series (1993–current) form a valuable archive over progress and development in this community. Additionally, a number of important books on specific topics within the scope of the community have been published. But until now, we have not had a book that surveys the entirety of the field. *Conceptual Structures in Practice* is the remedy for that, and it is our hope that the book will serve as a benchmark of research in Conceptual Structures as well as an important source of inspiration for students and researchers within this and related fields of scholarly effort.

Guided by fundamental research questions such as charting out the internal structure of a concept (Wille, Chapter 6), and the question of how to represent age-old knowledge about knowing (Sowa, Chapter 5), *Conceptual Structures in Practice* takes us through the basic but non-trivial task of establishing conceptual relations such as [T]−relation→[T] or the seemingly simplistic structure of a cross-table as the foundation for research that now reaches into the cutting edge of leading technology in knowledge representation and knowledge mining.

Conceptual Structures is now a mature field of research in Computer Science, which looks back at over 20 years of research. In the course of recent developments, Conceptual Structures find their way into applications including Semantic Web, Data Mining, Knowledge-Based Systems, Natural Language Processing, etc. The book captures the current state-of-the-art concerning such applications, by means of contributions by leading researchers in the field. It also contains brief introductions to the necessary formal backgrounds on Conceptual Graphs and Formal Concept Analysis, and is suitable for practitioners and as a basis for university courses. Each chapter ends with a list of hands-on projects for deepening the understanding of the covered topics.

Structure

The international community of Conceptual Structures can be seen as stemming from (at least) two traditions: The community studying Conceptual Graphs, in which John Sowa takes a prominent place, and the Darmstadt Group of mathematicians studying Formal Concept Analysis, in which Rudolf Wille and Bernard Ganter are the *sine non qua*. It is our belief that these two traditions over the years have been brought closer together to the present state of mutual respect, understanding, and gain. We are therefore very happy to present a truly remarkable selection of chapters stemming from both traditions, and for the first time outside the international conferences, presenting them in one volume. This historical background is quite naturally reflected in the organization of the chapters.

The book is composed of six parts, each of which could deserve a book of its own. Beginning with the minimally required formal background (Part I), we proceed to description of tools (Part II), visionary chapters by the initiators of the field (Part III) and applications to several areas that are currently very active and prominent in Computer Science (Parts IV–VI).

In Chapter 1, Markus Krötzsch and Bernhard Ganter provide an excellent introduction to Formal Concept Analysis. With great clarity, they take us through the basic definitions, and guided by well-chosen examples point to the potential of FCA. In addition, they also share a few concerns regarding future research, and they do so by applying the very tools they introduce. The second introductory chapter, by Frithjof Dau, offers the reader a broad introduction to classical Conceptual Graphs as well as some variations. Dau takes on the laborious task of dealing with the formal logics of CGs, a topic that over the years has taken up many pages in conference proceedings. This chapter therefore provides an overview of some of the most pertinent problems in this regard, and it also draws on other traditions, not least the work of Chein and Mugnier.

Part II on tools is also composed of two chapters. Peter Becker and Joachim Hereth Correia guide us through a number of FCA tools, bringing valuable insights and inspiration to both newcomers and more seasoned practitioners of Formal Concept Analysis. In her chapter, Galia Angelova presents a complex example of how graph operations and ontology-driven analysis can be employed over a knowledge base in a framework of Finite State Automata (FSA). Newcomers to the field of applying Conceptual Graphs will find much inspiration here, and may benefit from combining their reading of this chapter with the chapters in Part VI.

Part III contains two very important chapters in which John Sowa and Rudolf Wille, in a very dense, yet lucid manner, take us through various aspects of the background that led them to form the fields of Conceptual Graphs and Formal Concept Analysis, respectively. Bringing these important scholars together in this manner gives the readers an exceptional opportunity to witness seasoned scholars reflect upon a lifetime of work in representing and reasoning with knowledge. In a sense, much of the material covered in these chapters can be found elsewhere. But the composition and density presented in this context show perfectly the importance and also the beauty of the field of knowledge representation. We are also certain that core members of the KR communities will enjoy reading these chapters.

Parts IV–VI all deal with important aspects on ongoing research in Conceptual Structures, and in this capacity they could have been put under just one heading, but for the sake of clarity, we have chosen to arrange them after their main topics. Each of these chapters represents many years of research in their respective disciplines by either individuals or research groups. As such, chapters in these sections can be seen as benchmarks of Conceptual Structures in practice: they present state-of-the-art research, covering decades of material while reflecting on practices and their future paths.

Uta Priss demonstrates how Formal Concept Analysis can be used for complex linguistic data exploration, involving various nontrivial semantic relations and bilingual dictionaries. In a related approach, Philipp Cimiano discusses the notion of ontologies and then shows how to extract ontologies from text using FCA and other techniques. Sebastian Rudolph and Johanna Völker are also concerned with ontology learning, and combine FCA techniques with Description Logics. Jon Ducrou and Peter Eklund share important insights drawn from building FCA-based applications for faceted information retrieval. Here, interface concerns also play a role. In the next chapter, Aldo de Moor takes tools for handling conceptual structures into a social software context and shows how such tools and techniques are useful for communicating about complex topics. Michel Chein in his chapter reflects on decades of work on conceptual graphs in the LIRMM group, and introduces here the notion of banned type sets. Also Adil Kabbaj takes us through two decades of developing conceptual tools, leading to the impressive and comprehensive Amine system. In this chapter, you will also find a brief exposition of features in available CG systems. And finally, in the last chapter, Harry Delugach shows how conceptual graph theory can be utilized to handle dynamic information in active knowledge systems.

Acknowledgments

This impressive set of chapters written by long-term engaged researchers who share their struggles and visions provides an unprecedented overview of research on Conceptual Structures, and we are thankful that the authors have taken time from their busy schedules to write these chapters. We would like to mention that an important group of Canadian researchers is missing from the list. Specifically, Guy Mineau and Berhard Moulin, both of whom have been involved in the early editorial process, are not found in the list of authors. To their regret – and very much also to ours – it has not been possible for them to contribute to this volume.

We would also like to thank Randi Cohen at Chapman & Hall/CRC for her kind and competent support and encouragement throughout the process; and we extend our gratitude to Ulrik Sandborg-Petersen for his meticulous work with the LaTeX files, not least taking care of the technical notations.

Pascal Hitzler and Henrik Schärfe
Karlsruhe and Aalborg

About the Editors

Pascal Hitzler is assistant professor at the Institute for Applied Informatics and Formal Description Methods (AIFB) of the University of Karlsruhe in Germany. Prior to this, he graduated in Mathematics at Tübingen University, Germany, did his dissertation at the National University of Ireland in Cork, worked as a postdoctoral researcher at the Artificial Intelligence Institute at TU Dresden, Germany, and spent three months as a visiting research associate at Case Western Reserve University in Cleveland, Ohio. The focus of his research is foundations and applications of knowledge representation and reasoning, and his research record lists over 130 publications in such diverse areas as Semantic Web, neural-symbolic integration, knowledge representation and reasoning, lattice and domain theory, denotational semantics, and set-theoretic topology. He is co-author of the first German textbook on the foundations of the Semantic Web (*Semantic Web* – Grundlagen, Springer 2008). He is a steering committee member of the conference series Web Reasoning and Rules System (as vice-chair) and of the International Conference on Conceptual Structures, and also of the workshop series OWL – Experiences and Directions, and he is on the editorial board of several scientific journals. He has also been an organiser of and teacher at international enhancement programmes for highly skilled students in Mathematics and Computer Science, and has served as an editor for several books in this area. For more information, please see http://www.pascal-hitzler.de.

Henrik Schärfe is associate professor at the Department of Communication and Psychology at Aalborg University, Denmark where he has taught since 1999. He holds an MA in Humanistic Computer Science, and a PhD in Human Centered Informatics, both from Aalborg University. He is a very active educator, and has been part of developing study programs in Humanistic Informatics and Information Architecture. His favorite subjects to teach are the theory of science, categorization, and the relations between aspects of communication that can, and respectively cannot, be formalized. Furthermore, he has played an active role in developing teaching materials for basic and advanced studies in conceptual structures. His research centers on philosophical aspects of knowledge representation, especially concerning matters of time, cultural impact on theoretical constructions, epistemic narratology, and the history of ideas. Within these fields he has written extensively and has been published in Computer Science, the Social Sciences, as well as in the Humanities. His publications include a selection of papers on Conceptual Graphs and Formal Concept Analysis, typically applied to areas that traditionally belong to research conducted in the Arts. He has been an active member of the International Conference on Conceptual Structures (ICCS) since 2000, and chaired the 2006 conference in Aalborg together with Peter Øhrstrøm and Pascal Hitzler. He is also a member of the Editorial Board of ICCS. He serves as reviewer in various international communities and has edited a few books, and a small journal on formalization issues. For more information and a complete publication list, please visit http://www.hum.aau.dk/~scharfe.

Contributors

Galia Angelova
Bulgarian Academy of Sciences
Sofia, Bulgaria

Peter Becker
Nudgee, Australia

Michel Chein
C.N.R.S. and University Montpellier 2
Montpellier, France

Philipp Cimiano
Universität Karlsruhe (TH)
Karlsruhe, Germany

Joachim Hereth Correia
Technische Universität Dresden
Dresden, Germany

Frithjof Dau
SAP Research CEC Dresden
Dresden, Germany

Aldo de Moor
CommunitySense
Tilburg, The Netherlands

Harry Delugach
University of Alabama in Huntsville
Huntsville, Alabama, U.S.A.

Jon Ducrou
Amazon.com, Inc.

Peter Eklund
University of Wollongong
Wollongong, Australia

Bernhard Ganter
Technische Universität Dresden
Dresden, Germany

Adil Kabbaj
INSEA
Rabat, Morocco

Markus Krötzsch
Universität Karlsruhe (TH)
Karlsruhe, Germany

Uta Priss
Napier University
Edinburgh, UK

Sebastian Rudolph
Universität Karlsruhe (TH)
Karlsruhe, Germany

John F. Sowa
VivoMind Intelligence, Inc.

Johanna Völker
Universität Karlsruhe (TH)
Karlsruhe, Germany

Rudolf Wille
Technische Universität Darmstadt
Darmstadt, Germany

Part I

Introductions

Chapter 1

A Brief Introduction to Formal Concept Analysis

Markus Krötzsch

AIFB, Universität Karlsruhe (TH), Germany

Bernhard Ganter

Technische Universität Dresden, Germany

1.1   Introduction

Mathematics has, quite naturally, always been concerned not just with numbers and functions, but also with many other kinds of abstract objects and structures. Yet it became clear only in recent decades – especially with the advent of Computer Science – that approaches originating in algebra and discrete mathematics also find concrete practical use in many application domains where numerical methods are not satisfactory. A prominent example is the requirement to represent and process increasing amounts of computer-mediated "knowledge" in a formal way that allows for suitable algorithmic treatment.

Modern discrete mathematics supplies a number of tools for addressing this aspect, and previous decades saw the development of a rich theory of *structures*, comprising the theory of *ordered sets* and in particular the theory of *lattices*. Originally developed as a tool for describing mathematical structures, this field of research has since been found to be increasingly useful for much more applied disciplines. *Formal Concept Analysis* (FCA) provides a mathematical notion of *concepts* and *concept hierarchies* that is based on order and lattice theory.

The basis of FCA are remarkably simple data structures, so-called "formal contexts." To each such data set, FCA associates hierarchies of "formal concepts" which, as one can formaly show, carry the structure of complete lattices. In consequence, methods for processing and analyzing such data can draw from the extensive toolbox of lattice theory, which does provide both a rigorous formal foundation and insights for obtaining suitable algorithms. As we will see, one can also connect FCA with ideas from *formal logic*, which confirms the intuition that FCA is indeed an approach for representing and processing knowledge.

This chapter ist structured as follows. In Section 1.2, we begin our introduction with (formal) *contexts* and *concepts*, the central notions of FCA. We also introduce some basic properties, and relate formal concepts to the concept of a *closure operator*. *Concept lattices* and the related graphical representation of conceptual knowledge are presented in Section 1.3, before Section 1.4 takes a more logical viewpoint by introducing *attribute implications* and their relationships with formal contexts. Section 1.5 then gives an intuitive introduction to the method of *attribute exploration* that is widely used to support the construction of contexts and implications when no complete description is available yet.

1.2 Contexts and Concepts

The primary means of representing knowledge in Formal Concept Analysis are *(formal) contexts*[1], which describe binary relationships between a set of *objects* and a set of *attributes* from a domain of interest. More precisely, a formal context is given by the following components:

- a set G of *(formal) objects*,

- a set M of *(formal) attributes*, and

- a binary *incidence relation* I with $I \subseteq G \times M$.

One commonly denotes such a context by $\mathbb{K} := (G, M, I)$. The incidence relationship might be written in set notation ("$(g, m) \in I$"), or infix notation ("$g\ I\ m$"). As contexts essentially are binary relations, small contexts can conveniently be specified by *cross tables*, where each object is represented by a row, and each attribute is represented by a column. Figure 1.1 shows a simple context about planets in our solar system.

Although simple in structure, formal contexts turn out to be versatile tools for capturing information, and FCA provides numerous methods for processing, visualizing, and acquiring such information. An important basic notion

---

[1]The adjective "formal" that qualifies many notions of FCA is commonly omitted when the intended meaning is clear from the non-formal context.

	a	b	c	d	e	f	g
1 Mercury			×	×	×	×	
2 Venus				×	×	×	×
3 Earth	×		×		×	×	×
4 Mars	×		×	×	×		×
5 Jupiter	×	×	×		×		
6 Saturn	×	×	×	×	×		
7 Uranus	×	×					
8 Neptune	×	×	×				

FIGURE 1.1: A context about planets in our solar system; the attributes mean: a: "has some moon," b: "gas giant," c: "prograde rotation" (sun sets in the "West"), d: "average magnetic field weaker than on Earth," e: "was known as a planet in Ancient Rome," f: "average temperature above $0°C$," g: "stable atmosphere that contains nitrogen."

throughout FCA is that of a *(formal) concept*, which is introduced next. Concepts are based on the idea of considering certain sets of objects and attributes that are mutually related by means of the given incidence relation. Given a set O of objects of some context (G, M, I), it is interesting to consider the set of attributes common to all those objects:

- For any set of objects $O \subseteq G$, the set O' is defined as

$$O' := \{m \in M \mid m \, I \, g \text{ for all } g \in O\}.$$

Similarly, we can derive the set of all objects having all of attributes from a given set A:

- For any set of attributes $A \subseteq G$, the set A' is defined as

$$A' := \{g \in G \mid m \, I \, g \text{ for all } m \in A\}.$$

Note that we use essentially the same notation $\cdot'$ for two different *derivation operators*, one that maps sets of objects to sets of attributes, and another one that maps in the opposite direction. In cases where this simplification is overly ambiguous, it is common to use notations like $O^\uparrow$ and $A^\downarrow$ to distinguish the operators, or O^I and A^I to distinguish different incidence relations. Given a formal context (G, M, I), a pair (O, A) with $O \subseteq G$ and $A \subseteq M$ is a *(formal) concept* whenever we find that

$$O = A' \qquad \text{and} \qquad A = O'.$$

In this case O is called the *extent* and A is called the *intent* of the concept. Note how extent and intent of a concept obviously mutually determine each other. Given, for example, the context of Figure 1.1, we find that

($\{146\}, \{cde\}$) is a concept: Mercury, Mars, and Saturn are exactly the planets that have a weak magnetic field, propagate rotation, and were known to the Romans.

The concepts of a formal context are of specific interest, since they present combinations of objects and attributes that are "closed" in a certain sense: the intent includes *all* attributes common to the objects in the extent, and the extent includes *all* objects that support the attributes of the intent. In order to consolidate this somewhat vague intuition, it is useful to consider some further properties of the derivation operators. Given arbitrary sets O of objects and A of attributes, it is trivial to see the following: The attributes A are common to all objects in O *if and only if* every object in O has all attributes of A, i.e.,

$$A \subseteq O' \qquad \text{iff} \qquad O \subseteq A'$$

This simple property establishes an important relationship between the two derivation operators, asserting that they form what mathematics calls an *(antitone) Galois connection* with respect to the natural order of sets. There is a wealth of theorems about Galois connections, stating, e.g., that each operator of a Galois connection uniquely determines the other. In FCA, however, we are especially interested in the consecutive application of the two derivation operators. For example, given a set O of objects, O'' is the set of all objects that have all the attributes that the objects O have in common. One of the favorable properties of Galois connections is that this mapping is guaranteed to have the following features:

- Extensiveness: $O \subseteq O''$

- Monotonicity: if $O \subseteq P$ then $O'' \subseteq P''$

- Idempotency: $O'''' = O''$

The same could be said about the operator $A \mapsto A''$ on attributes. Functions with the above properties are known as *closure operators*, and allow for a rather intuitive understanding based on the idea of a *closed set*. Namely, applying the operator to a set *closes* it by adding certain hitherto missing elements (extensitivity). Once a set is closed, however, no further elements will be added (idempotency). In FCA, we therefore often speak of closed sets of objects, i.e. set of objects O for which we find that $O = O''$. Closed sets of attributes are defined dually.

Coming back to our above definition of formal concepts, we note that the defining properties $O = A'$ and $A = O'$ also imply that $O = O''$ and $A = A''$. In other words, concepts can be viewed as "matching" pairs of closed sets of objects and attributes. It is interesting to note that a single application of any derivation operator already suffices to obtain a closed set:

- For every set O of objects, O' is a closed set of attributes, and

- for every set A of attributes, A' is a closed set of objects.

This leads to a simple, albeit naive, method of computing all concepts of a finite context (G, M, I): it is enough to compute all closed sets of objects, and this can be done by applying the derivation operator to each available set of attributes. It is obvious that this is not an optimal strategy, since it is often the case that many different sets of attributes lead to the same closed set of objects. More information about the efficient generation of concepts can be found in [Ganter and Wille, 1999a, Section 2.1].

1.3 From Contexts to Concept Lattices

An important advantage of FCA is that it supports graphical (and algebraic) representations of knowledge based on so-called *concept lattices*. This is a special kind of order structures that is obtained by arranging the concepts of a formal context in a natural order: a concept (O_1, A_1) is smaller (or equal) than another concept (O_2, A_2) whenever $O_1 \subseteq O_2$. Note that $A_1 \supseteq A_2$ would be an equivalent condition. In this case we write $(O_1, A_1) \leq (O_2, A_2)$, and $\leq$ is referred to as the *(hierarchical) order* of concepts.

The *concept lattice* of a formal context (G, M, I) is the set of all concepts of (G, M, I) ordered by $\leq$, and is commonly denoted by $\underline{\mathfrak{B}}(G, M, I)$. As long as the formal context is reasonably small, it is possible to picture concept lattices in line diagrams, where greater elements are higher up in the picture, and neighboring elements are connected with a simple line. Figure 1.2 shows a possible visualization of the concept lattice for the context of Figure 1.1. An important structural property of concept lattices is that they are *complete lattices*, i.e. order structures in which every set of elements has both a greatest lower bound (infimum) and a least upper bound (supremum). For example, given the concepts $(\{46\}, \{acde\})$, $(\{34\}, \{aceg\})$, and $(\{234\}, \{eg\})$ of Figure 1.1, the supremum is given by $(\{12346\}, \{e\})$ and the infimum is given by $(\{4\}, \{acdeg\})$. Note that it is not difficult to read this information off the above graphical representation of the concept lattice, by just following upward and downward edges from the given set of elements.

Infimum and supremum thus are operations that are defined for every input set of concepts. In lattices of sets, these operations are often given by set intersection and set union. Only one of those can directly be applied in FCA: the intersection of any set of closed sets (of objects or attributes) is again a closed set. The union of closed sets, however, need not be closed, so that we must apply the closure operator again. This is reflected in the following theorem:

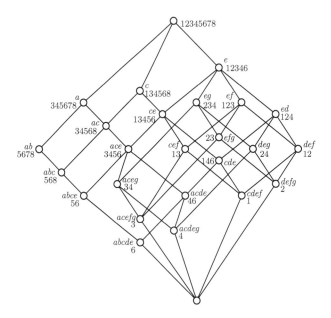

FIGURE 1.2: The concept lattice for the context in Figure 1.1, with full labeling.

The Basic Theorem on Concept Lattices. *A concept lattice* $\mathfrak{B}(G, M, I)$ *is a complete lattice where supremum and infimum of an indexed set* $(O_t, A_t)_{t \in T}$ *of concepts are defined as follows:*

- *Supremum:* $\bigvee_{t \in T}(O_t, A_t) = \left(\left(\bigcup_{t \in T} O_t \right)'', \bigcap_{t \in T} A_t \right)$

- *Infimum:* $\bigwedge_{t \in T}(O_t, A_t) = \left(\bigcap_{t \in T} O_t, \left(\bigcup_{t \in T} A_t \right)'' \right)$

Conversely, any complete lattice $(L, \leq)$ *is isomorphic[2] to some concept lattice. In particular, it is isomorphic to* $\mathfrak{B}(L, L, \leq)$.

For simplicity, some details on the possible encoding of complete lattices as formal contexts have been omitted here. It is apparent, however, that there is a close relationships between the algebraic structure of complete lattices, and the information encoded in formal contexts. Further details and the proof can be found in [Ganter and Wille, 1999a].

The labeling of the concept lattice in Figure 1.2 may still appear rather complicated, and it is worth looking for a simplified presentation. Indeed, a closer inspection reveals helpful regularities. Clearly, whenever an element is

---

[2]For us it suffices to translate "isomorphic" as "equal up to renaming of elements." Formal details are found in [Ganter and Wille, 1999a].

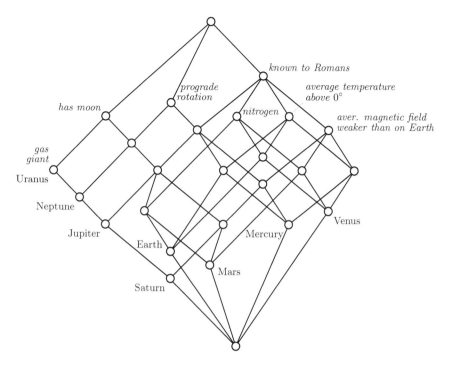

FIGURE 1.3:   The concept lattice for the context in Figure 1.1, with simplified labeling.

labeled with a certain object $g \in G$, all elements above it must also carry that label, this is a trivial consequence of the underlying subset ordering. Another simple observation is that the greatest lower bound of all elements that are labeled with g is also labeled with g, this follows from the intersection operation used in the Basic Theorem. Therefore, for every object g, there is a least element in the concept lattice labeled with g. This special element is the so-called *object concept* of g, and it can simply obtained as $(\{g\}'', \{g\}')$. Dually, every attribute $m \in M$ leads to an *attribute concept* $(\{m\}', \{m\}'')$ that constitutes the greatest element in the concept lattice which is labeled with m.

We can use these observations to simplify the visualization of concept lattices: instead of labeling elements with all their objects and attributes, we now merely label the object and attribute concepts with their generating objects and attributes. Hence, every object and attribute occurs exactly once as a label. Figure 1.3 shows how this method simplifies the visualization of the concept lattice from Figure 1.2. To read such a simplified concept lattice, one merely needs to recall that an object is contained in the extents of all concepts above its object concept, whereas an attribute belongs to the intent of all concepts below its attribute concept. Hence, in order to find, e.g., the attributes of Mercury in Figure 1.3, one merely needs to collect all attribute concepts above Mercury's object concept, yielding "weak magnetism," "positive average temperature," "known to Romans," and "prograde rotation." Conversely, the objects with the attribute "gas giant" are given by the object concepts below the attribute concept for "gas giant." Thus one can easily reconstruct the original formal context from the concept lattice.

Together with the fact that concepts can be completely characterized by either their extent or their intent, the above observations also ensure that there are always "enough" object and attribute concepts, respectively, to derive the labels of all other elements of the concept lattice. Indeed, the extent and intent of every element is given by the object concepts below and the attribute concepts above it, respectively. Another way of putting it is to say that every concept is the supremum of the object concepts below it, and the infimum of the attribute concepts above it. Object concepts thus are *supremum-dense* while attribute concepts are *infimum-dense* in the concept lattice.

1.4   The Logic of Attributes

Cross-tables and (the graphical presentations of) concept lattices are not the only ways of representing the content of a formal context. The closure operators introduced in Section 1.2 already hint at another option, since closure operators in computer science typically relate to logical deduction. Indeed, the deductive closure – the set of all logical consequences – of a logical theory satisfies the properties of closure operators as defined in Section 1.2. But deduction is concerned with sets of logical formulae, whereas we are merely dealing with sets of attributes or objects! It turns out that this does not inhibit the use of various ideas from formal logic, from which powerful tools can be derived for the use in FCA.

It is common to consider only sets of attributes in this approach. The situation for objects would be completely symmetric, but attributes often yield a more intuitive relationships to other logical approaches. Indeed, it is suggestive to say that a certain attribute m_1 *implies* another attribute m_2 if all objects that have m_1 also have m_2. This idea can be extended to sets of attributes, yielding statements of the form:

- "Every object that has all attributes of the set A_1 also has all attributes from the set A_2."

Formally, this statement can be written as an *implication between attributes* $A_1 \rightarrow A_2$. An attribute implication holds in a formal context whenever it is satisfied by all objects of that context. In the context of Figure 1.1, for example, the implication $\{a, d\} \rightarrow \{c, e\}$ holds: the planets that have some moon and a weak magnetic field (Mars and Saturn) both have a prograde rotation and were known to the Romans. Note that this is merely a material implication – something that is the case for the given data – but does of course not yield a causal relationships. Our simple definition of implications already leads to the sought logical representation of contexts: as we will see below, the set of all implications holding in a formal context suffices to reconstruct the structure of the concept lattice.[3] To this end, it makes sense to first consider the relevance of implications to arbitrary sets of attributes. A set of attributes A *respects* an implication $A_1 \rightarrow A_2$ if either $A_1 \not\subseteq A$ or $A_2 \subseteq A$. Moreover, A respects a set of implications $\mathcal{L}$ whenever it respects all implications in $\mathcal{L}$. We might be inclined to say that A is closed under the implications in $\mathcal{L}$ in the latter case, similar to a logical theory being closed under a set of derivation rules. Closed sets of attributes in FCA usually refer to concept intents, but the following theorem shows that logical closure and concept closure do indeed coincide.

Theorem   *Consider a context (G, M, I) and a set $\mathcal{L}$ of all implications between attributes holding in that context.*

- *Every concept intent of (G, M, I) respects all implications in $\mathcal{L}$.*

- *Every subset of M that respects all implications in $\mathcal{L}$ is a concept intent of (G, M, I).*

Therefore, the closure operator $A \mapsto A''$ may as well be seen as a logical closure that extends A by the consequences of all implications that apply to it in the given context. It is possible to represent contextual information by means of sets of implications: not just the implications of a given context but any set of attribute implications generates the intents of some concept lattice as in the theorem above. This can be handy in applications where the general rules governing the combination of attributes are more concise or more convenient than dealing with a possibly large set of different objects. Even if a context is already given, one is often interested in the implications that hold, either because of their informative value or because of the potential simplification of representation. Computing meaningful sets of implications can be a challenging task, since most of the implications of a context are trivial and of little interest. It is therefore necessary to single out the "relevant" implications from that set.

Such reduced sets of implications can still capture all semantic relationships holding in a context. Namely, an implication $A_1 \rightarrow A_2$ *follows* from a set $\mathcal{L}$

---

[3]Obviously, the attribute-based implications cannot provide detailed information about a context's objects, but they still fully specify the attribute-combinations that occur in objects.

$\{b\} \rightarrow \{a\}$	$\{f\} \rightarrow \{e\}$	$\{a,c,e,f\} \rightarrow \{g\}$
$\{a,e\} \rightarrow \{c\}$	$\{g\} \rightarrow \{e\}$	$\{a,b,c,e,g\} \rightarrow \{d,f\}$
$\{d\} \rightarrow \{e\}$	$\{c,e,g\} \rightarrow \{a\}$	$\{a,c,d,e,f,g\} \rightarrow \{b\}$

FIGURE 1.4: The stem base for the context of Figure 1.1.

of implications whenever each set A that respects $\mathcal{L}$ also respects $A_1 \rightarrow A_2$. Thus we may call a set of implications $\mathcal{L}$ *complete* for a given context (G, M, I) if all implications of (G, M, I) follow from $\mathcal{L}$. Complete sets of implications of course may still contain unnecessary elements, which could be derived from the other members. We therefore say that a set of implications is *non-redundant* if none of its implications follows from the others. Summing up, we are generally interested in complete but non-redundant sets of implications. Such sets are generally called *implications bases* of a context.

One particular way of constructing an implication base for contexts with finitely many attributes has been introduced in [Guigues and Duquenne, 1986], leading to what has been called the *Duquenne-Guigues base* or just *stem base*. A special virtue of this implication base is that it consists of the smallest possible number of implications: all other bases that might exist must have at least as many. The computation of the stem base can be computationally challenging for large contexts, but the basic algorithm is well-known and practical implementations are available, e.g., as part of the free *Concept Explorer (ConExp)*[4]. Figure 1.4 gives the stem base for the example of Figure 1.1. Note that the premises of the two last implications in the lower left of the table are not satisfied by any object. Those implications thus are vacuously true, and merely ensure closure for the bottom concept $(\{\}, \{a, b, c, d, e, f, g\})$.

1.5   Attribute Exploration

We have seen in the previous sections that contexts, concept lattices, and implication bases are alternative ways of representing essentially the same structural information. A context can be used to compute concept lattice and implication base, whereas the implication base defines possible intents from which a lattice structure and a suitable context can be derived. Thus one many freely choose the approach that seems most convenient to a particular application of FCA, and specify all input data in this way. There are, however, cases where neither a complete context, nor a lattice or a full implication base

---

[4]http://conexp.sourceforge.net/

is known, and we will now introduce a method for "exploring" incompletely specified information in an efficient way.

Assume, for example, that we wish to analyze the co-occurrence of certain keywords in journal publications of the year 2006. To simplify the example, we focus on the research topic *Semantic Web* and consider only six keywords: "ontology" (a general term for structured information ressources), "OWL" and "RDF" (common Semantic Web standards), and "Pellet," "Racer," and "KAON2" (three software tools for processing OWL). How could one construct a formal context that describes the logical relationships between those attributes? We do not have a database with all journal papers of 2006 available, so constructing a context directly is not possible. However, using an online search service such as Google Scholar[5], we can find out whether a certain combination of keywords (attributes) occurs in papers.

A naive approach thus would be to conduct a Web search for all 64 possible combinations of attributes, adding an object to the context for each existing combination. Obviously this would quickly become infeasible for larger numbers of attributes. Yet many of those combinations could easily be ruled out, e.g., if we know that there is no paper that mentions "Pellet" without mentioning "OWL" (which could be easily checked with a search request of the form "`Pellet -OWL`"). Implications like {Pellet} → {OWL} therefore can serve as shortcuts. The challenge now is to find an efficient strategy for checking possible implications and for adding relevant objects to the formal context.

FCA does indeed provide a method for doing this kind of incremental construction of formal contexts. This procedure is called *attribute exploration*, and is based on a modification of the algorithm for computing the stem base. The key point is that the stem base is computed step by step, deriving valid implications in a fixed order. In attribute exploration, one starts this computation with a partially specified or even empty context. Since the set of objects is still incomplete (just like the set of papers in our example), it might happen that invalid implications are computed. Therfore, the algorithm presents each computed implication to the user, who may then choose from two possible options:

- accept the implication as being generally valid for *all* objects of interest, or

- reject the implication by supplying an object for which the implication is not valid (a *counterexample*).

Step by step the algorithm thus completes both the logical specification and the relevant object set of the context, until the confirmed implications are a base for the constructed context. Normally this requires far less exploration steps (Web searches in our example) than a naive approach.

---

[5]`http://scholar.google.com`

For our example we start indeed with an empty context. The first question of the algorithm then is whether all objects have all the attributes, i.e. whether the empty set implies the set of all keywords. This is of course not the case, and a counterexample is needed. Indeed, we can easily find a paper that has none of the keywords.[6] The next question of the algorithm is whether the implication {Ontology} → {OWL,RDF,Pellet,Racer,KAON2} is valid. Again this can easily be rejected by seraching for a paper that contains "Ontology" but not "OWL." After adding a few more papers with various combinations of keywords, the questions of the algorithm quickly become more targeted, and implications such as {Pellet} → {OWL} are found valid. This rules out further questions, and after adding 17 objects the exploration is finished. Note that the number of necessary exploration steps is influenced by the counterexamples that are chosen. Usually the algorithm does not yield a minimal set of objects needed for determining the structure of the concept lattice, and so it is possible to further simplify the context. This *reduction* of a context, which is also implemented in standard tools, leads to a context of only 5 objects in our case. After completing the exploration, we obtain the following final set of implications:

$$
\begin{aligned}
\{\text{Ontology, Racer}\} &\rightarrow \{\text{OWL}\} \\
\{\text{OWL, Racer}\} &\rightarrow \{\text{Ontology}\} \\
\{\text{Pellet}\} &\rightarrow \{\text{Ontology, OWL}\} \\
\{\text{RDF, Racer}\} &\rightarrow \{\text{Ontology, OWL}\} \\
\{\text{Ontology, OWL, Pellet, Racer}\} &\rightarrow \{\text{KAON2}\} \\
\{\text{KAON2}\} &\rightarrow \{\text{Ontology, OWL, Pellet, Racer}\}
\end{aligned}
$$

The resulting concept lattice with object labels omitted is displayed in Figure 1.5. In this case the lattice is rather concise, and allows us to quickly capture essential relationships in our explored domain. For example, the position of the attribute concept for KAON2 tells us that there have been journal papers that mentioned this tool, and that all those papers also mentioned Racer, Pellet, OWL, and Ontology, but did not include a reference to RDF. Similarly, there was no paper that mentioned Pellet without mentioning OWL, whereas some paper(s) on Racer do not mention any of the other terms.

1.6   Summary and Outlook

In this chapter, we have provided a basic introduction to the central notions of FCA. We have seen that formal contexts, though simple in structure,

---

[6]However, we always add the search term "Semantic Web" in order to avoid unexpected uses of the chosen keywords in completely different research areas.

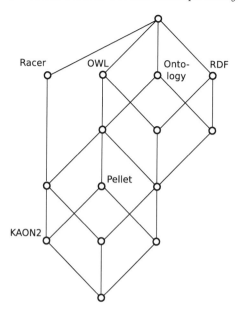

FIGURE 1.5: Relationships between keywords of Semantic Web research in 2006 journal papers, obtained by attribute exploration.

give rise to a number of algebraic and logical representations of information. The algebraic structure of a concept lattice, which emerges naturally from the basic notion of a formal concept, can also be used to generate helpful graphical representations of the relationships given by a certain context. The mathematical notion of closure operators finally builds a bridge between the concept closure induced by a formal context, and the logical closure of an implicational theory. The intimate relationship between both views on conceptual knowledge can be exploited in the method of attribute exploration. The practical importance of this approach is witnessed by various applications described in later chapters.

Naturally, many important topics in FCA could not even be mentioned within this short chapter. A relevant question, for example, is how to produce a readable visualization of a given concept lattice. This can be a challenging task even for comaratively small lattices, and various methodologies and algorithms have been proposed. Methods like the nesting of concept lattices provide further means of visualizing somewhat larger lattices more concisely. Another topic of practical significance is the extension of FCA towards attributes with more than two possible values. In so-called *many-valued contexts*, attributes such as "color" with possible vaules "red", "green", and "blue" are enabled via a method known as *conceptual scaling*. Finally, there is also a wealth of foundational results regarding the mathematical properties of structures encountered in FCA. For more information on these and other topics, and for additional details on the material presented here, the reader is referred to [Ganter and Wille, 1999a].

Chapter 2

Formal Logic with Conceptual Graphs

Frithjof Dau

SAP Research CEC Dresden, Germany

2.1 Introduction

This chapter aims to give an introduction into the formal theory of conceptual graphs (abbreviated by CGs), with an emphasis on how CGs can be understood as a diagrammatic approach to formal logic. Of course, as it can already easily be seen from the huge variety of topics covered in this book, formal logic is only one of many aspects of CGs. Anyhow, maybe not the majority, but at least a vast amount of CG research has focused on this specific facet, and even Sowa emphasizes the logic aspect of CGs.

CGs are based on diagrammatic representation of logic, namely Peirce's (1839-1914) existential graphs (for introductions into existential graphs, see [Zeman, 1964, Roberts, 1973, Shin, 2002]). In [Sowa, 1997a] Sowa writes that 'conceptual graphs (CGs) are an extension of existential graphs with features adopted from linguistics and AI. The purpose of the system is to express meaning in a form that is logically precise, humanly readable, and computationally tractable.'

The system of existential graphs is divided into three parts named *Alpha*, *Beta*, and *Gamma*. Alpha corresponds to propositional logic, and Beta corresponds to first-order predicate logic. Gamma is more complicated: It covers features of higher order and modal logic, the possibility to express self-reference, and other features. Due to its complexity, it was not completed by Peirce. Peirce's existential graphs and Sowa's CGs cannot be directly compared. Sowa adopted many ideas of existential graphs, even from Gamma, but CGs have a different and richer syntax, and they are tailored to better suit the needs of contemporary knowledge representation systems. Anyhow, Sowa's goal CGs to be "logically precise" and his reference to existential graphs clearly hints to the logical alignment of CGs. This is more explicitely expressed in [Sowa, 2000b], where Sowa states that 'CGs have been developed as a graphic representation for logic with the full expressive power of first-order logic and with extensions to support metalanguage, modules, and namespaces.'

In the following, we scrutinize CGs in terms of formal logic. First, in Section 2.2, we give a short introduction into CGs, as they have been introduced by Sowa. In this section, some core notations of CGs are defined. Next, in Section 2.3, we briefly investigate whether Sowa's CGs are indeed "logically precise," and we will see that they do *not* suit the needs of contemporary formal logic. For this reason, much research on the formal theory of CGs aims to fix the formal gaps and flaws of CGs. In Section 2.4, an overview over different approaches to turn CGs into a mathematically precise system of logic is given, and a core notation for a formal theory is provided. In different works, different fragments of CGs are elaborated in a precise manner. An overview of these fragments is given in Section 2.5.

A final remark to the definitions in the following sections has to be made. Different authors have used different notations, and the following definitions aim to select the most convenient ones among these. A term that is defined as it is used in this chapter is set in small caps, like VOCABULARY. If we refer to a notation that is not used in this volume, it is set in italics, like *support*.

2.2   Short Introduction to Conceptual Graphs

In this section, an introduction into the theory of CGs, as they have been invented by Sowa, is provided.

2.2.1   Simple Conceptual Graphs

In the following, the broad range of CGs is illustrated by several examples. The first one, given below, is probably one of the best known CGs.

$$\boxed{\text{CAT: Yoyo}} \!-\!\!\overset{1}{(}\,\text{on}\,\overset{2}{)}\!-\!\!\boxed{\text{MAT:}*}$$

The boxes in this graph are called CONCEPT BOXES. We have to stress that usually, the term *concept* istead of *concept box* is used, but this term is often used in a much broader, often semantical understanding in knowledge representation. Moreover, it can be confused with the *formal concepts* of Formal Concept Analysis. Finally, by using the term "box," we refer to some extent to the diagrammatic representations of CGs. For these reasons, this chapter sticks to the use of the term *concept box*. When in Section 2.4 the formalization of CGs by means of mathematical graphs is provided, we will speak of concept *vertices* instead. In each concept box, we have two entities: A TYPE LABEL and a REFERENT. In the graph above, the two concept boxes contain two different kinds of referents. The referent "Yoyo" in the left box is a name for an object. The referent "*" in the right box is called GENERIC MARKER, thus the right concept box is called GENERIC CONCEPT BOX. The generic marker denotes an unqualified object, i.e., it can be understood as an existential quantification. The type labels of the concept boxes are "Cat" and "Mat," respectively, and the concept boxes encode the information that the referents belongs to the type. Thus the left concept box encodes the atomar piece of information that Yoyo is a cat, and the right concept box encodes the information that there exists a mat. Type labels are ordered in a subtype-relation, according to their level of generality. For example, CAT is a subtype of ANIMAL. An ordered set of types is often called *support, type hierarchy, taxonomy,* or *alphabet.*

Besides the concept boxes, the graph contains an RELATION OVAL, labeled with "on." The concept boxes are LINKED with ARCS to the relation oval. The relation in the oval relates the referents of these concept boxes. In our example, the meaning of the relation oval is "Yoyo is on the (unknown) mat." So, the meaning of the whole graph is "Yoyo is a cat, there is a mat, and Yoyo is on that mat," or "the cat Yoyo is on a mat" for short.

Of course, the order of the arguments of the relations matter. In the diagrammatic representation, the order is represented by indexing the arcs that link the concept boxes to the relation oval with numbers. Another way to depict the order of the arguments is the use of arrows instead of arcs in the diagrams. The corresponding diagram is then

$$\boxed{\text{CAT: Yoyo}} \longrightarrow \left(\text{on}\right) \longrightarrow \boxed{\text{MAT: }*} \quad .$$

This approach works fine if only relations of arity 1 or 2 are used, but for relation with arities higher than 2 (e.g., "between," which is a ternary relation), this approach fails. So indexes are preferable.

Sowa uses various kinds of referents, for example,

- names: $\boxed{\text{PERSON: John}}$, $\boxed{\text{DOG: Lucky, Matula}}$,

- quantifiers: $\boxed{\text{MEAT: }*}$ (some meat), $\boxed{\text{DOG: }\forall}$ (all dogs),

- measure specifications: $\boxed{\text{TIME-PERIOD: @5 seconds}}$, $\boxed{\text{Money:@\$5}}$

- control marks ?, !, #: $\boxed{\text{DOG: ?}}$ (which dog?), $\boxed{\text{Dog: Lucky !}}$ (emphasis on Lucky), $\boxed{\text{MEATBALL: #}}$ (*the* meatball),

- generic plurals: $\boxed{\text{BONE: } \{*\}}$ (some bones) or $\boxed{\text{BONE: } \{*\}4}$ (four bones),

- prefixes: $\boxed{\text{PERSON: Dist}\{\text{Bill, Mary}\}@5}$ (five persons distributively including Bill, Mary, and others), or $\boxed{\text{LADY: Cum}\{*\}}$ (a set of Ladies).

(all examples taken from [Sowa, 1992]).

These various kinds of referents verify the comprehensive knowledge representation claim of CG. On the other hand, it is clear that most of them go beyond the expressiveness of first-order logic. For the elaboration of CGs in term of formal logic, we will restrict the referents to object names or the generic marker.

Below, another example is depicted, which covers more of the constituent elements of CG.

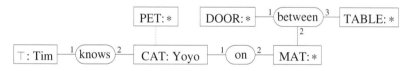

First of all, please note that concept boxes can be linked to more than one relation oval, and that we have a triadic relation in this graph. The type label 'T' in the leftmost concept box is the UNIVERSAL TYPE, which contains every object (of the respective universe of discourse) in its extension. As the type labels are usually ordered, T is the greatest element in this order. Finally note the dotted line between the concept boxes with the type labels Cat and Pet, respectively. This dotted line is a COREFERENCE LINK, expressing the identity of the referents in the linked concept boxes. Concept boxes that are linked with a coreference link are said to be COREFERENT. A COREFERENCE SET is a set of (pairwise) coreferent concept boxes. The two boxes with the type labels Cat and Pet are coreferent, and the subgraph consisting of these two boxes and the the coreference link expresses that Yoyo is a cat and there is a pet, which is the same as Yoyo, or for short: Yoyo is both a cat and a pet. The meaning of the whole graph is therefore Tim knows Yoyo, Yoyo is both a cat and a pet, and Yoyo is on a mat that is between a door and a table.

Although most practical examples are, CGs do not need to be connected. In the last example, we used two concept boxes, one of them being generic, and a coreference link to express that Yoyo is both a cat and a pet. Below, another graph is depicted that has exactly the same meaning as the last one. Now we use two non-connected, non-generic concept boxes to express that that Yoyo is both a cat and a mat.

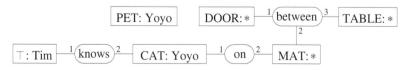

The CGs that can be constructed by means of concept boxes, relation ovals and coreference links, where the referents can be names for objects or the generic marker, are called SIMPLE CONCEPTUAL GRAPHS (SCGs). The class of SCGs has the expressiveness of the ∃, ∧-fragment of first-order logic (including identity). It is the most important and prominent fragment of CGs.

There are two kinds of SCGs that cannot be further decomposed. A SINGLETON GRAPH is a CG that consists only of a single concept box. A STAR GRAPH is a CG that contains a single relation oval, and concept boxes that are linked to that oval. Our first example is a star graph. Finally, even having no concept box or relation oval at all is a CG, the so-called BLANK GRAPH.[1]

2.2.2 Nested Conceptual Graphs

We have already seen a range of different referents in CGs, but one important referent, which vastly extends the expressiveness of CGs , is still missing. Sowa allows for whole CGs as referents in concept boxes. This construction is called NESTING. When CGs are nested, some CGs make statements and assertions about other CGs, i.e., the system of CGs offers the possibility of meta-level statements. Concept boxes whose referents are CGs are called *contexts*.[2] Besides concept boxes that contain CGs as referents, the complete area of the CG is a context as well, which is called the OUTERMOST CONTEXT. Below we present a well-known example of a nested CG.

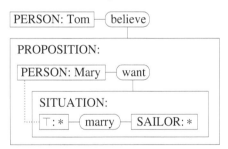

Besides the outermost context, this graph contains two further contexts, namely, the concept boxes of type PROPOSITION and SITUATION (which are common types of contexts). The graph can be read as follows: The person Tom believes a proposition, which is described by a graph itself. The proposition states that the person Mary wants a situation, which again is described by a graph. In this situation we have a concept box $\boxed{\top : *}$, which is connected with a coreference link to the concept box $\boxed{\text{PERSON:Mary}}$ in

---

[1]The blank graph corresponds to the empty *sheet of assertion* in Peirce's existential graphs.

[2]The term "context" occurs in several meanings and implementations in logics, linguistics, or artificial intelligence. An introduction into the various ideas of contexts is beyond the scope of this chapter. We refer to [Sowa, 2001] or Chap. 5 of [Sowa, 2000a] for an introduction and discussion of contexts in conceptual graphs.

the context above. So the situation is that Mary marries a sailor. The formal understanding of the whole graph is now: The person Tom believes the proposition that the person Mary wants the situation that Mary marries a sailor. In short, the person Tom believes that the person Mary wants to marry a sailor.

2.2.3   Negation in Conceptual Graphs

For the discussion of CGs in terms of mathematical logic, it is important to note that Sowa expresses negation with contexts, too. As Sowa states in [Sowa, 1997a]: "The EG [Existential Graph] negative contexts are a special case of the CG contexts. They are represented by a context of type NEGA-TION whose referent field contains a CG that states the proposition which is negated." Concept boxes of type NEGATION are introduced by Sowa as abbreviations for contexts of type proposition with an unary relation "NEG" attached to it (see [Sowa, 1992], where he says that 'Negation (NEG) is one of the most common relations attached to contexts'), and Sowa often abbreviates these contexts by drawing a simple rectangle with the mathematical negation symbol ¬ (e.g., in [Sowa, 1999]).

A well-known example for a CG with negations is presented below. The device of two nested negation contexts can be understood as an implication. So the meaning of the graph is "if a farmer owns a donkey, then he beats it."

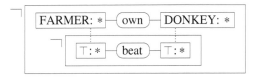

2.2.4   Conceptual Graphs and First-Order Logic

Sowa states in [Sowa, 1984] that "any formula in first-order logic can be expressed with simply nested contexts and lines of identity." (The term "line of identity" is adopted from Peirce's existential graphs and refers in this quotation to the coreference links of CGs). In particular CGs are designed to have at least the full power of first-order logic, that is first-order logic with identity, object names, and predicate names, but without function names. In the following, we will use the abbreviation FOL for this style of logic. Anyhow, as we have seen, the system of CGs is richer than FOL.

To show how CGs and FOL are related, Sowa defines a formal operator Φ that maps CGs to FOL. The definition of Φ can be found in [Sowa, 1984], and we find various comments and examples among the chapters of Sowa. The Φ-operator is useful for making the structural differences between CGs and the symbolic elaborations of FOL explicit, and it serves as a semantics for CGs as well. An example for the Φ-operator is the translation of the CG of

Section 2.2.3 to

$$\neg\exists x.\exists y.(FARMER(x) \wedge DONKEY(y) \wedge owns(x,y) \wedge \neg beat(x,y)).$$

The last formula is logically equivalent to the simplified formula

$$\forall x.\forall y.((FARMER(x) \wedge DONKEY(y) \wedge owns(x,y)) \rightarrow beat(x,y)).$$

2.2.5 Reasoning with Conceptual Graphs

When he developed the system of CGs, Sowa did not only adopt the iconicity and many syntactical and semantical elements of existential graphs. He also adopted the calculus of existential graphs. First of all, Sowa distinguishes between *equivalence rules*, which do not change the meaning of a CG, *specialization rules*, which (usually) specialize the meaning of a CG, and finally *generalization rules*, which in turn generalize the meaning of a CG. All rules transform a graph into a new graph. Equivalence rules can be carried out in both directions, whereas both specialization rules and generalization rules can be carried out only in one direction, and the specialization rules and generalization rules are mutually inverse. A simple example of a generalization rule is the generalization of a type or a referent in a concept box.

In [Sowa, 1984] Sowa provides a calculus, which is a one-to-one translation of Peirce's rules for existential graphs into the system of CGs. In [Sowa, 1997a] or in [Sowa, 2000b], this system is refined. Below, Sowa's rules from [Sowa, 1997a] are given:

Erasure: In a positive context, any graph u may be replaced by a generalization of u; in particular, u may be erased (i.e., replaced by the blank, which is a generalization of every CG).

Insertion: In a negative context, any graph u may be replaced by a specialization of u; in particular, any graph may be inserted (i.e., it may replace the blank).

Iteration: If a graph u occurs in a context c, another copy of u may be drawn in the same context c or in any context nested in c.

Deiteration: Any graph u that could have been derived by iteration may be erased.

Equivalence: Any equivalence rule (copy, simplify, or double negation) may be performed on any graph or subgraph in any context.

2.3 Conceptual Graphs from a Formal Point of View

In some sense the system of CGs is not fixed, but open-minded. It is designed to be used in fields like software specification and modelling, knowledge representation, natural language generation and information extraction, and these fields have to cope with problems of implementational, mathematical, linguistic, and even philosophical nature. Sowa addresses many of these problems in his landmark work [Sowa, 1984]. Due to the complexity of the system of CGs, it is nearly impossible and perhaps not even desirable to consider the overall system as an approach to formal logic. Consequently, all works including Sowa's which link CGs and formal logic only cover specific fragments of CGs. Anyhow, we have seen that one goal of CGs is to provide a humanly readable form of FOL.

In this section we scrutinize Sowa's CGs from the viewpoint of formal logic. It has to be acknowledged that in his works, Sowa provides core ideas for a formal elaboration of CGs. In the strict and narrow framework of mathematical logic, these core ideas are still ambigious and lack mathematical preciseness. This is the main reason why several authors elaborated fragments of CGs in minute mathematical detail. Before we come to an overview of these elaborations in the next section, this section first describes in which respects Sowa's ideas have to be refined from a mathematical point of view.

2.3.1 Sowa's Syntax

First of all, Sowa wanted to provide at least for a core of CGs a precise definition, termed *abstract syntax*, which was intended to result in an ISO-standard (see [Sowa, 1999, 2000b]). In the working draft [Sowa, 2000b] for the definition he writes: 'Informally, a CG is a structure of concepts and conceptual relations where every arc links a concept node [a concept box] and a conceptual relation node [a relation oval]. Formally, the abstract syntax specifies CGs as mathematical structures without making any commitments to any concrete notation or implementation.' The definition of the abstract syntax is not a mathematical definition, but it is written in common English. Furthermore, the definitions are incomplete, and many aspects of CGs that should be covered by the definition are only explained in the comments. So the first step for a formal elaboration of CGs has to provide mathematical definition for fragments of CGs. This is usually done by means of mathematical graph theory, as it is described in Section 2.4.1.

2.3.2 Sowa's Definition of Φ

Next we consider Sowa's definition of the Φ-operator, which is the "classical" approach to equip CGs with semantics. The fallacies in Sowa's definition of Φ have been addressed by several authors, for example, by Chein and Mugnier [Chein and Mugnier, 1992], Wermelinger [Wermelinger, 1995], and Dau (the author of this chapter) [Dau, 2003a].

Wermelinger writes in [Wermelinger, 1995]) that 'Sowa's original definition of the mapping (Φ) is incomplete, incorrect, inconsistent, and unintuitive, and the proof system is incomplete, too.' Although this judgment is formulated very harshly, it has somewhat to be acknowledged from a mathematical point of view. First of all, Wermelinger criticizes that the universal type ⊤, although it has a special meaning, is not treated in a special way. This can be considered a minor gap, which can be easily fixed. Another minor gap is that Sowa translates the blank graph into (), which is not a well-formed formula. More importantly, Wermelinger reveals that Sowa's algorithm to translate CGs into FOL formulas is not sufficiently specified. Depending on the the order in which graphs in a given context are translated, the algorithm yields different formulas that are not semantically equivalent, but the algorithm does not impose such an order. This is a surely a more serious gap.

Finally we consider how Sowa translates contexts into formulas. As CGs go beyond FOL, Sowa writes that the Φ-operator 'is only defined for features [of CGs] that can be represented in predicate calculus' [Sowa, 1992]. As Sowa does not clearly specify what features he talks about, different formal elaborations of CGs in fact cover different features. For example, different kinds of quantifiers are addressed. More important is the handling of nestings on formal elaborations. In [Sowa, 1992], Sowa provides the following graph for the proposition "Tom believes that Mary wants to marry a sailor" (this graph is more complicated than the first graph we provided for the same proposition, because Sowa explicates in this graph the linguistic roles of the verbs):

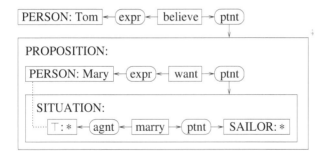

Afterward, he translates this graph to the following formula:

$$\exists x.Person(Tom) \wedge believe(x) \wedge expr(x, Tom) \wedge ptnt(x,$$
$$\exists y.\exists z.PERSON(Mary) \wedge want(y) \wedge SITUATION(z) \wedge$$
$$expr(y, Mary) \wedge ptnt(y, z) \wedge descr(z,$$
$$\exists u.\exists v.(marry(u) \wedge SAILOR(v) \wedge agnt(u, Mary) \wedge ptnt(u, v))))))$$

In this formula, whole formulas occur as arguments of relations, namely, of *ptnt* (patient) and *descr* (description), which Sowa uses to express nestings in FOL. But using formulas as arguments in relations is a feature beyond FOL, so this translation cannot be considered as translation to FOL. This gap in handling nestings is fixed by different formal elaborations in different ways.

2.3.3   Sowa's Understanding of Negation

We have seen that in CGs, negations are implemented as special contexts, namely as contexts of type "Proposition," linked to a relation oval labeled with "neg." Recall that contexts are used to draw meta-level propositions on CGs. Usually, negation is understood as a *logical* operator, and its properties have to be captured by the semantics and by any calculus. On the other hand, in CGs negation changes to a *meta-level* operator. This is stated quite explicit by Sowa in [Sowa, 2001]: 'To support FOL, the only necessary meta-level relation is *negation*.' Expressing negation as a meta-level operator might lead to difficulties in the formal treatment. To discuss this, we consider the following two CGs G_1 and G_2.

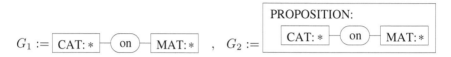

The meaning of the first graph is well known: It is "a cat is on a mat." The meaning of the second graph is, strictly speaking, "there exists a proposition, which states that a cat is on a mat." Quoting a proposition changes its character: There is a crucial difference between asserting and quoting a proposition, between using and mentioning a linguistic item. In particular these two graphs have different meanings. Now we consider the graph below:

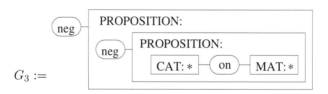

G_3 contains a double negation. In the usual understanding of negation as a logical operator, G_1 and G_3 are equivalent. On the other hand, in the CG understanding, G_3 formalizes a statement on a statment (first nesting) on a statement (second nesting), i.e., we have a meta-meta-statement. These points of view are conflicting. For this reason, formal elaborations deviate from Sowa's meta-level understanding of negation and implement negation on the logical level.

2.3.4   Sowa's Calculus

An example of Sowa's rules for reasoning with CGs has been given in Section 2.2.5. We have already seen that in different writings of Sowa, the set of rules is varied. Moreover, the rules are only informally described in common English. To obtain a precise and non-ambiguous understanding on how the rules act on CGs, a formal definition for the rules is needed.

Sowa's rules are basically Peirce's rules for existential graphs being adapted to the notations of CGs. Anyhow, first there are some crucial syntactical differences between existential graphs and CGs. In existential graphs, both existential quantification and identity are expressed by the same syntactical device, the so-called *line of identity*. In CGs, we have separate devices for these two logical functions: Existential quantification is expressed by the generic marker, and identity is expressed by coreference links. Moreover, in CGs, we have names for constants, a syntactical device that is missing in existential graphs. This leads to a higher expressiveness of CGs. The syntactical differences and the higher expressiveness have to be reflected by the calculus. To some extent, Sowa provides rules that cover these differences, but it is not clear whether these rules are sufficient. Sowa does neither prove the soundness nor the completeness of his set of rules for CGs.

In fact, it may be doubted that Sowa's rules are complete. A proof that Sowa's rules are not complete cannot be provided due to missing formal definitions of the syntax, semantics, and calculus for CGs. Anyhow, we provide a few examples of entailments between graphs, which are unlikely to be mirrored by proofs with Sowa's rules. First of all, the calculus has to cover the special meaning of the universal type $\top$. For example, it should be allowed to derive $\boxed{\top : \text{Tom}}$, if Tom is a valid object name, and it should be possible to derive $\boxed{\top : *}$, too. This seems to be impossible applying the calculus of Sowa.

Next, it seems that coreference links are not adequately treated by the calculus. For example, if the following graph on the left is a well-formed CG (which is not clear from the definition of CGs), it should be provably equivalent to the CG on the right:

$$\boxed{\text{DOG}: *} \qquad \boxed{\text{DOG}: *}$$

As identity is a transitive relation, the next two graphs should be provably equivalent, too:

$$\boxed{\text{DOG}: *} \cdots \boxed{\top: \text{Snoopy}} \cdots \boxed{\text{PET}: *} \quad \text{and} \quad \boxed{\text{DOG}: *} \cdots \boxed{\top: \text{Snoopy}} \cdots \boxed{\text{PET}: *}$$

Finally, given the graph

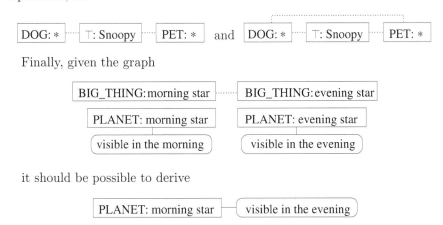

it should be possible to derive

$$\boxed{\text{PLANET: morning star}} - (\text{visible in the evening})$$

It is difficult to decide from Sowa's definition of the calculus whether these derivations are possible. A formal elaboration of the calculus allows addressing this question, and together with a formal elaboration of the syntax and the calculus, it is then possible to mathematically prove its soundness and completeness.

2.4   The General Approach for Formal Logic with CGs

Before we discuss different classes of CGs in more detail, this section provides a general overview of how formal logic can be carried out by means of CGs, and introduces some core notations.

Roughly speaking, the scrutiny of this chapter follows the common (model-theoretic) layered approach of formal logic, which can be outlined by the terms syntax, semantics, and calculus. That is, first CGs are introduced as purely syntactical structures. Particularly, the concept boxes and relation ovals in graphs are labeled with *names* for concepts and relations. These syntactical structures carry *per se* no meaning: They gain their meaning when some sort of formal semantics is defined. Thus after defining the syntax, the semantics is formally elaborated as well. Finally reasoning facilities, which are sound and complete w.r.t. the semantics, are provided.

Following this approach is not the sole possibility to formalize CGs. Particularly Wille and some scholars are interested in CGs as an extension of FCA for allowing judgments and conclusions in a *semantical manner*. That is, he does not strictly separate syntax and semantics of CGs. In his approach, he does not assign names to the concept boxes and relation ovals of CGs, but already (formal) objects, (formal) concepts and relations of a given power context family instead. That is, each CG is closely linked to a given power context family. Conclusions in Wille's approach are not drawn with the hep of inference rules of some adequate calculus, but via an algebraic investigation of the graphs. The approach following symbolic logic is called SYNTACTICAL or LOGICAL approach, the approach of Wille is called SEMANTICAL or ALGEBRAIC APPROACH. To some extent, the syntactical approach corresponds to the open-world paradigm, whereas the semantical approach correspond to the closed-world paradigm. This chapter focuses on the syntactical approach. For the semantical approach, please see Chapter 6.

Even for the syntactical approach to CGs, unfortunately there exists a variety of different formalizations with different notations. This chapter provides some references to these different notations, but due to space limitations, it does not provide the complete definitions. Instead, only the most convenient among the different formalizations is given. A more comprehensive comparison of the different formal approaches to SCGs can be found in [Aubert et al.,

2006a].

2.4.1 Syntax

The starting point in fixing the syntax is defining a set of names. In literature, there are different terms used for the set of names. The CG standard [Sowa, 1999] refers to this set as a *type hierarchy*. Authors like Prediger, Klinger, and Dau use the term *alphabet*. Other authors, e.g., Chein, Mugnier, Baget, Kerdiles, Simonet use *vocabulary* or *support* as well. Following the common notations of symbolic logic and knowledge representation, we will use the term VOCABULARY. Basically, a vocabulary consists of three ordered sets of the names for the objects, concepts (the types), and relations. Again, different authors use different notations here. The core notation of a vocabulary is fixed in the following definition.

DEFINITION 2.1 Vocabulary   *A* VOCABULARY *is a triple* $\mathcal{V} := (\mathcal{O}, \mathcal{C}, \mathcal{R})$ *of finite and pairwise disjoint sets* $\mathcal{O}$, $\mathcal{C}$, *and* $\mathcal{R}$. *The elements of* $\mathcal{O}$ *are called* OBJECT NAMES *or* INDIVIDUAL MARKERS. *The elements of* $\mathcal{C}$ *are called* CONCEPT NAMES. *The elements of* $\mathcal{R}$ *are called* RELATION NAMES.

Besides the object names, we assume to have a further sign "∗*." called the* GENERIC MARKER. *We assume to have a special concept name* ⊤, *called the* UNIVERSAL CONCEPT. *To each relation name* $R \in \mathcal{R}$, *we assign its* ARITY $ar(R)$ *(*$ar(R) \in \{1, 2, 3, \ldots\}$*). We set* $\mathcal{R}_n := \{R \in \mathcal{R} \mid ar(R) = n\}$.

Finally, the sets of object, concept, and relation names are ordered:

- *We have an order* $\leq_{\mathcal{O}}$ *on the set* $\mathcal{O} \cup \{*\}$, *with* $O \leq_{\mathcal{O}} *$ *for each* $O \in \mathcal{O}$, *and different elements of* $\mathcal{O}$ *are incomparable,*

- *We have an order* $\leq_{\mathcal{C}}$ *on the set* $\mathcal{C}$, *where* ⊤ *is the greatest element, and*

- *We have an order* $\leq_{\mathcal{R}}$ *on the set* $\mathcal{R}$, *where relation names with different arities are incomparable.*

"Conceptual graphs are also graphs" is the title of a 1995 report of Chein and Mugnier [Chein and Mugnier, 1995], nicely describing the fundamental approach to a formalization of CGs by means of mathematical graph theory. This approach has already been addressed by Sowa, who suggests formalizing CGs as labeled bipartite graphs (for this reason, Sowa sometimes speaks of concept *nodes* and relation *nodes*). The first formal definitions and theoretical results for SCGs have been provided in the landmark work [Chein and Mugnier, 1992]. In this approach, the underlying structure of a CG is a multi-bipartite graph (C, R, E), where C and R are disjoint sets of vertices, and E is a multiset, where each element $e \in E$ is a pair (c, r) with $c \in C$ and $r \in R$. Concept boxes of CGs correspond to concept vertices $c \in C$, relation ovals correspond to relation vertices $r \in R$, and the arcs of CGs correspond to edges $e \in E$. So each edge links a concept vertex (aka node) and a relation vertex (aka node). The bipartite graph (C, R, E) is augmented with additional mappings, which capture the full syntax of CGs. Labeling functions are provided that assign to each concept vertex a concept name (the type of the concept box) and an object name (the referent of the concept box), and to each relation vertex a relation name. The n edges incident to a given relation vertex r are labeled with $1, \ldots, n$ to indicate the order of the argument of the relation of r. Coreference links are modelled as a equivalence relation on C.

Another possibility is to formalize CGs as directed multi-hypergraphs. This approach has been first suggested by Wille in [Wille, 1997]. In this approach, the underlying structure of CGs is a directed multi-hypergraph (V, E, ν), where V is a set of vertices, E is a set of edges, and $\nu : E \to \bigcup_{k \in \mathbb{N}_0} V^k$ is a mapping. Now concept boxes correspond to vertices $v \in V$, and relation ovals correspond to edges $e \in E$. Note that, in contrast to the multi-bipartite approach, in this approach the order of the arguments of a relation has already been formalized. To ease the notation, in later works the term *directed multi-hypergraph* has been replaced by RELATIONAL GRAPH. Again, relational graphs have to be augmented with labelling functions to capture the object, concept, and relation names concept boxes resp. relation ovals are labeled with.

Roughly speaking, the "Montpellier-school," i.e., Chein and Mugnier and the scholars, follow the approach on bipartite graphs, and the "Darmstadt-school," i.e., Wille and his scholars, follow the relational graph approach. As multi-bipartite graphs, including the labelling function that captures the order of the arguments for the relations, and relational graphs can be mutually transformed into each other, it is a mere matter of taste which of these two approaches is taken. Anyhow, Baget (from the Montpellier-school) slightly advocates in [Baget, 2003] the relational graph approach for computational reasons. This chapter uses the relational graph approach.

It should be noted that the diagrams we presented in Section 2.2 are diagrammatic representations of an underlying abstract syntax. Already Sowa emphasizes in several places that the notation of CGs has to be independent of their diagrammatic representation. So, in contrast to symbolic notations of formal logic, we have two layers: A precisely defined *abstract syntax*, which is independent of any diagrammatic or topological properties of CGs, and the informally given diagrammatic representations of the abstract syntax. This might seem obvious, but is has to be acknowledged that some authors working on other versions of diagrammatic logic miss this important distinction. A thorough discussion of this issue can be found in [Howse et al., 2002, Dau, 2004].

To distinguish the more open computerscience-based theory of CGs to the formal mathematical theory, Wille coined the term *concept graph* for the mathematizations of CGs. Although this distinction is sensible, the use of these two different terms might lead to confusion, and it is not adopted by the whole CG community. For this reason, the term *concept graph* is, despite for referencing, not used in this chapter.

Unfortunately, even if we consider only SCGs, nearly all works differ in significant details. Care has to be taken on how in different works the labelling functions for object names and concept names is modelled, and how and coreference links are implemented. Sometimes the universal type ⊤ is required in the vocabulary, sometimes it is not. Most works assign *one* object name or the generic marker as referent to concept vertices, but some authors (for example, Klinger and Prediger) allow for *sets* of object names instead. Most works assign a *single* concept name to a concept vertex, but some works allow for a *set* of concept names, which is interpreted as the conjunction of the concept names (e.g., [Baget, 2003, Chein and Mugnier, 2004]). Coreference links are by some authors modelled as equivalence relations (e.g., Chein and Mugnier, or Prediger), sometimes as special edges, labeled with a name for identity (e.g., [Dau, 2003a,b, Baget, 1999]). Often coreference is allowed only between *generic* concept vertices (e.g., Chein/Mugnier, Prediger), sometimes it is allowed between arbitrary concept vertices (e.g., Dau). Some of these differences do not change the expressiveness of the system that is considered (for example, for SCGs, it does not make a difference whether we allow only single object names or sets of object names as referents), some of these difference can (slightly) change the expressiveness (for example, whether we allow coreference only between generic concept boxes or between arbitrary concept boxes).

2.4.2   Semantics

In literature, we basically find three kinds of semantics:

- The translation of CGs to FOL formulas by means of Φ.

- The evaluation of CGs in Tarski-style interpretations.

- The evaluation of CGs in power context families.

As mentioned in Section 2.2.4, Φ is the "classical" semantics for CGs. To some extent, it is not a "real" semantics, but a translation from some syntactical structures i.e., graphs to other syntactical structures i.e., formulas. As formulas of FOL have in turn a precisely defined semantics based on Tarski-style interpretations, Φ can be understood to provide an "indirect" evaluation of CGs in Tarski-style interpretations. The semantics via Φ is sometimes called LOGICAL SEMANTICS [Aubert et al., 2006a]. Unsurprisingly, some authors provide a *direct* evaluation of CG in Tarski-style interpretations, i.e., an EXTENSIONAL SEMANTICS. Finally, some authors (the authors of the Darmstadt-school) provide a called CONTEXTUAL SEMANTICS, where CGs are evaluated in power context families, which are contextual interpretations based on Formal Concept Analysis. For an introduction into FCA, please see Chapter 1.

All three types of semantics have to reflect the orders of concept names and relations names of the given vocabulary, and the special meaning of the universal concept as well. In the following three paragraphs, let $C_1, C_2 \in \mathcal{C}$ be arbitrary concept names with $C_1 \leq_{\mathcal{C}} C_2$ and $R_1, R_2 \in \mathcal{R}_n$ be arbitrary relation names with $R_1 \leq_{\mathcal{R}} R_2$.

For Φ, Sowa proposed to translate the orders into FOL-formulas. For two concepts C_1, C_2, the corresponding formula is $\forall x : C_1(x) \rightarrow C_2(x)$. Similarly for the relation names R_1, R_2, the corresponding formula is $\forall x_1, \ldots \forall x_n : R_1(x_1, \ldots, x_n) \rightarrow R_2(x_1, \ldots, x_n)$. Finally, the meaning of $\top$ is reflected by $\forall x : \top(x)$. The (finite) set of these formulas is denoted $\Phi(\mathcal{V})$.

An EXTENSIONAL INTERPRETATION is a pair (D, I), where D is a nonempty set, the UNIVERSE (OF DISCOURSE), and I is an INTERPRETATION FUNCTION, which maps object names to elements of D, concept names to subsets of D, and relation names of arity n to subsets of D^n. The special properties of the vocabulary are covered by additionally requiring that I is order-preserving, i.e., for two C_1, C_2 it holds $I(C_1) \subseteq I(C_2)$, and similarly for R_1, R_2 it holds $I(R_1) \subseteq I(R_2)$. Moreover, we require that $I(\top) = D$ holds.

Analogously to extensional interpretations, a CONTEXTUAL INTERPRETA-TION is a pair $(\vec{\mathbb{K}}, \lambda)$, where now $\vec{\mathbb{K}} = (\mathbb{K}_0, \mathbb{K}_1, \ldots)$ is a power context family and λ is a mapping that maps object names to elements of G_0, concept names to formal concepts of $\mathbb{K}_0$ and n-ary relation names to formal concepts of $\mathbb{K}_n$. Clearly, λ is the contextual counterpart of the extensional interpretation function I. Again, λ has to be order preserving, i.e., for C_1, C_2 it holds $\lambda(C_1) \leq \lambda(C_2)$, and for R_1, R_2 it holds $\lambda(R_1) \leq \lambda(R_2)$. Moreover, we require that $\top$ is mapped to the top concept of $\mathbb{K}_0$.

Given a CG G and an extensional or contextual interpretation, G is then EVALUATED in the interpretation to true or false. A precise definition of the evaluation depends of the kind and definition of the graph and the kind of the interpretation and is omitted here due to space limitations. If G evaluates to true in the interpretation, the interpretation is called an (extensional resp. contextual) MODEL of G, and we write $(D, I) \models G$ resp. $(\vec{K}, \lambda) \models G$. Finally, if we have a set of CGs $\mathcal{G}$ and a CG G such that whenever we have an interpretation that is a model for all CGs of $\mathcal{G}$, then it is a model for G as well, we say that $\mathcal{G}$ (SEMANTICALLY) ENTAILS G, and we write $\mathcal{G} \models G$. If $\mathcal{G}$ consists only of a single graph H, we write $H \models G$.

Contextual interpretations take benefit only of the order of the formal concepts of the power context family; its lattice structure is not used. The order of formal concepts in turn is based on the set-theoretical inclusion of the *extents* of formal concepts. For this reason, it is possible to assign to each contextual interpretation $(\vec{K}, \lambda)$ an extensional interpretation (D, I) such that we have $(\vec{K}, \lambda) \models G \iff (D, I) \models G$. Vice versa, we can assign to each extensional interpretation (D, I) a contextual interpretation $(\vec{K}, \lambda)$ such that we again have $(D, I) \models G \iff (\vec{K}, \lambda) \models G$. In the light of this observation, extensional and contextual interpretations can be considered equivalent. The details of this transformation can be found in [Dau, 2003a] or [Aubert et al., 2006a].

2.4.3 Reasoning Facilities

In literature, we find different kinds of reasoning facilities for CGs. The most important ones are projections, and sets of rules based on graph transformations. A thorough discussion of these kinds of reasoning facilities goes beyond the scope of this chapter; only an overview will be provided. Anyhow, some more details will be given in the next section. For all the following reasoning facilities, in the papers or treatises where they are described it has been proven that they are sound and complete.

Projections are graph-homomorphisms that respect the orderings of the names the vertices (and edges, if the relational graph approach is taken) are labeled with.[3] They are applied to SCGs. Roughly speaking, the information in an SCG G is the collection of the atomar information in the singleton graphs and star graphs the graph G is composed of, thus if we have a projection $\Pi : G_1 \to G_2$, then each piece of information in G_1 can be found in G_2 as well; i.e., G_2 contains all the information of G_1 (maybe more), which in turn is equivalent to $G_2 \models G_1$. Anyhow, projections have to cope with syntactically different, but semantically equivalent graphs, and with coreference links. More information on projections can be found in Section 2.5.1.

A set of graph transformations is the form of reasoning that is closest to the calculi for symbolic formalizations of logic. A graph transformation is a rule that transforms a graph G_1 into a graph G_2, and a proof is simply a sequence of CGs, where each CG in the sequence is obtained from its predecessor by one of the transformations. We can roughly distinguish between *simple* and *complex* graph transformations. A simple transformation transforms G_1 into G_2 by modifying small parts of G_1. Examples are the erasure of a relation edge, the merging of two concept vertices, which are coreferent (with some constraints for the concept names of these vertices), or the generalization of a object or concept name in a concept vertex resp. the generalization of a relation name in a relation edge. Complex graph transformations in turn modify a whole subgraph of a given CG. Examples are the erasure of a subgraph or the iteration (making a copy) of a subgraph. Complex transformations are in most cases inspired by Peirce's rules for existential graphs.

Roughly speaking, calculi for CGs that include full negation are based on the complete set of Peirce's rules, augmented with some simple graph transformation rules that are needed to cover the syntactical differences between existential graphs and CGs. The informally given calculus of Sowa is an example for such a calculus. A formally elaborated example is provided in [Dau, 2003a]. Calculi for CGs without full negation mostly or even totally consist of simple graph transformations. Examples for SCGs can be found in [Chein and Mugnier, 1992, 1995, Prediger, 1998a,b, Mugnier, 2000, Dau, 2003b], for

---

[3]The entailment relation between RDF graphs is similar to projections. See [Baget, 2004, 2005b, Dau, 2006] for papers where CG results are transferred to RDF.

CGs that allow for negation of concepts and relations in [Klinger, 2005], and for nested CGs without negation in [Prediger, 1998a, 2000].

Besides projections and graph transformations, other forms of reasoning facilities have been explored as well. First the notion of STANDARD MODELS has to be mentioned. The basic idea of standard models is to assign to a CG G a corresponding model $\mathcal{M}^G$ in which exactly the information represented by G is encoded. Once this is done, there are two possibilities to check whether a CG G_1 entails a CG G_2. First, we can check whether G_2 evaluates to true in the standard model $\mathcal{M}^{G_1}$ of G_1. The second is to check whether the information encoded in the standard model $\mathcal{M}^{G_1}$ of G_1 can be found in the standard model $\mathcal{M}^{G_2}$ of G_2. This can be verified by a meaning-preserving mapping from $\mathcal{M}^{G_2}$ to $\mathcal{M}^{G_1}$. This mapping is, roughly speaking, the semantical counterpart of the syntactical projection between G_2 and G_1. Standard models cannot be constructed for CGs that allow expressing of disjunction or full negation. They have been used for SCGs [Prediger, 1998a,b, Dau, 2003b] and CGs which allow for negation of concepts and relations [Klinger, 2001, 2002, 2005].

Other approaches to employ reasoning facilities can be based on tableaux algorithms (see [Kerdiles, 2001] for an tableaux algorithm for CGs with full negation), or resolution-like calculi combined with projections [Mugnier and Salvat, 1996, Baget and Salvat, 2006].

2.5 Different Forms of Conceptual Graphs

This section provides an overview of different formalizations of CGs that encompass some sound and complete reasoning facilities. This overview is not intended to be comprehensive. Instead, a (subjective) selection of important works is provided.

2.5.1 Simple Conceptual Graphs

Probably the most prominent and best investigated fragment of CGs is the class of simple conceptual graphs (SCGs). Nonetheless, as has already been mentioned in Section 2.4.1, there does exist a variety of different formalizations and even different understandings of SCGs. Basically, a SCG is a CG without contexts, which – due to Sowa's modelling of negation as a special context– excludes (full) negation. Anyhow, there are at least two finer subdivisions of SCGs: We will distinguish between NON-EXISTENTIAL and EXISTENTIAL SCGs, that is, SCGs without or with generic markers, and between SCGs without or with coreference links.

First we provide a possible formalization of SCGs by means of directed multi-hypergraphs, close to the definition in [Dau, 2003b].

DEFINITION 2.2 Simple Conceptual Graphs   A SIMPLE CON-
CEPTUAL GRAPH OVER $\mathcal{V}$ (WITHOUT COREFERENCE LINKS) *is a structure*
$G := (V, E, \nu, \kappa, \rho)$ *where*

- V *and* E *are finite sets of* (CONCEPT) VERTICES *and* (RELATION) EDGES,

- $\nu : E \to \bigcup_{k \in \mathbb{N}} V^k$ *is a mapping (we write* $|e| = k$ *for* $\nu(e) \in V^k$),

- $\kappa : V \cup E \to \mathcal{C} \cup \mathcal{R}$ *is a mapping such that* $\kappa(V) \subseteq \mathcal{C}$, $\kappa(E) \subseteq \mathcal{R}$, *and*
  all $e \in E$ *with* $|e| = n$ *satisfy* $\kappa(e) \in \mathcal{R}_n$, *and*

- $\rho : V \to \mathcal{O} \cup \{*\}$ *is a mapping.*[4]

The vertices v *with* $\rho(v) = *$ *are called* GENERIC VERTICES, *the vertices* v *with*
$\rho(v) \in \mathcal{C}$ *are called* OBJECT VERTICES. *We set* $V^* := \{v \in V \mid \rho(v) = *\}$.

A SIMPLE CONCEPTUAL GRAPH OVER $\mathcal{V}$ WITH COREFERENCE LINKS *is a*
structure $G := (V, E, \nu, \kappa, \rho, \Theta)$, *where* $(V, E, \nu, \kappa, \rho)$ *is a simple conceptual*
graph over $\mathcal{V}$ *and* Θ *is an equivalence relation on* V^*. *Two vertices* $v, w \in V^*$
with $v\Theta w$ *are said to be* COREFERENT.

Of course, the equivalence relation Θ can be canonically extended to V by
additionally setting $v\Theta w$ for two non-generic vertices v, w with $\rho(v) = \rho(w)$
(in some works, i.e., [Mugnier, 2000], the authors speak then of *coidentical*
instead of coreferent vertices). Note that with this understanding, even in
SCGs without coreference links, we can have different coreferent vertices, i.e.,
two object vertices labeled with the same object name.

Projections The notion of projections as a reasoning facility has already
been proposed by Sowa in [Sowa, 1984]. In [Chein and Mugnier, 1992], it has
been pointed out that projections are indeed graph-homomorphisms. They
respect the orders of the names, as well as coreference links. For the graphs
of Definition 2.2, they can be formally be defined as follows:

DEFINITION 2.3 Projection   Let $G_i := (V_i, E_i, \nu_i, \kappa_i, \rho_i, \Theta_i)$ $(i = 1, 2)$
be two SCGs. A PROJECTION *from* G_1 *to* G_2 *is a mapping* $\Pi : V_1 \cup E_1 \to$
$V_2 \cup E_2$, *which maps vertices to vertices and edges to edges, such that*

- $\nu_2(\Pi(e)) = (\Pi(v_1), \dots, \Pi(v_n))$ *for each edge* $e \in E_1$ *for which* $\nu_1(e) = (v_1, \dots, n_n)$,

- $\kappa_2(\Pi(x)) \leq \kappa_1(x)$ *for each* $x \in V_1 \cup E_1$,

- $\rho_2(\Pi(v)) \leq \rho_1(v)$ *for each* $x \in V_1$, *and*

- $v\Theta_1 w \Rightarrow \Pi(v)\Theta_2\Pi(w)$ *for all* $v, w \in V_1$

---

[4]The letter ρ is is intended to remind that ρ (rho) maps vertices to their referents.

Sowa argues that projection is sound, but does not prove its completeness. The soundness and completeness has been proven for the first time in [Chein and Mugnier, 1992]. But for the completeness, an additional constraint is needed. A SCG (without or with coreference links) is said to be in normal form, if no different concept vertices are coreferent. We then have for a vocabulary $\mathcal{V}$ and two SCGs G, H, where H is in normal form:

$$\Phi(\mathcal{V}), \Phi(H) \models \Phi(G) \quad \Longleftrightarrow \quad \text{there exists a projection } \Pi : G \to H \quad (2.1)$$

So projection relies on the target graph being normalized (this restriction has not been pointed out in [Chein and Mugnier, 1992]). A simple example for normal forms is given below. H is the normal form of G. Both G and H have the same meaning, but there exists only a projection from G to H, but not vice versa.

$$G := \boxed{\text{C: a}} \overset{1}{-} \bigcirc{R} \overset{2}{-} \boxed{\text{C: a}} \qquad H := \boxed{\text{C: a}} \overset{1}{\underset{2}{=}} \bigcirc{R}$$

A normal form of a graph can be easily computed by merging concept vertices that are coreferent. But this depends on whether it is possible to compute a concept name corresponding to the conjunction of the concept names of the vertices that are merged. This is not always possible. For this reason, in some works it is required that coreferent vertices are labeled with the same concept name, and other works consider SCG where vertices can be labeled with *sets* of concept names.

In [Chein and Mugnier, 2004], Chein and Mugnier consider existential SCGs with coreference links, and they overcome the need for normalforms in projections. To do so, they firstly allow concept vertices to be labeled with sets of concept names, which are interpreted as their conjunction. More importantly, they introduce the notion of COREF-PROJECTIONS. In their approach, for each vertex v they consider its COREFERENCE CLASS, that is, the set of all vertices that are coreferent to v. Coref-projections then map coreference classes to coreference classes instead of vertices to vertices.

Simonet investigates a different approach to eliminate the need for normalforms. In [Simonet, 1998, Chein et al., 1998], she provides a new translation Ψ from CGs to FOL, which covers not only the meaning, but also the syntactical structures of CGs. With this new semantics, projection is sound and complete without requiring that the target graph is normalized.

Calculi Sowa provided in [Sowa, 1984] a set of simple transformation rules for SCGs, but this set is not complete [Mugnier, 2000]. Several works have proposed different set of rules for SCGs. Prediger considers in [Prediger, 1998a,b] SCGs with a contextual semantics (in [Prediger, 1998b], she considers so-called *relational* power context families, the full contextual approach is provided in [Prediger, 1998a]). Prediger generally allows references with more than one object name. In [Prediger, 1998b], she considers non-existential SCGs without coreference links. For these graphs, a sound and complete calculus composed of ten simple graph transformations (except one rule that allows makeing a copy of the whole graph) is provided. In her (German) PhD-thesis [Prediger, 1998a], this calculus is extended to existential SCGs with coreference links between generic concept vertices, which then has eleven rules.

Chein and Mugnier provide for existential SCGs, both for graphs without and with coreference links, in [Chein and Mugnier, 1995, Mugnier, 2000] sets of sound and complete graph transformation rules. Similar to Prediger, coreference links only act on generic concept vertices. The calculus for SCGs without coreference links consists of five rules, called "relation duplicate," "unrestrict," "detach," "substract" (which corresponds to Peirce's erasure-rule), and "join." When coreference links are added, detach is replaced by a rule called "co-identical split," and a new rule "coreference deletion," which allows the erasure of coreference links, is added.

In [Dau, 2003b] existential SCGs are considered, where coreference links are modelled by special relation edges labeled with "=" (a similar approach can be found in [Baget, 1999], where the properties of the identity relation are modelled by special *rule graphs*, which are described in the next paragraph). In contrast to other approaches, coreference links between arbitrary concept vertices are allowed, even if both are non-generic. The set of rules he considers are exactly thse rules of his calculus for concept graphs with negation (see Section 2.5.4) that can be applied to CGs without negation. They are: erasure, iteration, deiteration, generalization, isomorphism, exchanging references, merging two vertices, splitting a vertex, ⊤-erasure, ⊤-insertion, identity-erasure and identity-insertion. This calculus could probably be simplified.

An "if-then"-statement cannot *ad hoc* be expressed with *one* SCG. Anyhow, Baget invented in [Baget, 1999] SCGs where the set of vertices is additionally divided into HYPOTHESIS VERTICES and CONCLUSION VERTICES. Moreover, FRONTIER-VERTICES are those hypothesis vertices having a conclusion-vertex as neighbor. Now, if such a rule-graph G_R is given, and a graph G such that the hypothesis-part of G_R projects to G. then the conclusion-part of G_R can be added to G, where the frontier-vertices of the conclusion are merged with the corresponding vertices in the projection of the hypothesis. With this idea, a knowledge base of SCGs can contain not only facts stated by graphs, but if-then rules as well (but as projection is involved, this approach still relies on the normalization of some graphs). Baget's idea can even be extended. For example, positive and negative constraints can be expressed with similarly augmented SCGs. This leads to a framework called *Constrained Derivation Model* [Mugnier, 2000, Baget and Mugnier, 2002b]. Within this model, already SCGs become a quite powerful knowledge representation language.[5]

---

[5]Besides the already mentioned contributions, the papers show that finding a projection between SCGs is equivalent to the problem to the conjunctive-query containment known from relational databases, and the constraint-satisfaction-problem, thus linking reasoning with SCGs to other formalisms, and it is argued for CGs instead of formulas both from a

Standard Models Standard models as a means for reasoning with CGs have been introduced by Prediger in [Prediger, 1998a,b]. In [Prediger, 1998b], she assigns to each non-existential SCG G without coreference links a standard model $\mathcal{M}^G$, which is a formal context (O^G, A^G, I^G) with an additional mapping λ^G, which maps n-ary relation names to n-ary relations over O, and she proves

$$G_1 \models G_2 \text{ iff } G_2 \text{ is valid in } \mathcal{M}^{G_1} \tag{2.2}$$

$$G_1 \models G_2 \text{ iff } I^{G_1} \supseteq I^{G_2} \text{ and } \lambda_{\mathcal{R}}^{G_1}(R) \supseteq \lambda_{\mathcal{R}}^{G_2}(R) \text{ for all } R \in \mathcal{R} \tag{2.3}$$

The so-called RELATIONAL CONTEXTS of [Prediger, 1998b] are in [Prediger, 1998a] replaced by power context families. For existential SCGs with coreference links, Prediger proves in [Prediger, 1998a] the direction "⇒" of (2.2), but does not provide a counterpart of (2.3).

In [Dau, 2003b] a translation into standard models is provided as well. It is proven that $G_1 \models G_2$ if and only if $\mathcal{M}^{G_1}$ entails $\mathcal{M}^{G_2}$. The entailment between $\mathcal{M}^{G_1}$ and $\mathcal{M}^{G_2}$ is expressed via the existence of a meaning-preserving mapping from $\mathcal{M}^{G_2}$ to $\mathcal{M}^{G_1}$, which is a sort of projection between models. Now, even in the existential case (with a slightly higher expressiveness of [Dau, 2003b] compared to [Prediger, 1998a]), we have in [Dau, 2003b] results that correspond to (2.2) and (2.3). So reasoning with SCGs fully carries over to reasoning with models. Consequently, Dau provides moreover a sound and complete set of four rules (removing an element, doubling an element, exchanging attributes, restricting the incidence relation) for entailment between models.

Other approaches Based on the PhD thesis of Salvat, approaches with resolution-style calculi (with implication as sole logical connective) are presented in [Mugnier and Salvat, 1996, Baget and Salvat, 2006]. Tappe investigates in [Tappe, 2000a,b] SCGs where the generic marker which represents existential quantification is replaced by the universal quantifier $\forall$. Tappe provides a contextual semantics and a sound and complete calculus.

2.5.2   Nested Conceptual Graphs

In contrast to SCGs, there does not exist a generally agreed upon semantics for nested CGs. Instead, different approaches exist. Some of them are presented in this section.

One critics given in Section 2.3 for Sowa's handling of nesting in his definition of the Φ-operator was his use of formulas as arguments in a predicate, as this goes beyond FOL. Preller et al. [Preller et al., 1998] have used this approach to define a non-classical logics with "nested" formulas, which is similar to the logic of contexts of [McCarthy, 1993]. Then they define a Gentzenstyle sequent formal system with nested formulas and prove the soundness

---

knowledge modelling and a computational point of view.

and completeness of this system with respect to the projection in simple and nested graph models.

A different approach has been taken by Simonet in [Simonet, 1998, Chein et al., 1998]. She inductively defines a nested CG by assigning to each concept vertex v an additional argument $Desc(v)$, where $Desc(v)$ is either the blank graph or a nested SCG. Then she provides a new Φ-operator for nested CGs by adding to each predicate a further argument called CONTEXT ARGUMENT, which denotes the context. So, n-ary relation names in nested CGs are translated to $n+1$-ary relation names in the corresponding FOL-formulas. Finally, she extends the notion of projection to these graphs. To obtain soundness and completeness of projection, the target graph has to be in normal form again, but the notion of normal forms has extended to cover nestings. First of all, each nesting of the target graph, understood as a SCG, has to be in normal form. Moreover, if v, v' are coreferent concept vertices, then $Desc(v')$ must be an *exact* copy $Desc(v)$, i.e., $Desc(v')$ is a copy $Desc(v)$ such that each generic vertex appearing in $Desc(v)$ is coreferent to its copy in $Desc(v')$. This is called STRONGLY NORMAL. Assumed that we have two graphs G, H where H is a strongly normal nested CG, we have a corresponding result to (2.1). The notion of being strongly normal can be weakened to a notion of inductively defined k-normality, which is, roughly speaking, the strong normality condition up the a level k of the depth of nestings. We omit the details due to space limitations. Finally, similar to her new semantics Ψ for SCGs, she provides a corresponding semantics Ψ for nested CGs, where projection is sound and complete without any need for normalizing the target graph.

In [Prediger, 2000], which is basically an excerpt of her PhD thesis [Prediger, 1998a], Prediger elaborates nested CGs similar to her approach to SCG. Her syntax is based on directed multi-hypergraphs, augmented with a mapping $\rho : V \to \mathfrak{P}(V)$ that models the nesting of graphs. First of all, she considers only non-existential nested CGs. In contrast to her approach to SCGs, she now assigns a single object name as referent to each vertex. She equips these graphs with a situation-based contextual semantics, which is now based on *triadic* power context families. Again, similar as for SCGs, she is able to assign to each nested CG a corresponding standard model. Her result (2.2) for SCGs can be lifted to the nested case, but she does not provide a counterpart for (2.3). Finally, she provides a set of eight inference rules (double a vertex, delete an isolated, non-complex vertex, double an edge, delete an edge, exchange a concept name, exchange a relation name, join vertices with equal references in the same nesting, copy a sub-concept graph into an equally referenced nesting) that is sound and complete.

2.5.3 Conceptual Graphs with Atomic Negation

Negation in CGs can be expressed on two levels. We can restrict negation to concept names and relation names only, or where can allow negating of whole subgraphs. The first kind of negation is called ATOMIC NEGATION and covered in this section, the second kind of negation is called FULL NEGATION and will be covered in the next section. These two levels lead to a significant difference in the expressiveness of the systems. When atomic negation is added to SCGs, the resulting system is still decidable, whereas full negation leads to the equivalence to full FOL, thus undecidability. Anyhow, if negation on the atomic level is implemented, to some extent it is possible to express disjunctive information. To provide an example from [Kerdiles, 2001]: Take the formulas $f_1 := P(a) \wedge R(a, b) \wedge R(b, c) \wedge \neg P(c)$ and $f_2 := \exists x \exists y : P(x) \wedge R(x, y) \wedge \neg P(y)$. Then f_1 entails f_2. To see this, we have to consider two cases: If $P(b)$ holds in an interpretation, then f_2 holds for $x := b$ and $y = c$, and if $\neg P(b)$ holds in an interpretation, then f_2 holds for $x := a$ and $y := b$. The conceptual graph formalism does not allow to explicitly express this kind of disjunctive information. The two approaches we present in this section have to cope with this problem, caused by the law of the excluded middle.

Kerdiles considers in [Kerdiles, 2001] the negation of relation names only. Syntactically, there are two approaches to implement atomic negation. It is either possible to augment a given vocabulary $\mathcal{V}$ by adding for each relation name $R \in \mathcal{R}$ a new relation name R^- to the vocabulary (which of course denotes the negation of R), or we can augment a SCG $G := (V, E, \nu, \kappa, \rho)$ with an additional function $sign : E \rightarrow \{+, -\}$, which divides the set of edges into positive and negative edges. These graphs are called POLARIZED. Obviously, both approaches work equally well and can mutually transformed into each other, which is even formally proven by Kerdiles in [Kerdiles, 2001]. Kerdiles extends the projections of SCGs to SCGs with atomic negation: In addition to the usual constraints, projection has to respect $sign$ as well. As usual, the target graph must be normalized. Besides this requirement, an additional constraint is put on the source graph: It has to be DISCRIMINATED. This basically means that the graph is the juxtaposition of two graphs, where one graphs contains only positive, and the other graph only negative relation edges. Not every SCG with atomic negation can be converted into an equivalent graph that is discriminated, thus Kerdiles considers only the fragment of graphs where this is possible. He then shows that for two SCGs with atomic negation G, H where G is in normal form and H is discriminated, that $G \models H$ holds iff there exists a projection from H to G.

The approach of Kerdiles has be extended in [Leclère and Mugnier, 2006, Mugnier and Leclère, 2007]. Similar to Kerdiles, Mugnier, and Leclère consider polarized CGs, where only relation names can be negated (but as they argue, concept names can be replaced by unary relation names, so this does not lead to a loss of expressiveness). They consider three different logical frameworks for interpreting atomic negation: The framework where they rely on the closed-world assumption CWA, the framework of classical logic (particularly, here the open-world assumption OWA is employed), and the framework of intuitionistic logic (again based on OWA). For the framework of CWA, to check whether a polarized CG H projects to a polarized CG G can be reduced to check whether the positive part of H projects (in the classical understanding) to some sort of completition of G. For the framework of intuitionistic logic, they argue that projection precisely captures the underlying intuition of projection (recall that the law of excluded middle, which causes the problem we mentioned at the beginning of this section, does not hold in intuitionistic logic). Finally, for checking whether a polarized CG H projects to a polarized CG G in the classical logic setting, on has to check whether H projects to maybe exponentially many completitions of G. The first part of [Leclère and Mugnier, 2006] elaborates this in detail for polarized CGs without coreference links. In the second part, coreference is added to polarized CGs by means of special edges. As identity is a relation that can be negated, they add a second kind of special edges that are used to express non-identity as well. Assuming that the unique name assumption hold, they then show that their results can be lifted to polarized graphs with identity and non-identity.

Klinger takes in [Klinger, 2001, 2002, 2005] a different approach. First of all, on the syntactical level, she assigns to each concept vertex two kinds of referents: a positive and a negative one (similar to Prediger, she allows sets of object names instead of single object names). To illustrate this, a simple example is provided below.

As it can be seen, Klinger uses variables instead of the generic marker for existential quantification. The referents on the left-hand side of the bar are the positive referents, the referents on the right-hand side of the bar are the negative referents. The intuitive meaning of this graph is "there exists a dog (namely, x), Yoyo is not a dog, Tim is a man, there exists something (namely y), which is not a man, x likes Tim, and Yoyo does not like y."

First of all, Klinger equips her graphs with a contextual semantics. If one demands that for a formal concept (A, B), the extent of the negation of (A, B) is the set-theoretic complement $G \backslash A$ of A, then this set does not have to be the extent of a concept itself. For this reason, Klinger does not assign only formal concepts to concepts and relation names. Instead, in her papers [Klinger, 2001, 2002], she considers contextual interpretations where *semi*-concepts of the contextual interpretations are assigned to the names of the vocabulary. In her PhD thesis [Klinger, 2005], she assigns the more general *proto*-concepts to the names (each semiconcept is a protoconcept; see Chapter 6).

In [Klinger, 2001], Klinger introduces so-called simple semiconcept graphs, which provide no means of quantification. Nonetheless, coreference can be expressed by relation edges labeled with the name "=." She provides the syntax and semantics of simple semiconcept graphs, and she assigns a standard model to each of them. [Klinger, 2002] is the extension of [Klinger, 2001] to the existential case (where existential quantification is expressed with variables, not with the generic marker). Besides the definitions and their discussion, no results are presented in [Klinger, 2002] and [Klinger, 2001].

Here full elaboration of CGs with atomic negation can be found in her PhD thesis [Klinger, 2005], where so-called (non-existential and existential) protoconcept graphs are fully elaborated. One of her main concerns is the satisfiability of graphs. As it is possible to express self-contradictory proposition with atomic negation, Klinger provides criteria when a protoconcept is satisfiable. For non-existential protoconcept graphs, standard models are introduced, and a result similar to (2.2) for these graphs is proven. The notion of standard models is dismissed for the existential case. Both for non-existential and existential protoconcept graphs, a sound and complete calculus is provided. For the non-existential case, it consists of 14 rules, that are besides one rule (iterating a subgraph) simple transformation rules. For the existential case, exactly these rules are adopted, and an additional rule, the "existential rule," is added. Unfortunately, this rule is quite complex: Its definition takes nearly three pages, and it is therefore hard to understand. As Klinger writes, the rule is needed in order to obtain the completeness for the existential case, but it "certainly is not a very practical one."

2.5.4 Conceptual Graphs with Full Negation

After discussing CGs with atomic negation, Kerdiles considers in [Kerdiles, 2001] full negation as well. He inductively defines CGs with negation, where each negated subgraph is a simple graph in normal form. Coreference is still employed as an equivalence relation on the generic vertices, with some syntactical restrictions. Kerdiles did not want to use coreference edges as representation of identity, but simply to mark the multiple occurrences of a variable in traditional textual languages. So if one wants to express identity (or its negation), this should be done in his system by using additional relation edges labeled with "=" (personal communication). For this system of CGs with (a slightly restricted) full negation, Kerdiles provides an extensional semantics, a translation of the graphs to FOL, and a sound and complete tableau algorithm.

A comprehensive approach to add full negation to CGs is provided by Dau in [Dau, 2000, 2001, 2003a]. In [Dau, 2000], Dau argues that to express full negation to CGs, a new syntactical entity to express negation has to be added, and he suggests using the negation ovals of Peirce's existential graphs (which are called "cuts" by Peirce for this purpose (recall the discussion of Section 2.3.3 – which is in fact taken from [Dau, 2003a] – that negation should not be implemented as meta-level operation, which is syntactically expressed by means of special contexts). Moreover, he argues that for CGs with full negation, it is convenient to express identity by relation edges labeled with "=" (between arbitrary concept vertices). The resulting graphs are called *concept graphs with cuts* (CGwCs). In the diagrammatic representation of CGwCs, to distinguish negation ovals from relation ovals, they are drawn bold. In contrast to the use of contexts, there is no reason to draw negation ovals only around "complete" subgraphs. Below, two CGwCs are depicted. The left one is the CGwC-counterpart of Sowa's CG provided in Section 2.2.3, the right one is a CGwC expressing "there are at least two things."

Dau's approach is in [Dau, 2000] carried out for non-existential CGwCs. A contextual semantics for CGwCs is provided, and a sound and complete calculus with twelve rules, based on Peirce's calculus for existential graphs, is given.

The full approach to both non-existential and existential CGwCs can be found in [Dau, 2003a], [Dau, 2001] is a rough overview of [Dau, 2003a] without proofs. For CGwCs, [Dau, 2003a] provides the syntax, both an extensional and contextual semantics, including mappings between these semantics, a sound and complete calculus, and translations from CGwCs to FOL (the Φ-operator) and vice versa, from FOL to CGwCs (called Ψ). The rules of the calculus are erasure, insertion, iteration, deiteration, double cuts, generalization, specialization, isomorphism, exchanging references, merging two vertices, splitting a vertex, $\top$-erasure, $\top$-insertion, identity-erasure, and identity-insertion. The first five rules correspond to Peirce's rules for existential graphs. The rules generalization and specialization capture the order of the names. The rules $\top$-erasure and $\top$-insertion capture the special properties of the concept name $\top$. The rules exchanging references, identity-erasure, and identity-insertion capture some special properties of the relation name $=$, and the rules merging two vertices, splitting a vertex rely on both the properties of $\top$ and $=$.

2.6 Further Literature

Conceptual graphs are a still a field of active research. This can be easily seen from this chapter, which basically cites research papers and several PhD theses. This section mentions some more important research works.

Both the Darmstadt school and the Montpellier school have generated a number of important PhD theses. The latter have not been cited yet, as they are due to the regulations in France published in French. Nonetheless, this section will give an rough overview of the (concerning the aim of this chapter) the most important ones. First of all, the Montpellier school developed a CG-based reasoner called "CoGITo." The first version has been designed and implemented in C++ by Ollivier Haemmerle in his thesis "Une plate-forme de développement de logiciels sur les graphes conceptuels" (1995).[6] He studied also the relationships between CGs and relational databases and developed a question/answering system. The thesis of Michel Leclère "Les connaissances du niveau terminologique du modéle des graphes conceptuels: construction et exploitation" (1995) focuses on type definitions, i.e., contraction, expansion and classification of simple conceptual graphs with defined types. The thesis of Eric Salvat "Raisonner avec des opérations de graphes: graphes conceptuels et règles de d'inférence" (1997) is devoted to the processing of "if-then" rules. He defined and implemented in CoGITo sound and complete graph-based forward and backward chaining mechanisms for 'if SCG then SCG' rules. Finally, the thesis of Jean-Francois Baget "Représenter des connaissanceset raisonner avec des hypergraphes: de la projection à la dérivation sous con-straintes" (2001) contains a lot of results. For example, he defined algorithms for projection based on constraint processing techniques, and he defined a family of formalisms based on SCGs, rules, and constraints and studied their complexity.

We already mentioned a pioneering paper of Wermelinger. To the best of my knowledge, he dedicated his MSc thesis "Teoria Básica das Estruturas Conceptuais" ('Basic Conceptual Structures Theory', New University of Lisbon, Portugal, 1995) to a formalization of conceptual graphs. But unfortunately, the thesis is published in Portugese.

Heaton's English thesis "Goal Driven Theorem Proving Using Conceptual Graphs and Peirce Logic" (Loughborough University of Technology, UK, 1994) focuses on conceptual graphs and existential graphs. He augments CG with constructs of Peirce's existential graphs, including Peirce's negation ovals. See http://myweb.tiscali.co.uk/openworld/index.html for an treatise summarizing Heaton's research on conceptual graphs so far.

Finally, I'd like to mention that in [Baader et al., 1998, 1999, Kerdiles, 2001], the decidable, so-called "guarded fragment" of conceptual graphs is investigated.

---

[6]CoGITo has later been extended to typed nested CGs and be renamed CogiTaNT. See http://cogitant.sourceforge.net/

Part II

Tools

Chapter 3

Software Tools for Formal Concept Analysis

Peter Becker

Nudgee QLD 4014, Australia

Joachim Hereth Correia

Technische Universität Dresden, Germany

3.1 Introduction

In this chapter we present some typical programs implementing methods from Formal Concept Analysis, each with some historic context and an overview of its main features. The idea is to give an impression of what the programs can do and how they are used; we are not attempting to replace the user manuals. The exercises for this chapter will apply these tools to concrete problems.

We start with three that are aimed at practitioners of Formal Concept Analysis: CONEXP is a generic, full-featured tool for editing formal contexts and creating line diagrams; CONIMP allows the investigation of implications in data, even in the case that the investigator does not have complete knowledge of the domain; and the TOSCANAJ suite allows the creation of conceptual information systems, which can be deployed for users without experience with Formal Concept Analysis.

The fourth program is an example of an application using Formal Concept Analysis but aimed at a much more general audience: DOCCO allows querying a local store of documents, supporting the retrieval of searched items by line diagrams. We conclude with a brief list of related programs and tools that we could not present in detail here.

3.2   The CONEXP **Concept Explorer**

CONEXP[1] is a graphical tool for Formal Concept Analysis that implements several aspects of Concept Analysis. Serhiy Yevtushenko wrote it in Java and presented it originally in [Yevtushenko, 2000]. For some time it was developed in parallel with the tools from the TOSCANAJ suite and developers exchanged many ideas. Therefore some aspects like manipulating the context or the diagram or the highlighting features are similar. We will describe them in detail in this section but the information applies to TOSCANAJ and its editors, too.

By default, CONEXP starts with a view on the context editor, which can be controlled by mouse or keyboard. The default context has 15 objects and attributes that can be renamed if desired.

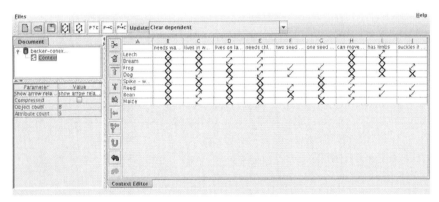

FIGURE 3.1:   CONEXP overview: menus and toolbars at top, overview and available parameters left, context editor right.

Figure 3.1 shows a screenshot of CONEXP after editing the context table to represent the data from the example "Living Beings and Water" (see Example 1 in [Ganter and Wille, 1999a]). In the vertical toolbar left from the context are buttons to add an object, purify, or reduce the objects. Objects can be removed by right-clicking on the object name and selecting "Remove object(s)." Other buttons fulfill the corresponding function for attributes, removing an attribute can be done in an analogous manner. Also, the user can reduce the whole context (not only objects or attributes) or transpose the context (exchanging objects and attributes) with a button click. Finally, most edit actions can be undone and redone, also with buttons from this toolbar.

---

[1]http://conexp.sourceforge.net/

When the mouse hovers over any button CONEXP gives a short description of the function in a tool tip.

On the left hand CONEXP provides an overview of available information (in the screenshot in Figure 3.1 only the context) and available parameters and information. The value for "Show arrow relation" has been changed to "show arrow relation." CONEXP then calculates the up-, down-, and doublearrows and displays them in the context table. Any change to the incidence relation has an automatic effect on the displayed arrows.

The last two lines indicate the number of objects and attributes. These numbers are editable to quickly change the size of the context.

3.2.1 Concept Lattices

The horizontal toolbar at the top of the window provides access to basic functionality and the central features, for instance, starting a new context (and thus closing the currently edited data), opening a file or saving the current data to a file. CONEXP will warn about the possible data loss when the user selects to start a new context or to load another file after having changed some data after the last saving.

The two buttons labeled with a lattice diagram look similar but have different functions. The left one only calculates the number of formal concepts for the current context (19 in our example). This is useful if the context is rather large and one wants to estimate the usefulness of looking at the concept lattice or other of CONEXP s advanced features.

The other one of these two buttons creates a diagram of the concept lattice for the current context and shows it in the editor for lattice line diagrams. In the following we will describe CONEXP's features with regard to diagram generation and manipulation. The other functions available from the horizontal toolbar will be presented in the subsequent sections.

One of the reasons to develop CONEXP was the investigation of different algorithms to draw (semi-)automatically diagrams of concept lattices (see [Yevtushenko, 2004]). Therefore the program has many drawing and layout options and allows the choice of one of several algorithms for an automatic drawing. The user can also change some parameters for these algorithms to investigate their influence.

The diagrams of concept lattices are drawn as usually, that is, concepts are represented by circles and lines indicate the ordering of the concepts; the object and attribute labels are attached to the concepts with a line. Figure 3.2 shows CONEXP with a diagram derived from our example data. The program uses some graphical hints to provide more information to the user. In the grayscale rendering these hints are difficult to discern, therefore we will describe them in more detail.

The circles may be empty, half, or completely filled. Those where the upper half is filled blue have an attribute attached, those with the lower half filled

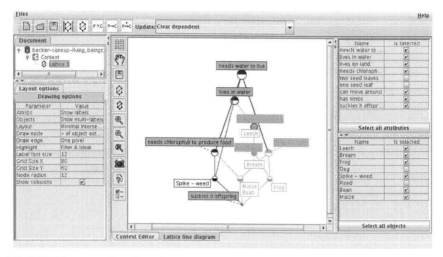

FIGURE 3.2: CONEXP lattice line diagram editor, hinting problems with the node positions and highlighting to ease reading the diagram.

black an object and those with both fillings combined naturally object–and attribute–attached.

The user can interact with the diagram by dragging nodes and labels to other places to improve the layout and the readability. These changes or the layout calculated automatically from the context can lead to wrong overlappings where a line crosses a node in the representation but the corresponding concept is not in relation to the nodes attached to the intersecting line. In such a case, the line and the node not belonging on the line are marked in red. In the grayscale version of the screenshot in Figure 3.2 these are only visible as slightly thicker lines: the node with the label "lives in water" does not belong to the line between the topmost and the one with the label "needs chlorophyll to produce food" on the left side. Likewise on the other side, the line from the topmost node to the grayed out one and the gray node with the label "Leech" are marked. These do not intersect but are too near to each other and are thus marked, too.

The gray nodes and lines belong to another feature helping the user to read and interpret the diagrams. Selecting a node with a left click, the node and all nodes above and below are rendered with blue lines and borders but all nodes incomparable to the selected node are grayed out and also all lines between these grayed out nodes. Additionally, the text in all labels belonging to objects not in the extent and to attributes not in the intent of the concept corresponding to the selected node are displayed in lighter colors.

The buttons left to the diagram allow various modes to control the interaction with the diagram. From top to bottom these are "snap to grid," "pan," "export diagram" (allowing image formats as well as formats from other FCA software, a textual and also the Graphviz[2] format), "toggle drag mode" (single concept or all concepts below the selected concept at once), "fit to screen," "zoom in," "zoom out," "no zoom," "store lattice as view" (allows to keep the current state of the diagram until the context is changed), "lattice statistics" and "save preferences." The features for view modifications work as in usual graphics programs, the diagram interaction features are better explored directly than explained here.

In the panel at the right side the user can select or deselect objects and attributes, thus restricting the context to the corresponding subcontext, as done in Figure 3.2. The diagram is then recalculated to adapt to the new context. Note that CONEXP does not remember previous changes to the diagram when changing the selection temporarily (use the "store lattice as view" feature for that).

3.2.2 Attribute Exploration

The next interesting feature of CONEXP is attribute exploration. For a long time, CONIMP (see the following section) has been the only software tool to implement this technique of Formal Concept Analysis (see Section 2.3 from [Ganter and Wille, 1999a]). In CONEXP, the user can start attribute exploration clicking the corresponding button labeled "P?C" in the top toolbar. CONEXP starts asking questions that the user has to answer. In our example the first question is "Is it true, that all objects that have attribute(s) x needs water to live?" As indeed all living beings need water, we answer this question with "yes." Please note that "objects" refer to the objects of the context in the more general sense, that is, "living beings," not only those already listed as (formal) objects in the context.

The next question is "Is it true, that when object has attribute(s) needs water to live, suckles it offspring, that is also has attribute(s) lives on land, can move around, has limbs?" Here the answer is "no." Whales need water to live and suckle their offspring, but while they can move around and have limbs (the flippers) they do not live on land. After answering "no" CONEXP presents the user with a mini-context editor with only one object. The attributes are those from the context and the crosses are set for all attributes in the premise (that is "needs water to live" and "suckles it offspring" in our example). The user can then add the lacking attributes ("lives in water." "can move around" and "has limbs"). After selecting "provide counterexample," the new object is added to the context. At any point, the user can stop the exploration.

---

[2]http://www.graphviz.org/

3.2.3   Implications and Association Rules

The two buttons in the top toolbar labeled with "P" and "C" and an arrow between them (the second having a "p" above the curly arrow) provide a list of implications and of association rules. In our example, the implication button (straight arrow) calculates a list of ten implications, with the attributes of the premise left and the attributes from the conclusion right from the arrow (==>). Of course, these are not all implications valid in the context, but only a generating base, the so-called *stem base* or *Duquenne-Guiges base*. At the beginning of each line stands the number of the implication (for easier reference) and the number of objects having all attributes of the premise (and consequently of the conclusion) in angle brackets. Some implications are true, because no object fulfills the premise (for instance no object has all the attributes "needs water to live," "lives in water," "lives on land," "can move around," and "has limbs suckles it offspring,") and therefore all objects with these attributes have also all other attributes (in our example "needs chlorophyll to produce food," "two seed leaves," and "one seed leaf"). These implications seem irritating at first and have to be interpreted with special care. Therefore CONEXP marks implications where no object fulfills the premise in red, all others in blue.

Partial implications, investigated by Luxenburger in [Luxenburger, 1991] and [Luxenburger, 1993], are a generalization of implications. Today, they are usually called *association rules* in the literature. An implication is only valid if all objects having the attributes of the premise have the attributes of the conclusion. For an association rule it is sufficient if there are "enough" objects with the attributes of the premise that also have the attributes from the conclusion. Formally, "enough" is defined as a percentage (the so-called *confidence*). For instance, we find that not all objects who "need water to live" and "can move around" also "have limbs" (the leech does not); therefore this would be wrong as an implication. But all beings except the leech with the first two attributes also have limbs. Therefore, if the required percentage is 75% or lower we accept this as an association rule in our context.

Obviously, the number of association rules can become rather large. The higher the confidence level the smaller the number of resulting rules. Besides the confidence level, there is a second parameter to reduce this number. One can restrict the presentation to only those rules where there are a certain number of objects fulfilling the premise. This is the so-called *support*. If the support is higher than 0, some implications may not appear as association rules (for instance, those that have been marked red in the list of implications). To simplify identification, CONEXP marks those association rules that are implications blue or red (as in the list of implications), all proper association rules are green.

3.3   ConImp

ConImp is the software in the realm of Formal Concept Analysis with the longest-running development history. Since its first versions in 1986 (under the name BAmn) it evolved continually to improve and to adapt to new theoretical findings. The main developer of ConImp as well as the principal researcher in the theory of concept analysis with partial knowledge is Peter Burmeister. You can download ConImp as executable for Windows and for Linux from his homepage[3] where the detailed ConImp documentation [Burmeister, 2003] can be also found.

Resulting from its early origins, ConImp has not a graphical user interface like the other software tools presented in this chapter. Its interface is textual and keyboard-controlled. As the menu structure grew organically and is today up to three levels deep, it is sometimes difficult to find the feature one looks for. In the meantime, some features of ConImp have been reimplemented in other tools like ConExp.

Therefore we will not give an overview of all features of ConImp, some of which originate from specific research interests. Instead we will focus on a feature that sets ConImp apart from other available tools of Formal Concept Analysis: attribute exploration with background implications and incomplete knowledge. We will give a step-by-step walkthrough how this works. It is best if you can repeat all steps directly with your own copy of ConImp.

As a preparation you should save a context prepared with ConExp in the CXT format that was introduced by ConImp originally (Choose Files->Save as and then select "Cxt Files" as file type). Then start ConImp (under Linux, execute "./conimp" in the directory where the executable is stored, under Windows just double-click the downloaded executable). After starting ConImp you are presented with a screen giving some information about development and usage restrictions, and after a keystroke the main menu appears. With "L" you can load a context you previously saved. After an additional keystroke you can enter the directory where the file is saved (keep it empty if the context is in the same directory as the executable) and then the file name. ConImp will ask you if the path derived from this information is correct and after your confirmation it tries to read the file.

You will then see the main menu expanded by additional options. In the lower part of the menu you find several options to calculate some information about the context. ConImp does not automatically compute all these because this may take some time for large contexts and is not always necessary. On the other hand it means you have to specifically start the calculation of the information you want.

---

[3]http://www.mathematik.tu-darmstadt.de/~burmeister/

As said above, we will not investigate all features of CONIMP in detail. Feel free to explore the features and possibilities. Be careful to have a backup of interesting data because CONIMP is changing some data (for instance, the context) during some of the operations (it usually warns about this before doing so, but you should copy your data beforehand).

We will now focus on the attribute exploration with incomplete knowledge. From the main menu we choose "I – Implication menu" and are presented with several options. You may at a later stage save or retrieve discovered implications. We continue with "I – Next menu."

3.3.1   Background Implications

In this menu you can start the most important tasks for attribute exploration. In particular, you can enter *background implications*: these are valid implications that you do not want to be asked about during the exploration process. A typical example of background implications is the knowledge that some attributes cannot appear together. In our example about living beings and water, no plant can have one seed leaf and two seed leaves at once. To enter such knowledge we type "H – Editor of background implications" and then "E – Extend present list," which opens CONIMP's context editor. Due to the textual interface the presentation of attributes is restricted; the first nine letters are shown in columns. This usually suffices to identify what attribute is meant but in contexts with long attribute names, this display can be difficult to read.

For our example, we mark the two columns with the seed leaf attributes with the letter "P" for premise and either type in every other column "C" for conclusion or just *, which is a shortcut (it prints "K" instead of "C" in the remaining columns but it means the same thing). The semantics entering such a premise is the same as the implications with no extent in CONEXP's implications list: There are no objects having all attributes; therefore saying that every object that has the two seed leaf attributes also has all other attributes is equivalent to saying that this cannot happen.

Another piece of background knowledge is that beings with limbs can move. To enter this second implication we move the cursor downward, which replaces the line with the previous implication by an empty one. We enter "P" in the column for "has limbs" and "C" in the column for "can move." We might also consider the column "needs water." Probably every being with limbs that moves also needs water to live, therefore it would be correct to add "need water" to the conclusion. But this is not necessary, even without this additional attribute in the conclusion the implication is valid. Having the background implications entered we press the escape key to return to the implication menu.

3.3.2 Attribute Exploration with Partial Knowledge

Of course, entering background implication is not necessary to start the attribute exploration, but it helps to concentrate on important questions. We now select "I – interactive calculation." If you want to recall later the questions posed by CONIMP and your answers select "Y" when asked about the protocol file and enter directory and filename. The first question (or "proposal of implication") consists of an empty premise and "needs wat(er)" as conclusion, that is, if every living being needs water to live. The answer is yes. We also see other possible answers. The second one is no, then we have to provide a counter-example. This is similar to CONEXP. The interesting feature of CONIMP are the next two possibilities that you can choose if you are not certain. Usually you should then select the fourth "O – I am uncertain and want optimal strategy," which tries to complete the exploration in as few steps as possible (that is optimally). This part is not yet extensively tested, therefore the old-style exploration with uncertainty (third answer), which is reliable but not always optimal is still available.

The second question reads "needs water, **suckles it offspring**" ⇒ "**lives on land**, can move, **has limbs**." All attributes considered in the question are displayed but you have only to investigate those printed in bold. The others follow from already known implications. As in the CONEXP exploration, "whale" is a counter-example that we can enter after answering no. In the following context editor we enter crosses and blanks according to the attributes. All question marks have to be replaced by crosses or blanks. We can answer the next proposal "needs water, **suckles it offspring**" ⇒ "can move, **has limbs**" with yes since all mammals have limbs. Now, CONIMP informs us that it derived from the entered information to accept the implication "needs water, has limbs" ⇒ "can move."

The fourth proposal is "needs water, **one seed leaf**" ⇒ "**needs chlorophyll**." Most plants (including those with one seed leaf) need chlorophyll, so the answer could be yes. On the other hand, there are plants that do not need chlorophyll, for instance, fungi. But because we are uncertain if some of those plants that do not need chlorophyll nevertheless have one seed leaf, we answer "O." This is the first time we include uncertainty in the exploration process, therefore CONIMP informs us that it switches to 3-valued logic. Then CONIMP continues with the next proposal. What it has done in the background is to add a fictitious counter-example for the uncertain implication.

3.3.3 Transforming Contexts with Partial Knowledge to Normal Contexts

We will stop the exploration at this point and return using the escape key twice to return to the the main menu. Using "S – Save context" we save the current data under a new file name. CONIMP stores the context in the cxt format, but due to the uncertainty the other programs have difficulties interpreting the data. CONEXP complains about a syntax error due to the question marks in the last line of the file.

CONIMP allows updating the data to transform the context into a usual two-valued one (this of course erases some of the information). We select "I" twice to return to the implication menu and then press the return key, which calculates the valid implications in the context (respecting the partial knowledge represented by question marks), with return and escape we go to the main menu and select "U – Transform" and choose "A" to update the context. The question marks are replaced by crosses if there is any implication in the context that implies this cross ("C" is similar but restricts itself to implications that do not include fictitious objects). We return with escape to the main menu and save the context under a new name. Now we can process the context with CONEXP or other software; for instance, look at the diagram of the now extended context.

3.4 The TOSCANAJ **Suite**

The history of TOSCANAJ starts in the early 1990s, when a first prototypical version of its predecessor TOSCANA was developed at the Darmstadt University of Technology. The first officially released version, TOSCANA 2, was implemented shortly afterward based on a Formal Concept Analysis library written by Frank Vogt ([Vogt, 1996]), who also authored the editor used together with TOSCANA: ANACONDA. The two programs were first presented at the CeBIT'93 and the Graph Drawing Conference 1994 ([Vogt and Wille, 1995]).

After years of partly academic and partly commercial development of TOSCANA, the KVO workgroup around Peter Eklund started reimplementing a new version of the former C++ product in Java in 2001 — the old version had accumulated a number of issues over time; most noticeably it had been built on an outdated and by then only partially supported GUI library. The result of this is the TOSCANAJ project, which is hosted as open source project on Sourceforge[4]. The TOSCANAJ suite of tools consists of TOSCANAJ itself and two editors: ELBA for Conceptual Information Systems on relational databases and SIENA for creating Conceptual Information Systems without requiring a database engine.

---

[4]http://toscanaj.sourceforge.net

3.4.1 The TOSCANAJ **Workflow**

TOSCANAJ displays a predefined set of diagrams that has been created with one of the editor components. This is then combined with information from a data source to create a "TOSCANA System" or more generally a "Conceptual Information System" (often abbreviated to "Conceptual System"). The predefined set of diagrams together with the information of how to retrieve and display the data is called the "schema" for the Conceptual System or "Conceptual Schema" for short.

Crucial to understanding a TOSCANA system is the notion of concrete and realized scales. A *concrete scale* is a formal context whose objects are queries against the underlying many-valued context, usually expressed in terms of the attributes of the many-valued context. When the user requests a diagram, TOSCANAJ will evaluate the queries and display the diagram defined for the concrete scale, but replacing the queries with the actual objects of the many-valued context that match each query. The result of this replacement is called the *realized scale* (see [Kollewe et al., 1994] for details and a more formal description). In a typical system on top of a relational database the objects of the concrete scale will be SQL clauses and the objects of the realized scale will be objects from the database.

The distinction between concrete and realized scales allows TOSCANA systems to be split into two phases: first creating the Conceptual Schema and later displaying the diagrams of the realized scales. Creating the schema is done by a person trained in creating such systems who is called a "conceptual system engineer," usually in cooperation with an expert from the domain to be analyzed. Analyzing the realized scales with regard to the current set of data is then a task that any domain expert can do — the separation of the two phases allows using the Conceptual System without much experience in Formal Concept Analysis and it also allows efficient updates of the diagrams on current data as long as the structure of the concept lattice is not changed.

Figure 3.3 shows an overview of the different roles and tools involved in this workflow. The following subsections will discuss the different tools individually, starting with TOSCANAJ itself.

3.4.2 TOSCANAJ **as Conceptual Information System Browser**

Figure 3.4 shows the TOSCANAJ program during a normal usage session. A Conceptual Schema was loaded in the XML format used by the TOSCANAJ suite (CSX) and a number of diagrams have been selected to be used during the analysis. The top left list shows all available diagrams, the list below shows the diagrams selected for analysis and the main part shows the current diagram, calculated based on the data found in the underlying data source and the current context defined by the user.

During an analysis session a TOSCANAJ user can use a number of functions that are described in the following.

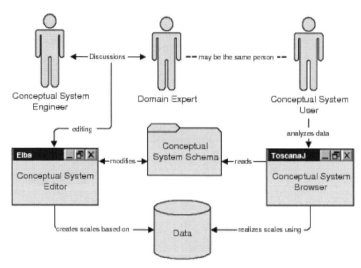

FIGURE 3.3:   The TOSCANAJ workflow.

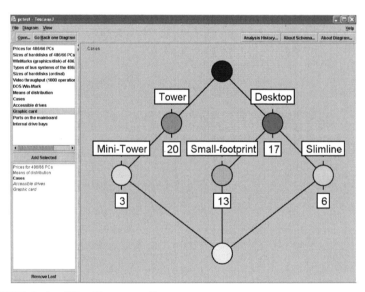

FIGURE 3.4:   The TOSCANAJ program.

Displaying Basic Diagrams

A diagram to display can be selected from the list of all diagrams available. It will be added to the list of diagrams selected for analysis and if no diagram is currently shown it will be displayed. Selected diagrams will be visited in the selected order after each drill-down operation (see below). With the default settings the color used in the nodes denotes how large a concept's extent is compared to the set of all objects in the diagram. This means darker nodes represent concepts with large extents — a feature useful in larger diagrams where it can help to draw attention to unusually small or large extents.

In a diagram the user can move the labels but not the nodes. While technically possible, changing the diagram layout is not offered to avoid users creating confusing layouts; diagram layout is considered the sole responsibility of the conceptual system engineer. The basic positions of the labels are defined in the schema but since the label contents and in consequence the label sizes can vary, the user needs to be able to move the labels.

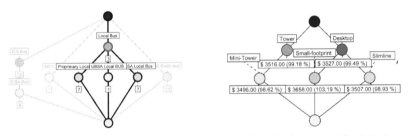

FIGURE 3.5:   (a) Highlighting a node; (b) System-specific label contents.

Another operation available to the user is "highlighting": when a node is clicked upon, the node and its upset and downset are highlighted, together with all labels relevant for the concept's intent and extent. This is similar to the feature in CONEXP and shown in Figure 3.5a.

Changing Label Contents

The contents of the object labels can be changed to at least three different options: the absolute number of objects, the relative number of objects (as percentage of all objects in the diagram) and the list of the objects; either of these can be displayed based on the extent of a concept or the object contingent (the objects that are in the extent of a concept but not in any extent of a proper subconcept).

Specific TOSCANAJ systems may have additional queries defined that can be used in the labels such as the one shown in Figure 3.5b: here the average price for the object set is shown, including how it relates to the average of all objects in the diagram (e.g., the average price of a small-footprint PC is 103.19% of the overall average price for PCs). It is also possible to move the labels around, although this change is not persistent across sessions: it will be reset when the Conceptual Schema is opened again.

Opening a More Detailed View

In many TOSCANAJ systems a more detailed view can be opened as popup window. This can be a database report (if the label is an aggregate) or a detailed view of a single object (if the label is a list of objects). Each TOSCANAJ system can define multiples of these views customized for the application. A variety of tools are available to implement such views, although the most common one is to use a component that renders HTML with the database information in it. Additional viewing components can be easily provided for specific system, e.g., to provide graphical representations.

Drilling-Down

The set of objects displayed in further diagrams can be restricted to an extent or object contingent of a concept, an operation called "drill-down" (also known as "filtering" or "zooming"). This is done by double-clicking on a node in a diagram: if a further diagram is available in the list of selected diagrams the view will change to that diagram with the set of objects restricted to the selected concept's extent or object contingent, depending on a setting in the menu.

Nesting Diagrams

Diagrams can be nested, which means in each node of the first diagram (the "outer diagram") a copy of the second diagram is displayed (the "inner diagrams"). The objects for the inner diagrams are restricted to the object contingent of the concept that is represented by the node of the outer diagram. Nesting can be turned on or off; if turned on the program will use the current diagram as outer diagram and the following one as inner diagram as long as there is still another diagram selected.

The overall diagram uses a formal context defined as apposition of the two original contexts: the attributes are joined into one set, the object set is kept (it has to be the same on the two inputs) and an incidence relation is used such that an object is incident with an attribute if it did so in either of the inputs (cf. [Ganter and Wille, 1999a]). Nesting can be particularly useful to get an impression of how two different feature sets correlate on a given set of data.

Figure 3.6 shows a nested diagram displayed by TOSCANAJ.

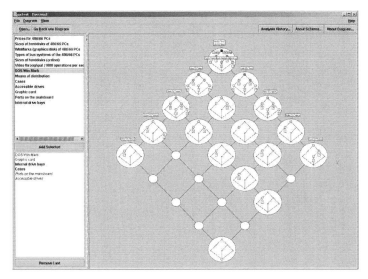

FIGURE 3.6: TOSCANAJ displaying a nested line diagram.

Printing and Diagram Export

Diagrams can be printed or exported into a number of graphic formats. The available options for graphic export vary depending on the installation: TOSCANAJ itself supports JPG and PNG export, but a number of plugins are available that add support for formats such as PDF or SVG. The size of the export defaults to the current size on screen, but a user can define a larger size to gain better quality for the bitmap formats such as JPG or PNG — with PDF or SVG this option will just change the default size when displaying the diagram since these formats allow scaling a graphic anyway.

3.4.3 Conceptual OLAP

The main application area for TOSCANA systems is the analysis of relational databases, an approach referred to as "Conceptual Online Analytical Processing" (Conceptual OLAP). The term "online" is used here to distinguish the approach from batch processing: in online processing the analysis is done live on the database, while batch processing uses exported data from a database.

TOSCANAJ has a number of optimizations to allow this approach on larger databases. Most noticeably TOSCANAJ will never query any lists of realized objects unless explicitly asked by the user. To display a diagram showing either object contingent or extent sizes, TOSCANAJ will query only the number of objects in each object contingent. The extent sizes are never queried but inferred as the sum of all object contingent sizes in the downset of a concept. This approach reduces the database load to at most one SQL COUNT query per node as long as the user does not request a list. Since a COUNT query is usually very efficient even for large data, simple line diagrams do scale very well with the database size.

In other respects TOSCANAJ has to rely on the setup of the database. If the database is not efficient in evaluating the query clauses used by the conceptual system engineer, performance will not be good on larger data sets. A typical example are range queries such as a "between size 3 and 4" — databases are often not very efficient in answering these, special index structures ("spatial indexes") are required to make these efficient.

If the data is fixed and the database can be modified by the conceptual system engineer, another option is to create specific index structures for the diagrams — this is the most efficient way and ELBA offers a tool to do this: for each diagram a new column is created that contains a number as identifier denoting the object contingent an object belongs to (this is always well defined). These columns are then indexed and the diagrams will contain only clauses checking these columns for particular identifiers. If the same approach is modelled into the databases using triggers or similar technology this scheme can also be used on the live data.

A typical Conceptual OLAP System will contain a number of diagrams modelling different aspects ("facets") of the data. The queries will be optimized for the data and the database engine in use and additional indexes or structural extensions might have been added into the database to optimize performance. Often the display options of TOSCANAJ are extended by defining new label contents and specialized database views as popups. These latter options still require manually editing the CSX files, although the tools ensures that these manual editions are kept during other changes.

Setting up such a Conceptual OLAP System can be quite a bit of work, but the approach allows the creation of a system that is highly optimized for ease of use and speed for the given domain; allowing the final user of the TOSCANAJ system to analyze the data efficiently without any technical knowledge.

3.4.4 Creating Conceptual OLAP Systems with ELBA

The tool with which Conceptual OLAP Systems for TOSCANAJ are created is called ELBA. Figure 3.7 shows the main window of ELBA with a diagram displayed. The look is similar to TOSCANAJ itself – in fact a lot of the components are reused, but the way they are combined is quite different: ELBA allows editing the diagram in various aspects, it allows adding new diagrams and it contains a number of features to help with setting up a Conceptual OLAP System. On the other hand ELBA does not offer features such as nesting or creating the realized scales; it solely operates on one concrete scale at a time.

ELBA is optimized for the purpose of creating Conceptual OLAP Systems; it connects to a relational database and then uses the information from that database to guide the user in creating a Conceptual OLAP System. Most

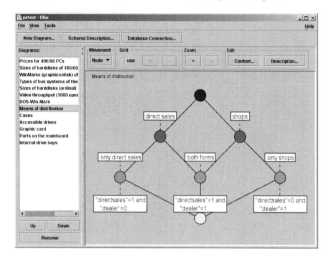

FIGURE 3.7: ELBA, the editor for Conceptual Systems on databases.

noticeably ELBA will query the database schema for the tables and columns available in the database to guide the user when creating scales.

To connect to a database the user can choose from a number of connection options, from specific file-based types such as using SQL scripts with the embedded database engine of TOSCANAJ or using MS Excel or MS Access files to the generic option of specifying a connection using the options available for a JDBC connection[5]. After the database connection is established, the user will then be asked to specify a column in one of the available database tables that is considered to contain the set of objects for the conceptual system.

After this process a new but empty Conceptual OLAP System has been established. Now diagrams can be added using a number of different assistants to create different types of scales. For example the ordinal scale generator will allow the user to define a name for the new scale (and thus diagram), a column of the database table which represents the many-valued attribute (restricted to the columns with ordinal datatypes), a direction of the scale (increasing/decreasing or both; bounds can be included or excluded in each step) and a list of bounds that will be used to define the steps of the ordinal scale. The dialog will show the current minimum, maximum, and average of the data in the database to give some direction on which values might constitute useful bounds for the scale. Once all these parts have been entered, an ordinal scale can be generated that will contain the appropriate objects and attributes of a concrete scale for the system.

In a similar fashion the other scale generators create other types of scales,

---

[5]http://java.sun.com/javase/technologies/database/

often by using information from the database to guide the user. There is a scale generator for nominal scales and another for cross-ordinal scales (the cross-product of two ordinal scales). There are also two generic scale generators: one that lets the user define the context of the concrete scale manually, the other lets the user define a list of attribute names and matching queries — a concrete scale will be generated that contains all possible combinations of those attributes.

A number of additional tools help a conceptual system engineer with specific tasks. The database content can be exported as SQL: this is useful to migrate smaller external databases into SQL script to be used with the embedded engine in TOSCANAJ, which allows creating more portable systems. The realized scales can be exported into a simple XML format, either containing only the object counts or the full lists of all realized objects for each scale. This format can be useful for specific analysis or debugging. Another tool extends the database with extra information to allow faster queries as described in 3.4.3 and it is also possible to import conceptual systems in the format used by the older TOSCANA versions and ANACONDA (CSC).

On the level of a single context the usual editing operations are available as well as export options and a consistency check that makes sure that the concrete scale is valid for the current database (all queries work and every object is matched exactly once). The diagrams can be edited by dragging nodes and labels. Dragging of the nodes can be done using different approaches: either just the dragged node is moved; or the node is moved in combination with other nodes, such as its upset or downset. The default movement works in a way that the additivity of the diagram is always ensured (c.f. [Becker, 2005c]).

3.4.5 SIENA

The third member in the TOSCANAJ suite is a tool similar to ELBA, but used to create TOSCANAJ systems that are independent of relational databases. It is called SIENA and at its core it allows editing the diagrams and contexts in the same way ELBA does — most of the GUI elements are just reused. SIENA does not offer the features relating to relational databases but instead it allows importing formal contexts from text-files in the CXT format used, e.g., by CONIMP or the object-attribute list format (OAL) used by CONEXP. Multiple contexts can be imported into one system and the diagrams will be generated for each. If the system is saved, SIENA will ensure that the object sets are the same for each diagram to allow the use of filtering and nesting in TOSCANAJ. If an object is not in a diagram, it will be added assuming that it does not match any attribute — a new top concept will be created if the former top concept has a non-empty intent.

This approach allows easily creating diagrams from contexts exported by other tools, e.g., allowing SIENA to be used as front-end for scripted generation of diagrams. It also allows creating whole TOSCANA systems by just generating a number of contexts and then importing them into one system in SIENA; a feature that has been proven to be quite useful in *ad-hoc* analysis of data not stored in relational databases or not even in the structure of a single many-valued context to start with.

SIENA is also intended to allow for the editing of many-valued context and conceptual scaling to create the diagrams similar to CERNATO ([Becker, 2001]), but this functionality has not been completed at the time of writing.

3.4.6 Further Reading

A more in-depth description of the TOSCANAJ suite can be found in [Becker and Hereth Correia, 2005]. There are also manuals available for TOSCANAJ and ELBA. [Becker, 2004] introduces another aspect of TOSCANAJ that has not been discussed here: the combination of Conceptual OLAP with quantitative approaches. [Becker, 2005c] discusses the two-step layout approach used to create the diagrams in a way that moving nodes does not easily break structure.

3.5 DOCCO

While the tools presented so far address users familiar with Formal Concept Analysis, a number of other tools exist that use Formal Concept Analysis to create programs addressing a larger audience without prior knowledge of Formal Concept Analysis. One example of such a tool is the desktop-search tool DOCCO[6] which is a spin-off of the TOSCANAJ project.

DOCCO creates a full-text index over a set of documents found on a computer's file system. That means it opens all known documents and creates a structure that can then be used to quickly find documents containing certain words. Additionally information about the files, such as a document's author or the position in the file system can be used to retrieve them. The idea is to make collections of documents easier accessible — where a user previously had to browse through a number of folders to find a document, with DOCCO finding that document can be faster if specific keywords can be provided.

There are a number of other tools that take a similar approach, for example, GOOGLE DESKTOP[7], COPERNIC DESKTOP SEARCH[8] or tools built into an op-

---

[6]http://tockit.sourceforge.net/docco/,

[7]http://desktop.google.com

[8]http://www.copernic.com

erating system such as SHERLOCK[9] in MacOS X, BEAGLE[10] on UNIX/Linux
or the Windows Indexing Service[11], which allows faster queries using the nor-
mal search interface. DOCCO distinguishes itself mostly through the Formal
Concept Analysis-based visualization; but it also differs by being Open Source
Software (true only for Beagle in the list above) and being cross-platform as
Java application.

3.5.1   Basic Usage

For a typical usage scenario consider a paper presented in a workshop on
a conference. When retrieving such a document from a collection in a file
system one has to ask if it was classified as belonging to the conference or
the workshop or both? Were acronyms or full names used for the conference
or workshop? Was the paper also classified against author(s), year, or other
facets such as the topic of the paper? How is a paper that has multiple authors
classified? What does the person retrieving the document do if the file system
wants them to choose the conference first, but they know only author and
year?

DOCCO is written under the assumption that for most document collections
on a computer many of these questions can not be answered — at least not
all the time. Instead DOCCO proposes an approach where document can be
searched by using queries that can be answered quickly and can be overspec-
ified, which means that queries can be used that do not match the target
document or even any document completely.

Take the example shown in Figure 3.8: here a query was used that contains
four keywords: `retrieval`, `dimension`, `apposition` and `lattice`. In this
example none of the documents contains all four of these words. If the query
would be submitted normally against the indexing backend, the result set
would be empty.

Here DOCCO behaves differently from other search programs but rather
follows an approach similar to what TOSCANAJ does: the query is broken
down into its parts and each part is considered a separate query against the
underlying many-valued context with the documents. A Boolean scale is
created using the query parts as objects, which then is used as concrete scale
that can be realized using the indexing engine. The resulting lattice is then
turned into the diagram shown to the user.

The fact that none of the documents matches all four keywords can be seen
by the fact that the bottom element of the lattice has no extent: its extent
is the set of all documents matching all the query parts. Nodes higher up

---

[9]http://developer.apple.com/macosx/sherlock

[10]http://beagle-project.org

[11]http://msdn2.microsoft.com/en-us/library/ms689644.aspx

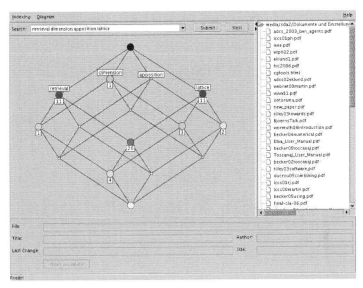

FIGURE 3.8: Example query in DOCCO.

in the diagram represent partial matches, allowing the user to identify which combinations of keywords exist and how common they are.

On an existing index this result can be produced by just entering the keywords into the search fields and submitting the search by clicking the button to the right or just hitting enter. After the diagram has been created, a node can be selected to restrict the query results to the extent of the matching concept and/or a document in the treeview on the right can be selected to show additional information on that particular document. If no node is selected, the list of all documents matching any query part will be shown — this is the extent of the top concept. Documents can be opened either by double-clicking their name or using the button at the bottom of the screen.

The indexing engine used for DOCCO is LUCENE[12] from the Apache project, written in Java. It allows storing the mapping between a set of objects ("documents") to keywords in a way that is very efficient for querying. It also comes with a query system that allows a number of operations to define more complex queries, such as Boolean operations, quotation to query phrases instead of single words and wildcards. Also queries can directed against different parts of the documents; by default the body of a document is queried, but it is also possible to query the title, author's names, or the file's name or location. A full list of the operations and fields is available in DOCCO's help system.

---

[12]http://lucene.apache.org/

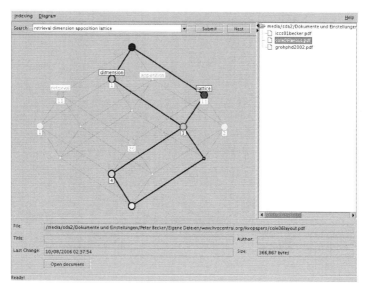

FIGURE 3.9:   A document has been selected.

3.5.2   Other Features of the User Interface

In the default settings and usage, DOCCO will show a diagram representing
a Boolean lattice with the lattice generated from the queries embedded into
it. This is similar to the distinction between concrete and realized scales
used in TOSCANA and TOSCANAJ, only that in this case we do not have
theoretical background provided through a schema, so a Boolean lattice is
used to structure the display. This approach of embedding the actual lattice
into a Boolean lattice can be changed to just displaying the actual lattice
instead. Additionally the user has the choice between seeing the contingent of
the concepts (documents matching just a given combination of query parts,
but none of the others) or the extent of the concepts (documents matching
the given combination of query parts and maybe some or all of the others).

All diagrams can be edited by dragging nodes. The way the diagram
changes matches the "Additive" mode in the TOSCANAJ tools, which means
the additivity of the diagram never gets broken. To override this behavior the
user can hold the shift key while dragging, in which case only the node under
the mouse cursor will be moved.

If queries are submitted not by using the Submit button (or by pressing
enter), but instead through the Nest button, a nested diagram will be generated by using the already existing lattice for the outer diagram and the lattice
created by the new query for the inner diagram. This feature is intended for
a structural analysis of larger document collections.

3.5.3 Indexing a Document Collection

To create an index of a document collection, DOCCO has to scan all relevant documents and process them to extract the keywords and meta-information. DOCCO is able to index a number of document types, the basic installation supports extracting information from plain text, HTML, XML, Open Office 1.x, Open Document Format and RTF files. Plugins can be installed to support Word, Excel, PowerPoint, and PDF documents.

The user can create multiple separate indexes by providing different start directories to recurse in, different mappings from file types to the document processors (e.g., ".dot" might be Word templates in one part of the file system, but GraphViz graph descriptions in another) as well as different text analyzers. The text analyzers are responsible for dropping common words such as "the" and "but" ("stopwords") as well as reducing words to their stem, so different inflections of the same word are not distinguished anymore ("stemming"). Once different indexes are created they can be activated separately or in combination to allow for different retrieval purposes.

The indexing is done in the background, so the query part of the tool is accessible while indexing is running. Additionally an option exists that allows running an update of all indexes from the command line, which can be used to run automatic updates without user interaction.

Note that requiring the user's initiative to update an index is a major difference to other tools with similar functionality. The drawback of this approach is that an index is not always up-to-date. Additionally updating requires scanning a whole section of the file system, while other tools use operating system callbacks to get notified of changes to files or folders. On the other hand letting the user request the updates means that there is no need for a constantly running system process and that the work required for indexing is done whenever the user requests it. Depending on the tool used having a constantly running indexing service can noticeably affect system performance.

3.5.4 Further Reading

DOCCO is described in [Becker, 2005a] with a similar scope to this section. Further detail about DOCCO is presented in [Becker and Cole, 2003], including references to similar work in the area of email collections. The tool itself comes with an online-help system and the Web site gives a little bit of information.

3.6   Related Work

Besides the tools presented in this chapter there are several others you can use to apply Formal Concept Analysis. In this section we hint to some of these, without claiming completeness.

Conceptual Software Library COLIBRI

For a long time the FASTCONCEPT library for Fast Concept Analysis (see [Lindig, 2000]) by Christian Lindig was the most often used and cited software library for Formal Concept Analysis software. The library and its OCaml successor have been revived in the COLIBRI project. Now there is a Java-Version,[13], too, and all of them are Open Source. They are not a tool per se, but a library providing the data structures and efficient algorithms for FCA.

Integrated Analysis Environment GALICIA

GALICIA is a project jointly developed by several universities in Canada and France. It is intended to become an integrated environment for the whole process of Formal Concept Analysis (opposed to the TOSCANAJ-workflow where several separate tools are needed). Besides features for context manipulation and lattice visualization it also includes the possibility to operate directly on lattices. Also support for so-called *relational context families* is planned to support data that is stored across several relations. On the project home-page[14] you can find a more detailed description of the software, including links to binaries, source, and a French user manual.

CAMELIS, Browser for Logical Information Systems

CAMELIS[15] is primarily developed by Sebastien Ferré and now by the Team LIS at the IRISA in France. LIS stands for Logical Information Systems, the idea to extend, for instance, file systems by logical attributes and make these available. CAMELIS is a prototype realizing such an LIS on a file system. You can therefore navigate not only by directory or file name, but also by date or other attributes. It is also possible to open bibtex-files (bibliography files used together with the typesetting system LaTeX, also often available online) and navigate these by author, year, etc.

---

[13]http://code.google.com/p/colibri-java/ also references the other two projects
[14]http://www.iro.umontreal.ca/~galicia/
[15]http://www.irisa.fr/LIS/ferre/camelis/

Fast Visualization of Relational Data with TupleWare/CASS

TupleWare[16] is a second spin-off of the ToscanaJ project. The basic idea behind TupleWare is to use operations of the relational algebra instead of conceptual scaling: a formal context is generated from an arbitrary relation, which can be imported from text files (tab-delimited, CSV) or queried from data sources such as SQL databases or RDF stores. In the basic case Tuple-Ware generates a number of diagrams from this relation by using the values of one column as object set and the values of other columns as attribute set for a formal context each. The incidence relation used is the projection of the original relation onto the two sets.

This approach allow the creation of lattice diagrams from relational data in a quick and flexible way, with most of the power coming from the query languages used (e.g., SQL for the relational databases). Since the object set is selected only once the resulting diagrams can be saved as one ToscanaJ system and nesting/filtering can be used in ToscanaJ for further analysis. As an additional feature the object and attribute sets can also be projections onto two or more columns, in which case cross-product values can be analyzed. Tupleware is described in [Becker, 2005b].

One larger application of the TupleWare approach is the CASS toolkit.[17] CASS stands for "Conceptual Analysis of Software Structure": it takes program source code and generates a number of relations from that, partly directly from the abstract syntax tree, partly by inferring higher abstractions such as a generic notion of "depends" between program parts. The resulting database is then analyzed using the TupleWare approach combined with a notion of "navigation spaces" that allow browsing between different views of the data ([Cole and Becker, 2005]).

3.7   Student Projects

3.7.1   Creating a Conceptual Information System with Elba

1. Discuss how a CIS can be set up for a data source such as the Internet Movie Database (IMDB)[18]. Address the following questions:

   - Which parts are suitable as objects?
   - Which parts are suitable as attributes?

---

[16]http://tockit.sourceforge.net/tupleware/

[17]http://tockit.wiki.sourceforge.net/CASS, a different version without TupleWare is available at http://griff.sourceforge.net/cass_project.html

[18]http://www.imdb.com

- How can a many-valued context be modelled from the relational data?
- What additional queries and views would be useful for such a CIS?

2. Identify other data sources suitable for a TOSCANAJ system. Discuss how they compare to the IMDB data.

3. Select one data source available for download and import it into a relational database.

4. Create a TOSCANAJ system based on this database using ELBA. Try to include:

  - views for the individual objects;
  - reports for object sets;
  - a specific query for the labels;
  - some online help for the system.

Note that some parts might require editing the CSX file.

3.7.2 Creating a Conceptual Information System with SIENA

1. Create a program/script that reads part of a file system and exports a number of different contexts in CXT or OAL format, including:

  - the size of the files found;
  - the owner of the files;
  - the extension of the files;
  - access rights on the files.

  Each of these aspects should be in a separate context file, but all should share the same set of input files.

2. Import the files generated into SIENA and generate diagrams. Discuss which diagrams are useful and which ones are not. Identify approaches to avoid some of the issues encountered.

3. Create a TOSCANAJ system based on multiples of these diagrams. Use TOSCANAJ to create some nested line diagrams and discuss their usefulness for this application.

Chapter 4

Efficient Computation with Conceptual Graphs

Galia Angelova

Institute for Parallel Processing, Bulgarian Academy of Sciences
25A Acad. G. Bonchev Str., 1113 Sofia, Bulgaria
galia@lml.bas.bg

4.1 Introduction

This chapter overviews an idea to treat the Simple Conceptual Graphs (SCGs) with binary conceptual relations as a regular language, defined over the support symbols. These considerations are motivated by the need to provide fast conceptual search in run-time, a challenge that is viewed as one of the knowledge-processing bottlenecks in the emerging semantic systems. Our approach suggests an internal representation of the SCGs with binary conceptual relations, which is based on Finite State Automata (FSA), and provides extraordinary run-time speed in calculations of injective projections. The choice of FSA as supporting formalism is natural since the minimal deterministic FSA are known for their operational efficiency in many kinds of applications, including compact encodings of morphological dictionaries and ultra-fast text search. Actually the suggested idea is *(i)* to encode *off-line* the knowledge base of SCGs and all injective generalizations of all the subgraphs as a single, minimal acyclic FSA, thus preparing a compact conceptual archive of all possible SCGs having injective projections onto the given knowledge base, and *(ii)* to intersect in *run-time* the projection query with the minimal FSA, finding the projection answers in linear time that depends on the query length only. This two-stage approach to the calculation of injective projection enables performing off-line the most time-consuming computations of subgraphs and their injective generalizations according to the knowledge base support. The critical point in such a scenario is to find suitable data structures, which store all resulting injective generalizations as a compact conceptual resource; fortunately the minimal acyclic FSA encode efficiently finite languages of words. To achieve our goals, we have to answer the following questions given a knowledge base of SCG with binary conceptual relations and its support:

- How to enumerate all injective generalizations of all the subgraphs in the knowledge base, keeping links to the original subgraphs for which the generalizations are computed?

- How to interpret this (finite but rather long) list of generalizations as a list of finite words over certain finite alphabet? Moreover, how to design an internal representation of all generalizations, which provides their unique off-line encoding in a way suitable for encoding of the future projection graphs in run-time – to enable easy matching of the run-time query onto the off-line pre-computed generalizations?

- How to organise efficiently all the words into a minimal FSA?

- How to implement the run-time projection calculation, to ensure maximal speed of query mapping onto the minimal FSA?

Obviously all the generalizations in a finite conceptual world form a finite set as well, but are these amounts of symbol strings manageable at all, by some feasible and scalable approach? Answering the above-listed questions requires in-depth understating of the logical and graphical nature of SCGs. This chapter summarizes the findings of our research work, which are discussed in more detail in [Angelova and Mihov, 2008], and presents the original results of a recent experiment in conceptual information compression.

4.2   Definitions

4.2.1   Support

Conceptual graphs have been proposed in [Sowa, 1984] as a knowledge representation formalism, which is founded both on logic and graph theory. The underlying ordinary graphs reflect the identity of concept types in the corresponding logical formulas: a graph node, which is common for two different graph edges, corresponds to a common argument in two different conceptual relations. Many scientists intensively studied the CG formalism and have made essential steps toward its elaboration; for instance, the notion of support was formally introduced in the paper [Chein and Mugnier, 1996]. The support fixes the background ontological knowledge and enables the representation of domain facts as separate assertions. As we consider SCGs with binary conceptual relations, in this chapter we define the support with binary conceptual relations.

Definition 1. A **support** is a 4-tuple $S = (T_C, T_R, I, \tau)$ where:

- T_C is a finite, partially ordered set of distinct concept types. The partial order defines the hierarchy of concept types: for $x, y \in T_C$, $x \leq y$ means that x is a subtype of y. We say that x is a specialization of y and y is a generalization of x; also that y subsumes x. The universal type $\top$ (top) subsumes all types in T_C. All types in T_C subsume the absurd type $\perp$ (bottom);

- T_R is a finite, partially ordered set of distinct relation types. The partial order defines the hierarchy of relation types. $T_C \cap T_R = \emptyset$. Each relation type $R \in T_R$ has arity 2 and holds between two different concept types $x, y \in T_C$ or between two distinct instances x_1, x_2 of a concept type $x \in T_C$. A pair of concept types $(c_{1R}, c_{2R}) \in T_C \times T_C$ is associated to each relation type $R \in T_R$; (c_{1R}, c_{2R}) define the greatest concept types that might be linked by the relation type R. R holds between the concept types $x, y \in T_C$ if $x \leq c_{1R}$ and $y \leq c_{2R}$. All pairs (c_{1R}, c_{2R}) are called *star graphs*.[1] If $R_1, R_2 \in T_R$ and $R_1 \leq R_2$, then it holds that $c_{1R1} \leq c_{1R2}$ and $c_{2R1} \leq c_{2R2}$. The lattice of relation types also has the universal type $\top$ as top node and the absurd type $\perp$ as the bottom node;

- I is a set of distinct individual markers (*referents*) that refer to specified concept instances. $T_C \cap I = \emptyset$ and $T_R \cap I = \emptyset$. The *generic marker* $*$, where $* \notin (T_C \cup T_R \cup I)$, refers to an unspecified individual of some specified concept type x. The members of I are not ordered but $i \leq^* $ for all $i \in I$;

- τ is a mapping from I to T_C and associates individuals to concept types. In this way the concept types have instances in contrast to the relations types; τ defines the *conformity* of individuals to concept types. □

Example 1. Let us define a sample support S_1 with eleven concept types, four relation types and two individuals:

---

[1]The "star graph" in Definition 1 is a more specific notion than the one defined in Chapter 2. Here a star graph $[C1] \rightarrow (R) \rightarrow [C2]$ fixes the greatest concept types $C1$ and $C2$, for which the conceptual relation R holds. Thus the star graphs restrict the generalizations that can be computed for each conceptual relation.

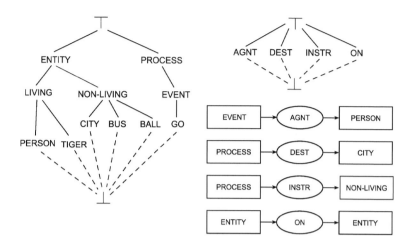

FIGURE 4.1: Type hierarchies of concept and relation types and star graphs for all relation types.

- T_C = {ENTITY, PROCESS, LIVING, NON-LIVING, EVENT, PERSON, TIGER, CITY, BUS, BALL, GO} which are ordered as shown in Figure 4.1;

- T_R = {AGNT, DEST, INSTR, ON} with partial order and star graphs shown in Figure 4.1;

- I = {John, Boston} which are not ordered;

- τ(John) = PERSON, τ(Boston) = CITY.

Chapter 2 discusses the variety of SCGs definitions; here we consider the connected, simplified SCGs in normal form without coreference links. So we shall deal with SCGs that contain no duplicated relation nodes linked to the same instances and no duplicated individual concept nodes. The duplicated generic concept nodes refer to different unspecified instances of the concept type in question. Moreover, we are interested in the finite, bipartite, directed graphs, which underlie the CGs; these graphs contain two kinds of vertices: *c-vertices* (corresponding to the concept boxes) and *r-vertices* (corresponding to the relation ovals). Each vertex is labeled by the name of the corresponding type. In the ordinary graphs, which underlie the SCGs, the *c*-vertices alternate with the *r*-vertices. Graph edges encode the links between concept and relation types, when the concepts are arguments of the relations. Usually the SCGs are multi-graphs, when they are defined for n-ary conceptual relations. But we consider binary conceptual relations only and no multi-edges and loops appear because each relation type R holds between two different concept types $x, y \in T_C$ or between two distinct instances x_1, x_2 of a concept

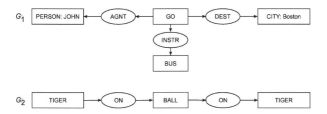

FIGURE 4.2: Two SCGs G_1 and G_2, defined over the support S_1.

type $x \in T_C$. In this way our SCGs respresent connected directed graphs with at most one edge between each two vertices. For binary conceptual relations, we can talk about the *first* and *second* argument of every conceptual relation. So we have an *incoming arc* from vertex c_1 to vertex r and an *outgoing arc* from vertex r to vertex c_2 when $r(c_1, c_2)$ holds – in other words, when c_1 is the first argument of the conceptual relation r and c_2 is the second argument of the conceptual relation r.

Example 2. Figure 4.2 shows two SCGs with binary conceptual relations defined over the support S_1, which will be used as a sample knowledge base (called KB_1) in this chapter. The concept nodes [PERSON:John] and [PER-SON:Mary] in G_1 are individual nodes as they refer to specified instances of the concept type [PERSON]. According to the default convention, the generic marker * is omitted in the generic concept nodes [GO], [BALL], [BUS], and [TIGER]. There are two distinct instances of [TIGER] in G_2 because the two c-vertices with label TIGER are distinct ones.

Chapter 2 discusses the translation of CGs to first-order logic, considers the definition of the formula operator ϕ in [Sowa, 1984] and lists its drawbacks: the treatment of the universal type and the blank graph as well as the insufficient specification of the CG contexts' translation to logical formulas. So in general, the formula operator ϕ itself needs a more elaborated definition. But we are interested in the logical formulas, corresponding to non-blank SCGs without contexts, and especially in the way how the support symbols (labels of concept types, relation types, and individuals) are placed within such a logical formula. Therefore we keep in mind the formula operator ϕ in [Sowa, 1984] and recall the main principles of juxtaposing support symbols to the elements of a given SCG G:

- If G contains k generic nodes, we assign a distinct variable $x_1, x_2, \ldots, x_k$ to each one;

- For each concept node c of G, let *identifier*(c) be the variable assigned to c if c is a generic node, which refers to an unspecified individual of type c, or *referent*(c) if c is an individual node. Each concept node c of G is represented as a monadic predicate whose name is the same as *type*(c) and whose argument is *identifier*(c);

- Each binary relation node r of G is represented as a binary predicate whose name is the same as $type(r)$. Let the first argument of the predicate be the identifier of the concept node linked to the incoming arc of r and the second argument of the predicate be the identifier of the concept node linked to the outgoing arc of r;

- The logical formula, corresponding to G, consists of a quantifier prefix $\exists x_1 \exists x_2 \ldots \exists x_k$ and the conjunction of all predicates for the concept nodes and relation nodes of G.

As seen from the above items, the logical interpretation of an SCG with binary conceptual relations contains binary predicates, one per each binary relation r and its concept-arguments c_1 and c_2 such that r holds from c_1 to c_2. All binary predicates correspond to triples *concept1-relation-concept2* where the concepts are first and second arguments of the relation; we will call these binary predicates *elementary conjuncts* of the logical interpretation and will use essentially this notion in our considerations. The monadic predicates define the domain of the variables $x_1, x_2, \ldots, x_k$ and the correspondences between individual markers and concept types.

Example 3. The sample SCG G_1 and G_2 are translated to logical formulas as follows:

- Graph G_1: *John is going to Boston by bus.*

$$\exists x \exists y \; \text{GO}(x) \,\&\, \text{BUS}(y) \,\&\, \text{PERSON}(John) \,\&\, \text{CITY}(Boston) \,\&$$
$$agnt(x, \text{PERSON}(John)) \,\&\, dest(x, \text{CITY}(Boston)) \,\&\, instr(x, y)$$

- Graph G_2: *A tiger is on a ball, which is on (another) tiger.*

$$\exists x \exists y \exists z \; \text{TIGER}(x) \,\&\, \text{BALL}(y) \,\&\, \text{TIGER}(z) \,\&\, on(x, y) \,\&\, on(y, z)$$

There are three elementary conjuncts in the formula of G_1, which correspond to the three conceptual relations *agnt*, *dest*, and *instr*, and two elementary conjuncts in the formula of G_2, which correspond to the two conceptual relations $on(x, y)$ and $on(y, z)$. Please note that the linear order of the conjuncts in the logical formulas is arbitrary. Also note that two distinct variables x and z are assigned to the two concept nodes with label TIGER in graph G_2.

4.2.2 Projection

Chapter 2 introduces the projection operation, a kind of graph morphism between two SCG. Let G be a query graph; the projection finds all its specializations in the knowledge base by searching mappings of the query to every

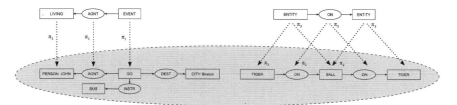

FIGURE 4.3:   Injective projections of SCGs onto the knowledge base KB_1.

SCG in the knowledge base. Obviously, the query graph has to be defined over the same support, which ensures the mapping of its labels onto the knowledge base graphs. From practical perspective, the projection is the basic CG instrument supporting the conceptual search. Here we are interested in conceptual patterns, which are isomorphic to the query graph, so we shall introduce the notion of injective projection.

Definition 2. [Mugnier, 1995] An **injective projection** π is a kind of projection defined as follows. Let G and H be two SCG, defined over the same support. Then $\pi : G \to H$ is a graph πG, where $\pi G \subseteq H$ and πG is isomorphic to G. $\square$

Example 4. Given a query SCG G and a knowledge base, the projection extracts all the subgraphs in the knowledge base which are specialisations of G. Figure 4.3 shows two query graphs and their three injective projections π_1, π_2, and π_3 onto G_1 and G_2. Obviously the query

$$[\text{ENTITY}] \to (\text{ON}) \to [\text{ENTITY}]$$

has two different injective projections π_2 and π_3 onto G_2.

The efficient projection calculation has been a challenging task for many years. The projection algorithms, dealing with the logical forms, usually rely on Prolog inference mechanisms. The approaches, based on graph theory, integrate algorithms for subgraph search and mapping. Given two CGs G and H defined over the same support, it is NP-complete to decide whether there is a projection $\pi : G \to H$. However, polynomial algorithms for projection are proposed for SCGs when the underlying ordinary graphs are trees [Mugnier and Chein, 1992, Mugnier, 1995]. Thus the most efficient approaches to projection calculation are influenced by algorithms solving equivalent problems in graph theory. Another interesting problem ("given an integer k, is the number of injective projections $\pi : G \to H$ smaller than k?") is also NP-complete [Mugnier, 1995]. In all cases, the projection of a query graph G onto a given knowledge base is calculated graph by graph, i.e., the query graph is mapped independently on each particular graph.

We recall below further basic definitions and certain facts concerning finite state automata.

4.2.3   Finite State Automata

Definition 3. A **deterministic finite state automaton** A is a 5-tuple $A = \langle \Sigma, Q, q_0, F, \Delta \rangle$, where Σ is a finite alphabet, Q is a finite set of states, $q_0 \in Q$ is the initial state, $F \subseteq Q$ is the set of final states, and $\Delta \subseteq Q \times \Sigma \times Q$ is the *transition* relation. The transition $\langle q, a, p \rangle \in \Delta$ *begins* at state q, *ends* at state p and has the *label* a. □

Definition 4. Let A be a FSA. A **path** c **in** A is a finite sequence of $k > 0$ transitions: $c = \langle q_0, a_1, q_1 \rangle \langle q_1, a_2, q_2 \rangle \ldots \langle q_{k-1}, a_k, q_k \rangle$, where $\langle q_{i-1}, a_i, q_i \rangle \in \Delta$ for $i = 1, \ldots, k$. The integer k is called the *length* of c. The state q_0 is called the *beginning* of c and q_k is called the *end* of c. The string $w = a_1 a_2 \ldots a_k$ is called the *label* of c. The null path of $q \in Q$ is 0_q, beginning and ending in q with label ε, where ε is the empty symbol. □

Definition 5. Let $A = \langle \Sigma, Q, q_0, F, \Delta \rangle$ be a FSA. Let Σ^* be the set of all strings over the alphabet Σ, including the empty symbol ε. The **generalized transition relation** Δ^* is the smallest subset of $Q \times \Sigma^* \times Q$ with the following closure properties:

- For all $q \in Q$ we have $\langle q, \varepsilon, q \rangle \in \Delta^*$;

- For all $q_1, q_2, q_3 \in Q$ and $w \in \Sigma^*$, $a \in \Sigma$: if $\langle q_1, w, q_2 \rangle \in \Delta^*$ and $\langle q_2, a, q_3 \rangle \in \Delta$, then $\langle q_1, w \cdot a, q_3 \rangle \in \Delta^*$. □

Definition 6. The **formal language** $L(A)$ accepted by an FSA

$$A = \langle \Sigma, Q, q_0, F, \Delta \rangle$$

is the set of all strings, which are labels of paths leading from the initial to a final state:

$$L(A) := \{ \ w \in \Sigma^* | \exists q \in F : \langle q_0, w, q \rangle \in \Delta^* \ \}.$$

These strings will be called **words** of the language $L(A)$. □

Languages accepted by FSA are **regular** languages. Every finite list of words over a finite alphabet of symbols is a regular language. Among the deterministic automata that accept a given language, there is a unique automaton (excluding isomorphisms) that has a minimal number of states. It is called the **minimal** deterministic automaton of the language.

Definition 7. Let $A = \langle \Sigma, Q, q_0, F, \Delta \rangle$ be a FSA. Let Σ^+ be the set of all strings w over Σ, where $|w| \geq 1$. The automaton A is called **acyclic** iff for all $q \in Q$ there exist no string $w \in \Sigma^+$ such that $\langle q, w, q \rangle \in \Delta^*$. □

A necessary and sufficient condition for any deterministic automaton to be acyclic is that it recognises a finite set of words. There are algorithms which construct acyclic deterministic automata given their regular languages - finite sets of words.

Definition 8. A **deterministic finite state automaton with markers at the final states** A is a 7-tuple $A = \langle \Sigma, Q, q_0, F, \Delta, E, \mu \rangle$, where Σ is a finite alphabet, Q is a finite set of states, $q_0 \in Q$ is the initial state, $F \subseteq Q$ is the set of final states, $\Delta \subseteq Q \times \Sigma \times Q$ is the *transition* relation, E is a finite set of markers, and $\mu : F \to E$ is a function assigning a marker to each final state. $\square$

4.3 Construction of Compact FSA-Based Encoding of the Knowledge Base

As we have said earlier, we aim at the off-line pre-computing of all subtasks in the calculation of injective projection, which do not depend on the run-time query graph. This two-stage treatment is feasible if we manage to encode efficiently the pre-processing results. As it goes about off-line computations, we have (enough) time to extract and analyse the knowledge base subgraphs, to calculate all their injective generalizations and to store all the information in a carefully elaborated construction – a minimal acyclic FSA with markers at the final states. We define some important notions below and then summarise the FSA construction steps.

4.3.1 Conceptual subgraphs and encoding of the concept nodes identity

In contrast to the usual approaches to projection, which deal with the logical formulas or the underlying graph structures, we are focused on the elementary conjuncts of the SCGs.

Definition 9. [Angelova and Mihov, 2008] Let G be a SCG with binary conceptual relations. A **conceptual subgraph** of G is a connected graph G_{cs} such that:

- as with ordinary graphs, $G_{cs} \subseteq G$ and

- G_{cs} consists only of nodes and edges of G, which correspond to triples *concept1-relation-concept2*, i.e., G_{cs} is a SCG too.

In the remaining part of this chapter, the conceptual subgraphs will be called **subgraphs**. $\square$

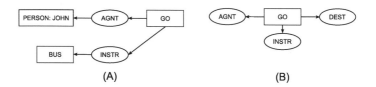

(A)   (B)

FIGURE 4.4:   Conceptual vs. ordinary subgraphs of the SCG G_1.

Example 5. The notion of subgraph is essential for our off-line constructions. The conceptual subgraphs are the patterns of all SCGs which have
injective projection onto the given knowledge base, so it is important to consider especially them and to filter out the other subgraphs. Figure 4.4(A)
shows a connected ordinary graph which is a subgraph of G_1 according to
definition 9, as it contains only nodes corresponding to elementary conjuncts
of G_1. Figure 4.4(B) shows another connected subset of G_1 nodes, which has
no conceptual interpretation according to the support S_1. Obviously, there
exist numerous connected bipartite graphs which cannot be considered as conceptual graphs in any support; Figure 4.4(B) exemplifies such a graph. These
kinds of subgraphs are eliminated during our off-line computations.

We want to interpret the SCGs with binary conceptual relations as a regular
language, which means to encode them as words over certain finite alphabet.
The proposal is to simplify and normalize every SCG from the knowledge
base and then to represent it as a finite string of support labels, by writing
down its elementary conjuncts triple by triple in linear order, replacing the
variables by their concept types and explicating the types and instances of
the individual concept nodes [Angelova and Mihov, 2008]. Then G_1: *John is
going to Boston by bus* can be encoded as a string of nine support labels, where
the information from the monadic predicates of the logical interpretation is
stored within the binary predicates. Each triple is given in the linear order
argument1-relation-argument2:

GO AGNT PERSON:John GO DEST CITY:Boston GO INSTR BUS (1)

Similarly, G_2: *A tiger is on a ball, which is on (another) tiger* is represented
by the string:

TIGER ON BALL   BALL ON TIGER   (2)

It is obvious that the linear sequences of labels (1) and (2) fail to encode
the identities between the concept types, which correspond to the structural,
topological connections of the underlying ordinary graphs. In fact we need to
define classes of equivalence among the arguments of the conceptual relations,
in order to express their identity, and this task is similar to the task of set
elements partitioning into disjoint equivalence classes. So the set partitioning
is a suitable framework to model the equivalence of SCGs' concept nodes.

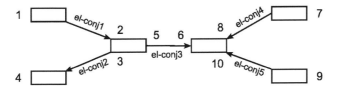

FIGURE 4.5: Concept nodes of a SCG with five elementary conjuncts, numbered from 1 to 10.

The number of set partitions is growing too fast, as it reflects the combinatorial nature of the partitioning, and it is given by the so-called Bell numbers [Weisstein, 2007]. The paper [Angelova and Mihov, 2008] considers in detail the subclasses of set partitioning for four elements (there are 15 classes, i.e., $B_4 = 15$) and analyzes them as potential models of the structural links between the four arguments of two SCGs' elementary conjuncts. It turns out that nine classes are irrelevant to our considerations and the number of irrelevant classes is growing with the graph length. For brevity we do not present here all details of these considerations; we only stress that each label linearization of a knowledge base SCG – (1), (2) and so on – will be juxtaposed a "topological" type that describes the argument identity. Further discussion of these topological types is given in Example 6. As G_1 and G_2 are very simple to illustrate the proposed solution, Example 6 presents an SCG with five elementary conjuncts and ten concept nodes, six of them distinct ones.

Example 6. Let us consider the topological patterns for a graph with five elementary conjuncts, given at Figure 4.5, and all its subgraphs. For convenience, only the concept nodes are displayed with focus on the elementary conjuncts *el-conj1*, *el-conj2*, *el-conj3*, *el-conj4*, *el-conj5*, which correspond to certain binary conceptual relations r_1, r_2, r_3, r_4, and r_5. We notice immediately that the graph connectedness is due to the equivalent arguments of the different conceptual relations, therefore we focus on the concept identity. The concept nodes are numbered from 1 to 10, conjunct per conjunct, according to the direction of the relation arcs, no matter that some nodes are equivalent; e.g., *el-conj1* has arguments 1 and 2, where argument 1 is the concept type at the beginning of the incoming arc to the respective conceptual relation r_1 and argument 2 is the concept type at the end of the outgoing arc of the conceptual relation r_1. Table 4.1 lists all the subgraphs of the graph given at Figure 4.5 and their structural relations, if we assume that the elementary conjuncts are to be linearized as symbol strings in order corresponding to their indices *el-conj1*, *el-conj2*, *el-conj3*, *el-conj4*, *el-conj5*.

Let us discuss in more detail the six subgraphs of three elementary conjuncts in Table 4.1. They have six types of concept nodes identity since the latter are listed using the graph-specific indices of the concept nodes. But we can consider them as structural patterns of linkages among the six relation

Subgraphs enumerated by elementary conjuncts	Encoding the identity of the concept nodes		Subgraphs enumerated by elementary conjuncts	Encoding the identity of the concept nodes
Subgraphs of 1 el. conj.	-		**Subgraphs of 3 el. conj.**	
el-conj1	-		el-conj(1&2&3)	2=3=5
el-conj2	-		el-conj(1&3&4)	2=5|6=8
el-conj3	-		el-conj(1&3&5)	2=5|6=10
el-conj4	-		el-conj(2&3&4)	3=5|6=8
el-conj5	-		el-conj(2&3&5)	3=5|6=10
			el-conj(3&4&5)	6=8=10
Subgraphs of 2 el. conj.				
el-conj(1 & 2)	2=3		**Subgraphs of 4 el. conj.**	
el-conj(1 & 3)	2=5		el-conj(1&2&3&4)	2=3=5|6=8
el-conj(2 & 3)	3=5		el-conj(1&2&3&5)	2=3=5|6=10
el-conj(3 & 4)	6=8		el-conj(1&3&4&5)	2=5|6=8=10
el-conj(3 & 5)	6=10		el-conj(2&3&4&5)	3=5|6=8=10
el-conj(4 & 5)	8=10			
			Subgraphs of 5 el. conj.	
			el-conj(1&2&3&4&5)	2=3=5| 6=8=10

TABLE 4.1:   Enumeration of Subgraphs and Encoding of Concept Types Identities.

arguments. Let the three elementary conjuncts be enumerated in the particular linear order as $r_1(x_1, y_1)\&r_2(x_2, y_2)\&r_3(x_3, y_3)$. Then the subgraph *el-conj*(1&2&3) has arguments identity $y_1 = x_2 = x_3$, the two subgraphs *el-conj*(1&3&4) and *el-conj*(1&3&5) have arguments identity $y_1 = x_2|y_2 = y_3$, the two subgraphs *el-conj*(2&3&4) and *el-conj*(2&3&5) have arguments identity $x_1 = x_2|y_2 = y_3$, and the subgraph *el-conj*(3&4&5) has arguments identity $y_1 = y_2 = y_3$. Thus there are only four types of concept nodes equivalences for subgraphs of three elementary conjuncts wrt the particular order of label linearization in Table 4.1. The number of all equivalence classes for a set of six elements is much larger, as $B_6 = 203$.

4.3.2   Construction of the Minimal FSA

The construction is described using the fact, that every acyclic automaton can be defined by the finite list of words belonging to the automaton language. The automaton minimisation in general has complexity $O(n \log(n))$, where n is the number of states of the initial automaton [Hopcroft, 1971]; however it

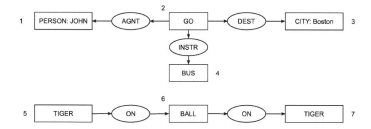

FIGURE 4.6:   Indexing of all concept instances in the knowledge base.

is performed iteratively and requires construction of intermediate automata, which are rather large compared to the output minimal automaton. For the simpler case of acyclic automata, the thesis [Mihov, 2000] and paper [Daciuk et al., 2000] propose an algorithm for direct construction of the minimal acyclic automaton, recognizing a finite list of words sorted in lexicographic order. This is done by incremental construction of automaton paths word by word; at every step the algorithm constructs an automaton which is "minimal except for the last word." So in our considerations we aim primarily at the construction of a sorted list of words, encoding all the injective generalizations, and then we employ results from the automata theory to ensure the construction of the minimal FSA.

The automaton construction can be done off-line in several steps [Angelova and Mihov, 2008], which are summarized here and illustrated by computations for the graphs G_1 and G_2.

Step 1, ordering the support labels: Define lexicographic order for all the support symbols (items of T_C, T_R, and labels $x : i$ such that $x \in T_C$, $i \in I$ and $\tau(i) = x$). For S_1, let the order be:

AGNT < BALL < BUS < CITY < CITY:Boston < DEST < ENTITY < EVENT < GO < INSTR < LIVING < NON-LIVING < ON < PERSON < PERSON:John < PROCESS < TIGER

Step 2, indexing all concept nodes: Juxtapose indices to all the distinct concept nodes in the knowledge base, to ensure their treatment as distinct instances (which is particularly important for the generic concept types). For KB_1, let the indices be given as shown in Figure 4.6.

Step 3, computation and encoding of subgraphs: Calculate all (conceptual) subgraphs of all SCGs with binary conceptual relations in the knowledge base, keeping track of the source conceptual graphs. For each subgraph, encode linearly its labels as symbol strings, by interpreting each triple *concept1-relation-concept2* as a string of three support labels CONCEPT1 RELATION CONCEPT2. Sort the strings triple by triple, using the lexicographic order defined at step 1, taking into consideration the indices of the knowledge base concept nodes as they are assigned at step 2. For the sorted strings, for each subgraph, encode the concept nodes identity and assign these strings as annotation to the linear sequence of subgraph labels.

All the subgraphs of KB_1 are listed below (there are 10 different subgraphs with length of 1-3 elementary conjuncts). As a preparation for the future construction steps, certain markers M1, M2, ... are assigned to these subgraphs. The original graphs G_1 and G_2 are included, too – e.g., M6 is G_2 with triple labels, sorted using the lexicographic order defined at step 1. The identity of the concept nodes is encoded for each subgraph after a separator "/," similarly to the encoding shown in Table 4.1 where the separator '|' enables the encoding of different equivalence classes of concept nodes. We also keep track of the original graph – G_1 or G_2.

M1 = {GO_2 AGNT PERSON:John$_1$} (comes from G_1)

M2 = {GO_2 DEST CITY:Boston$_3$} (comes from G_1)

M3 = {GO_2 INSTR BUS$_4$} (comes from G_1)

M4 = {TIGER$_5$ ON BALL$_6$} (comes from G_2)

M5 = {BALL$_6$ ON TIGER$_7$} (comes from G_2)

M6 =′{BALL$_6$ ON TIGER$_7$ TIGER$_5$ ON BALL$_6$} / 1=4 (comes from G_2)

M7 = {GO_2 AGNT PERSON:John$_1$ GO_2 DEST CITY:Boston$_3$} / 1=3 (comes from G_1)

M8 = {GO_2 AGNT PERSON:John$_1$ GO_2 INSTR BUS$_4$} / 1=3 (comes from G_1)

M9 = {GO_2 DEST CITY:Boston$_3$ GO_2 INSTR BUS$_4$} / 1=3 (comes from G_1)

M10={GO_2 AGNT PERSON:John$_1$ GO_2 DEST CITY:Boston$_3$ GO_2 INSTR BUS$_4$} / 1=3=5 (comes from G_1)

Step 4, computation and encoding of all injective generalizations: For all subgraphs found at Step 3, calculate all possible injective generalizations according to the support. To every generalization G, juxtapose the subgraph G' for which G is computed in order to remember the particular knowledge base fragment G' such that $\pi : G \to G'$ if G is given as an injective projection query in run-time. Record each generalization G as a word of triples and sort it triple by triple using the lexicographic order defined at step 1. Consider every generalization G and its subgraph G'. Compare the triple order of G' to the triple order of G. If the triples are not correspondingly linearised, a new subgraph G'' has to be juxtaposed to G. G'' is constructed from G' by reordering its triples to correspond to the triples in G, thus encoding the exact projection of each concept and relation type from G to G''. When the triples of a subgraph G' are re-ordered to build G'', encode again the information of G'' concept nodes identities. In case no marker for G'' was designed so far, create a new marker that presents another linearisation of the G' triples. Finally, when all generalizations are computed and the correspondingly sorted subgraphs' labels are assigned to them as annotation markers, sort all the generalizations in a single list of words, using the lexico-graphic order defined at step 1. When one generalization is assigned several different subgraphs $G_{s1}, G_{s2}, \ldots, G_{sk}$, create a new marker for the union of all markers corresposnding to $G_{s1}, G_{s2}, \ldots, G_{sk}$.

Tables 4.2 to 4.4 list all 104 injective generalizations for the subgraphs of KB_1, sorted in lexicographic order, with original subgraphs (markers) juxta-posed to each generalization. We see that the subgraph with marker

$$\text{M6} = \{\text{BALL}_6 \text{ ON TIGER}_7 \text{ TIGER}_5 \text{ ON BALL}_6\} \ / \ 1{=}4$$

had to be duplicated but with reordered triples, so it becomes
$$\text{M11} = \{\text{TIGER}_5 \text{ ON BALL}_6 \text{ BALL}_6 \text{ ON TIGER}_7\} \ / \ 2{=}3$$

This duplication is due to the fact that in some injective generalizations of M6 – e.g.

$$\text{ENTITY ON NON-LIVING NON-LIVING ON LIVING}$$
$$\text{ENTITY ON NON-LIVING NON-LIVING ON TIGER}$$

the triples, which generalize the M6 triples, are reordered after the lexico-graphic sorting, so we can lose track of the concept instances linearization in the original subgraph to where the injective generalizations are to be pro-jected. Therefore the triples of M6 are reordered and recorded once again as M11 with proper encoding of the concept nodes identity. Tables 4.2 to 4.4 contain 18 generalizations of G_2, which are projected to M6, and 8 general-izations of G_2, which are projected to M11.

One generalization:

$$\text{ENTITY ON ENTITY ENTITY ON ENTITY}$$

can be projected to both M6 and M11, so a new marker M12 = { {M6 / 1=4}, {M11 / 2=3} } is created. Similarly, M13 = {M4, M5} is designed to encode that ENTITY ON ENTITY has injective projections on both M4={TIGER$_5$ ON BALL$_6$} and M5={BALL$_6$ ON TIGER$_7$}.

Please note that marker duplications, due to reordered triples, do not hap-pen for all subgraphs. For instance in G_1, all three elementary conjuncts have identical first arguments and the conceptual relations AGNT, DEST, INSRT cannot be generalized, so the triples order in all sorted generalizations always corresponds to the triple order in the initially linearized subgraph at step 3. So no marker duplications are needed for all the subgraphs of G_1.

At step 4, there is a real danger that too many subgraphs' versions are created – by recording subgraphs several times to reflect the lexicographic order of triples in all their generalizations, and this in principle might cause combinatorial explosion during the process of brute-force enumeration and sorting. However the experimental results prove that the data size and the number of nodes identity types remain manageable. The tests also confirm the expectation that the resulting FSA has a rather compact structure and the space needed for the final conceptual archive is relatively small. Last but

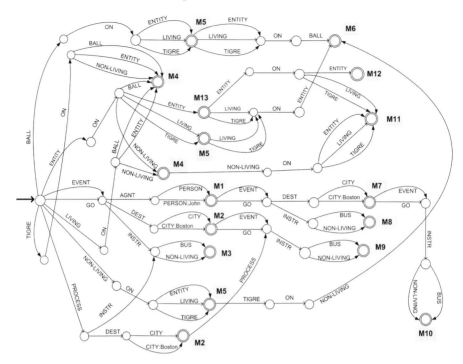

FIGURE 4.7: Minimal acyclic FSA with markers at the final states, which encodes all injective generalizations of KB_1.

not least, we should always remember that all these computations are to be performed off-line, so the operational efficiency is not of primary importance in this case.

Step 5: Consider the generalizations' labels as *words* over the finite alphabet of support symbols and the knowledge base subgraphs as *markers* attached to the generalizations. Build the minimal acyclic FSA with markers at the final states, which recognizes the language L of all words (sequences of generalizations' labels) and juxtaposes every word of the marker, which is associated with it. We will call this automaton the minimal knowledge base FSA.

Figure 4.7 contains the minimal knowledge base FSA for the finite language L_1 listed in Tables 4.2 to 4.4. The automaton has 49 states and 86 transition arcs, which encode 104 generalizations with the length of 3-9 alphabet symbols. All beginning symbols, which are common for several words, are encoded only once. The common ending symbols for words with the same marker are also encoded only once.

Labels of injective generalization as sorted strings and markers of corresponding subgraphs
BALL ON ENTITY, M5
BALL ON ENTITY ENTITY ON BALL, M6
BALL ON ENTITY LIVING ON BALL, M6
BALL ON ENTITY TIGER ON BALL, M6
BALL ON LIVING, M5
BALL ON LIVING ENTITY ON BALL, M6
BALL ON LIVING LIVING ON BALL, M6
BALL ON LIVING TIGER ON BALL, M6
BALL ON TIGER, M5
BALL ON TIGER ENTITY ON BALL, M6
BALL ON TIGER LIVING ON BALL, M6
BALL ON TIGER TIGER ON BALL, M6
ENTITY ON BALL, M4
ENTITY ON ENTITY, M13
ENTITY ON ENTITY ENTITY ON ENTITY, M12
ENTITY ON ENTITY ENTITY ON LIVING, M11
ENTITY ON ENTITY ENTITY ON TIGER, M11
ENTITY ON ENTITY LIVING ON ENTITY, M6
ENTITY ON ENTITY TIGER ON ENTITY, M6
ENTITY ON LIVING, M5
ENTITY ON LIVING LIVING ON ENTITY, M6
ENTITY ON LIVING TIGER ON ENTITY, M6
ENTITY ON NON-LIVING, M4
ENTITY ON NON-LIVING NON-LIVING ON ENTITY, M11
ENTITY ON NON-LIVING NON-LIVING ON LIVING, M11
ENTITY ON NON-LIVING NON-LIVING ON TIGER, M11
ENTITY ON TIGER, M5
ENTITY ON TIGER LIVING ON ENTITY, M6
ENTITY ON TIGER TIGER ON ENTITY, M6
EVENT AGNT PERSON, M1
EVENT AGNT PERSON EVENT DEST CITY, M7
EVENT AGNT PERSON EVENT DEST CITY EVENT INSTR BUS, M10
EVENT AGNT PERSON EVENT DEST CITY EVENT INSTR NON-LIVING, M10
EVENT AGNT PERSON EVENT DEST CITY:Boston, M7
EVENT AGNT PERSON EVENT DEST CITY:Boston EVENT INSTR BUS, M10
EVENT AGNT PERSON EVENT DEST CITY:Boston EVENT INSTR NON-LIVING, M10
EVENT AGNT PERSON EVENT INSTR BUS, M8
EVENT AGNT PERSON EVENT INSTR NON-LIVING, M8
EVENT AGNT PERSON:John, M1
EVENT AGNT PERSON:John EVENT DEST CITY, M7

TABLE 4.2: Enumeration of all Possible Injective Generalizations at Step 4 (Part 1/3).

Labels of injective generalization as sorted strings and markers of corresponding subgraphs
EVENT AGNT PERSON:John EVENT DEST CITY EVENT INSTR BUS, M10
EVENT AGNT PERSON:John EVENT DEST CITY EVENT INSTR NON-LIVING, M10
EVENT AGNT PERSON:John EVENT DEST CITY:Boston, M7
EVENT AGNT PERSON EVENT DEST CITY:Boston EVENT INSTR BUS, M10
EVENT AGNT PERSON EVENT DEST CITY:Boston EVENT INSTR NON-LIVING, M10
EVENT AGNT PERSON EVENT INSTR BUS, M8
EVENT AGNT PERSON EVENT INSTR NON-LIVING, M8
EVENT AGNT PERSON:John, M1
EVENT AGNT PERSON:John EVENT DEST CITY, M7
EVENT AGNT PERSON:John EVENT DEST CITY EVENT INSTR BUS, M10
EVENT AGNT PERSON:John EVENT DEST CITY EVENT INSTR NON-LIVING, M10
EVENT AGNT PERSON:John EVENT DEST CITY:Boston, M7
EVENT AGNT PERSON:John EVENT DEST CITY:Boston EVENT INSTR BUS, M10
EVENT AGNT PERSON:John EVENT DEST CITY:Boston EVENT INSTR NON-LIVING, M10
EVENT AGNT PERSON:John EVENT INSTR BUS, M8
EVENT AGNT PERSON:John EVENT INSTR NON-LIVING, M8
EVENT DEST CITY, M2
EVENT DEST CITY EVENT INSTR BUS, M9
EVENT DEST CITY EVENT INSTR NON-LIVING, M9
EVENT DEST CITY:Boston, M2
EVENT DEST CITY:Boston EVENT INSTR BUS, M9
EVENT DEST CITY:Boston EVENT INSTR NON-LIVING, M9
EVENT INSTR BUS, M3
EVENT INSTR NON-LIVING, M3
GO AGNT PERSON, M1
GO AGNT PERSON GO DEST CITY, M7
GO AGNT PERSON GO DEST CITY GO INSTR BUS, M10
GO AGNT PERSON GO DEST CITY GO INSTR NON-LIVING, M10
GO AGNT PERSON GO DEST CITY:Boston, M7
GO AGNT PERSON GO DEST CITY:Boston GO INSTR BUS, M10
GO AGNT PERSON GO DEST CITY:Boston GO INSTR NON-LIVING, M10
GO AGNT PERSON GO INSTR BUS, M8
GO AGNT PERSON GO INSTR NON-LIVING, M8

TABLE 4.3:   Enumeration of all Possible Injective Generalizations at Step 4 (Part 2/3).

Labels of injective generalization as sorted strings and markers of corresponding subgraphs
GO AGNT PERSON:John, M1
GO AGNT PERSON:John GO DEST CITY, M7
GO AGNT PERSON:John GO DEST CITY GO INSTR BUS, M10
GO AGNT PERSON:John GO DEST CITY GO INSTR NON-LIVING, M10
GO AGNT PERSON:John GO DEST CITY:Boston, M7
GO AGNT PERSON:John GO DEST CITY:Boston GO INSTR BUS, M10
GO AGNT PERSON:John GO DEST CITY:Boston GO INSTR NON-LIVING, M10
GO AGNT PERSON:John GO INSTR BUS, M8
GO AGNT PERSON:John GO INSTR NON-LIVING, M8
GO DEST CITY, M2
GO DEST CITY GO INSTR BUS, M9
GO DEST CITY GO INSTR NON-LIVING, M9
GO DEST CITY:Boston, M2
GO DEST CITY:Boston GO INSTR BUS, M9
GO DEST CITY:Boston GO INSTR NON-LIVING, M9
GO INSTR BUS, M3
GO INSTR NON-LIVING, M3
LIVING ON BALL, M4
LIVING ON ENTITY, M4
LIVING ON NON-LIVING, M4
LIVING ON NON-LIVING NON-LIVING ON ENTITY, M11
LIVING ON NON-LIVING NON-LIVING ON LIVING, M11
LIVING ON NON-LIVING NON-LIVING ON TIGER, M11
NON-LIVING ON ENTITY, M5
NON-LIVING ON ENTITY TIGER ON NON-LIVING, M6
NON-LIVING ON LIVING, M5
NON-LIVING ON LIVING TIGER ON NON-LIVING, M6
NON-LIVING ON TIGER, M5
NON-LIVING ON TIGER TIGER ON NON-LIVING, M6
PROCESS DEST CITY, M2
PROCESS DEST CITY PROCESS INSTR BUS, M9
PROCESS DEST CITY PROCESS INSTR NON-LIVING, M9
PROCESS DEST CITY:Boston, M2
PROCESS DEST CITY:Boston PROCESS INSTR BUS, M9
PROCESS DEST CITY:Boston PROCESS INSTR NON-LIVING, M9
PROCESS INSTR BUS, M3
PROCESS INSTR NON-LIVING, M3
TIGER ON BALL M4
TIGER ON ENTITY M4
TIGER ON NON-LIVING M4

TABLE 4.4: Enumeration of all Possible Injective Generalizations at Step 4 (Part 3/3).

(A) EVENT AGNT PERSON EVENT INSTR BUS   **(B)** BALL ON TIGER   TIGER ON BALL
                    1=3   1=4|2=3

FIGURE 4.8:   Encoding projection queries as words over the alphabet of support labels.

4.3.3 Run-Time Calculation of Injective Projection

The minimal knowledge base FSA provides efficient searching of injective projection. In run-time, when a SCG projection query G arrives to the system, the query will be normalized and then turned into a linear record of sorted triple labels, plus encoding of the identity of its concept nodes. There are several popular formats to represent conceptual graphs—linear, CGIF, graphical etc.—so the translation into sequence of support labels depends on the input form. As the normalization and sorting of labels is an obligatory task, we can estimate the complexity as $O(n \log(n))$, where n is the number of G symbols. Assigning the annotation, which encodes the identity of concept nodes in the query graph, can be done for constant time. Actually every query is turned to a word, which may (or may not) belong to the regular language, encoding all possible injective generalizations. Figure 4.8 illustrates such words; the word at Figure 4.8(A) belongs to the language L_1 but the one at Figure 4.8(B) does not belong to it.

The check whether a query word belongs to the language is performed for n steps, when n is the number of projection query symbols. If the word does not belong to the language, the query has no projection onto the knowledge base, which is the case of the query at Figure 4.8(B). If the word belongs to the language; i.e., the automaton is traversed to a final state and its unique marker is found, the marker content is compared to the topological structure of the query, and the answer (i.e., all the answers) are delivered at once if their structural links are identical to the query's one. So the answer to the query at Figure 4.8(A) will be the subgraph in M8. The computations at this step concern the query only while the minimal knowledge base FSA is simply traversed n times along its transition arcs.

At the end let us stress further important features of our approach:

1. The lexicographic order (whatever it is) and the conventions for description of the concept nodes identity (whatever they are) provide means for unique encoding of all SCGs with binary conceptual relations over the given support, which enables the comparison and mapping of projection queries to knowledge base generalizations;

2. All graph-specific labels, which index the distinct concept instances, are stored in the FSA markers. Thus the alphabet consists of the support

symbols only, which facilitates the intersection between the minimal knowledge base FSA and the query graphs to arrive at a future moment;

3. All subgraphs with the same injective generalizations are grouped into one marker, so all the information regarding multiple projection answers will be pre-computed off-line (keeping the source graph identifier) and will be delivered at once in run-time, as a set of injective projections;

4. Ahe proposed solution is rather efficient regarding the computational complexity. The cyclic and acyclic SCGs are treated in the same way and a whole range of interesting problems—e.g., "are two given SCGs equivalent"—can be easily solved if the minimal knowledge base FSA is integrated in the computations.

4.4 Experimental Assessment of Conceptual Data Compression

The minimal FSA at Figure 4.7 displays encouraging features. Each word has either common beginning symbols or common ending symbols with other words, or both. The number of markers is much smaller compared to the number of words and this fact also facilitates the compression. However, KB_1 is too simple; therefore in order to assess a realistic conceptual resource, we have made a test experiment with a larger support with 1025 concept types and 10 relation types (Figure 4.9). The concept type hierarchy uses the famous top level categories of John Sowa [Sowa, 2000a][2] and has another 1000 artificially designed concept types, associated to the top level in a lattice with maximal depth 20 levels and maximal breath 98 types (average breath 51 types). There are 10 artificially designed relation types with star graphs, which are built using the uppermost concept types. To generate numerous generalizations, the average number of parent supertypes is 2.00029 per type (maximally 8 parents, minimally 1 parent).

The experimental knowledge base consists of 329 SCGs with binary conceptual relations, each with a length of 3-12 elementary conjuncts, and average length of 5.65 elementary conjuncts. These SCGs are constructed using randomly selected concept and relation types. The number of (conceptual) subgraphs in the knowledge base is 11146. These subgraphs have 140031027 different injective generalizations with some 3618 different topological structures in the lexicographically sorted injective generalizations.

---

[2]See also http://users.bestweb.net/~sowa/ontology/toplevel.htm

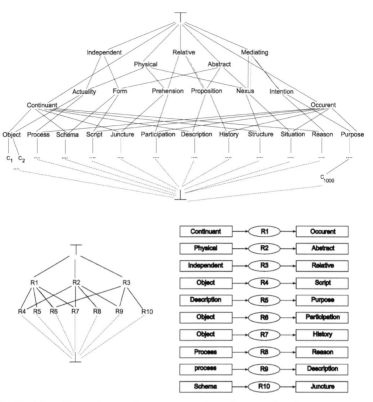

FIGURE 4.9: Experimental support using the top-level concept types of [Sowa, 2000a].

Table 4.5 presents the number of topological structures (strings encoding the identity of concept nodes), which are computed separately for all subgraphs with the same length (i.e., with the same number of elementary conjuncts). Their number is relatively restricted (3618), especially if it is compared to the number of all possible equivalence classes given by the respective Bell number.

The automaton, compressing all the injective generalizations of the experimental knowledge base, is built off-line using the algorithm described in [Mihov, 2000, Daciuk et al., 2000]. The input text file contains all injective generalizations sorted in lexicographic order, each followed by the description of the concept nodes identities and the marker-index, which for the experimental data contains the identifier of the SCG and its subgraph. For instance,

$$(1,1,1002)(2,3,10)(1002,2,5)(1002,3,10)\&1,4,6|3,7\&92\&13$$

actually means

$$\text{C1 R1 C1002 C2 R3 C10 C1002 R2 C5 C1002 R3 C10}$$
$$1{=}4{=}6|3{=}7 \text{ (subgraph 13 of } G_{92}\text{)}$$

Number of elementary conjuncts in the injective generaliza-tions	Number of arguments to be par-titioned in equiv-alence classes	Number of different topological structures in the ex-perimental knowledge base	Corresponding Bell number (maximal number of possible equivalence classes)
2	4	5	B_4=15
3	6	35	B_6=203
4	8	139	B_8=4 140
5	10	342	B_{10}=115 975
6	12	461	B_{12}=4 213 597
7	14	761	B_{14}=190 899 322
8	16	769	B_{16}=10 480 142 147
9	18	271	B_{18}=682 076 806 159
10	20	513	B_{20}=51 724 158 235 372
11	22	281	B_{22}=4 506 715 738 447 320
12	24	41	B_{24}=445 958 869 294 805 000
	TOTAL:	**3618**	**450 518 001 943 669 000**

TABLE 4.5: Kinds of Topological Structures in the Experimental Knowledge Base, Compared to the Numbers of Set Partitions for the Corresponding Number of Arguments (set members).

Number of states	Number of transitions	Number of words in the FSA language	FSA Size Bytes / MB	Size of the input text file
23 956 007	43 347 641	140 031 027	642 490 216 B 612,73 MB	~ 13 GBytes

TABLE 4.6: Compression of Injective Generalizations by a Final State Automaton.

The minimal knowledge base FSA is built using UNICODE and the support alphabet is considered as a subset of UNICODE. It contains the support labels and the digits 0,1,...,9, which encode the concept nodes' identities. Only two separators are needed after switching to UNICODE – the symbol "&," which separates the elementary conjuncts, the enumeration of the identical concept nodes and the markers, and the symbol '|', which separates the different equivalent classes of identical concept nodes. In this way the automaton is built over an alphabet of 1047 symbols.

Table 4.6 describes the final minimal acyclic FSA, which recognizes all different injective generalizations of the knowledge base subgraphs (their number is 140031027). As shown in the last two columns, the input file is compressed about 21,2 times.

4.5 Conclusion

This chapter summarizes an approach for internal SCGs representation which is tailored to off-line data preprocessing. It includes a brute-force enumeration of all possible injective projections to a given knowledge base at a certain particular moment, which is a computationally intensive task. However, the resulting encoding of the finite list of all generalizations as a minimal acyclic FSA with markers at the final states substantially reduces the space, needed to store this conceptual resource, and radically improves the run-time calculation of injective projection. As the whole knowledge base is turned into a single deterministic automaton, the run-time speed depends on the query length only, no matter how large the initial knowledge base is. At the same time the final archive is rather compact, it is only 2-3 times bigger than the zipped-versions of the input text file, and still can be kept in the RAM during the run-time processing.

It is important to stress that the suggested encoding is an alternative knowledge base representation and the graphs can be kept in other formats, too, to facilitate other CG operations and visualization for different purposes. We believe that effective processing of very large conceptual resources will be a

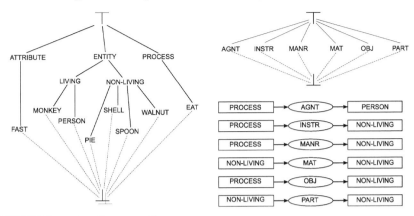

FIGURE 4.10: Support S_2.

must in the foreseeable future; therefore new methods for knowledge processing should be invented to cope with the future challenges in advanced knowledge-based applications.

Acknowledgments

The author is grateful to Stoyan Mihov for the fruitful collaboration in the development of the ideas presented here as well as to Ognian Kalaydjiev and Pavlin Dobrev for their support in the experimental tests.

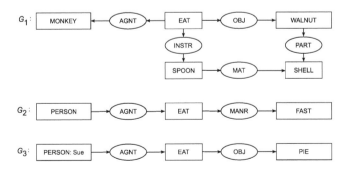

FIGURE 4.11: Knowledge base KB_2 of SCGs that encode facts about the process EAT.

4.6   Student Projects

4.6.1   Project 1

This chapter introduces an encoding of all SCGs with binary conceptual relations, which has to be maintained off-line. Some SCGs might be deleted or added to the knowledge base and other SCGs could be updated. Update KB_1 by adding to it the cyclic graph shown at Figure 4.8(B). What changes need to be made in the FSA with markers at the final states, which is shown at Figure 4.7?

4.6.2   Project 2

Consider the support S_2 and the knowledge base KB_2 presented respectively in Figures 4.10 and 4.11. One of the SCGs in KB_2 contains five elementary conjuncts and numerous subgraphs. However, the flat type hierarchies and the star graphs in S_2 restrict the number of all injective generalizations, so the resulting data set is still easy to maintain even manually. Let the support S_2 contains thirteen concept types, six relation types and one individual:

- T_C = {ATTRIBUTE, ENTITY, PROCESS, LIVING, NON-LIVING, EAT, FAST, MONKEY, PERSON, PIE, SHELL, SPOON, WALNUT} with partial order shown at Figure 4.10;

- T_R = {AGNT, INSTR, MANR, MAT, OBJ, PART} with partial order and star graphs shown at Figure 4.10;

- I = {Sue};

- τ(Sue) = PERSON.

For S_2 and KB_2, perform steps 1-4 from the algorithm sketched in section 3, by building and ordering all subgraphs and their injective generalizations. Try to construct the minimal FSA with markers at the final states, which encodes the list of all sorted generalizations, by creating states according to the lexicographic order of the word symbols. Construct the automaton word by word. For every new word, produce a new branch for the beginning symbols, which are not present as prefix of the previous word. Join the common ends of words with equivalent markers, producing an automaton that is similar to the one shown at Figure 4.7. Actually the acyclic automaton at Figure 4.7 is build by following the lexicographically sorted list of words in Tables 4.2 to 4.4 and the "order" of its arcs is clearly seen at Figure 4.7. What would happen, if the final automaton is a not minimal one? Is it easier to build a trie with markers at the final states, in order to model the list of all injective generalizations?

Part III

Foundations

Chapter 5

Conceptual Graphs for Representing Conceptual Structures

John F. Sowa

VivoMind Intelligence, Inc.
sowa@vivomind.com

Abstract   A conceptual graph (CG) is a graph representation for logic based on the semantic networks of artificial intelligence and the existential graphs of Charles Sanders Peirce. CG design principles emphasize the requirements for a cognitive representation: a smooth mapping to and from natural languages; an "iconic" structure for representing patterns of percepts in visual and tactile imagery; and cognitively realistic operations for perception, reasoning, and language understanding. The regularity and simplicity of the graph structures also support efficient algorithms for searching, pattern matching, and reasoning. Different subsets of conceptual graphs have different levels of expressive power: the ISO standard conceptual graphs express the full semantics of Common Logic (CL), which includes the subset used for the Semantic Web languages; a larger CG subset adds a context notation to support metalanguage and modality; and the research CGs are exploring an open-ended variety of extensions for aspects of natural language semantics. Unlike most notations for logic, CGs can be used with a continuous range of precision: at the formal end, they are equivalent to classical logic; but CGs can also be used in looser, less disciplined ways that can accommodate the vagueness and ambiguity of natural languages. This chapter surveys the history of conceptual graphs, their relationship to other knowledge representation languages, and their use in the design and implementation of intelligent systems.

5.1   Representing Conceptual Structures

Conceptual graphs are a notation for representing the conceptual structures that relate language to perception and action. Such structures must exist, but their properties can only be inferred from indirect evidence. Aristotle's inferences, with later extensions and qualifications, are still fundamentally sound: the meaning triangle of symbol, concept, and object; logic as a method of analyzing reasoning (*logos*); and a hierarchy of psyches ranging from the vegetative psyche of plants, the psyche of primitive animals like sponges, the locomotive pysche of worms, the imagery of psyches with sight and hearing, to the human psyche of an animal having logos (*zôon logon echein*). The medieval Scholastics extended Aristotle's logic, linguistics, and psychology, but Locke, Condillac, and the British empiricists developed more loosely structured theories about associations of ideas. Kant introduced schemata as tightly structured patterns of concepts and percepts, which became the foundation for many later developments. Peirce integrated aspects of all these proposals with modern logic in constructing a theory of signs that he called *semeiotic*.

In the 20th century, behaviorists tried to avoid hypotheses about conceptual structures; fortunately, many psychologists ignored them. Otto Selz [1913, 1922], who was dissatisfied with the undirected associationist theories, adapted Kant's schemata for a goal-directed theory he called *schematic anticipation*. Selz represented each schema as a network of concepts that contained empty slots, and he asked subjects to suggest appropriate concepts to fill the slots while he recorded their verbal protocols. Figure 5.1 shows a schema that Selz used in his experiments.

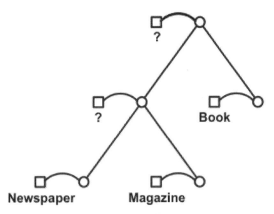

FIGURE 5.1:   A schema used by Otto Selz.

The expected answers to the question marks in Figure 5.1 are generalizations of the words at the bottom: the supertype of Newspaper and Magazine is Periodical, and the supertype of Periodical and Book is Publication. After analyzing the methods subjects use to solve such puzzles, Selz proposed a theory of goal-directed search that starts with a schema as an anticipation of the final result and propagates the question marks to subsidiary schemata. Selz's theories have a strong similarity to the backtracking methods developed by Newell and Simon [1972]. That similarity is not an accident. Newell and Simon learned Selz's theories from one of their visitors, the psychologist Adriaan de Groot, who used Selz's methods to study the thinking processes of chess players. One of their students, Quillian [1968], cited Selz as a source for his version of semantic networks. For computation, Quillian designed a *marker passing* algorithm, inspired by Selz's ideas for propagating question marks from one schema to another.

Another source for semantic networks was the *dependency grammar* developed by Lucien Tesnière [1959]. Figure 5.2 shows a dependency graph for the sentence *L'autre jour, au fond d'un vallon, un serpent piqua Jean Fréron* (The other day, at the bottom of a valley, a snake stung Jean Fréron). At the top is the verb *piqua* (stung); each word below it depends on the word above to which it is attached. The bull's eye symbol indicates an implicit preposition (*à*).

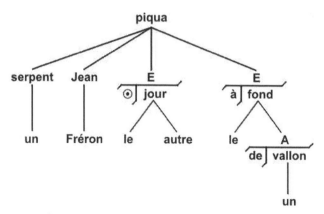

FIGURE 5.2:   A dependency graph in Tesnière's notation.

Tesnière had a major influence on linguistic theories that place more emphasis on semantics than syntax. Hays [1964] proposed dependency graphs as an alternative to Chomsky's notation, and Klein and Simmons [1963] developed a related version for machine translation. Those systems influenced Schank [1975], who adopted dependency graphs, but shifted the emphasis to concepts rather than words. Figure 5.3 shows a *conceptual dependency graph* for the sentence *A dog is greedily eating a bone*. Instead of Tesnière's tree notation, Schank used different kinds of arrows for different relations, such as ⇔ for the agent-act relation and an arrow marked with *o* for object or *d* for direction. He also replaced the words *eat* and *greedily* with labels that represent the concept types `Ingest` and `Greedy`. The subscript 1 on `Dog` indicates that the bone went into the same dog that ingested it.

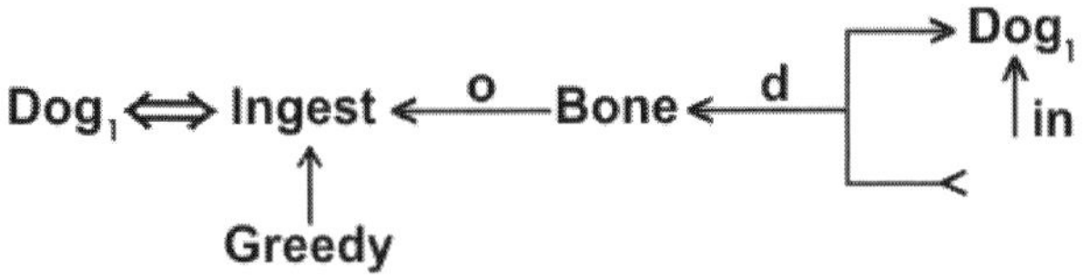

FIGURE 5.3:   Schank's notation for conceptual dependencies.

The early semantic networks were used for machine translation and question answering, but they could not represent all the features of logic. The first publication on conceptual graphs [Sowa, 1976] combined semantic networks with the quantifiers of predicate calculus and labeled the links between concepts with the *case relations* or *thematic roles* of linguistics [Fillmore, 1968]. That paper also presented a graph grammar for conceptual graphs based on four *canonical formation rules*. As an application, it illustrated CGs for representing natural language questions and mapping them to *conceptual schemata*. Each schema contained a declarative CG with attached *actor nodes* that represented functions or database relations. For computation, it proposed two kinds of marker passing for invoking the actors: backward-chaining markers, as in the networks by Selz and Quillian, and forward-chaining markers, as in Petri nets [Petri, 1965]. As an example of the 1976 notation, Figure 5.4 shows a conceptual graph for the sentence *On Fridays, Bob drives his Chevy to St. Louis*.

The rectangles in Figure 5.4 are called *concept nodes*, and the circles are called *conceptual relation nodes*. An arc pointing toward a circle marks the first *argument* of the relation, and an arc pointing away from a circle marks the last argument. If a relation has only one argument, the arrowhead is omitted. If a relation has more than two arguments, the arrowheads are replaced by integers $1,...,n$. Each concept node has a *type label*, which represents the type of entity the concept refers to: `Friday`, `Person`, `Drive`, `City`, `Chevy`, or `Old`. One of the concepts has a *universal quantifier* ∀ to represent *every Friday*; two concepts identify their referents by the names `Bob` and `"St. Louis"`; and the remaining three concepts represent the existence of a Chevy, an instance of driving, and an instance of oldness. Each of the six relation nodes has a label that represents the type of relation: agent (`Agnt`), point-in-time (`PTim`), destination (`Dest`), possession (`Poss`), theme (`Thme`), or attribute (`Attr`). The CG as a whole asserts that on every Friday, the person Bob, who possesses

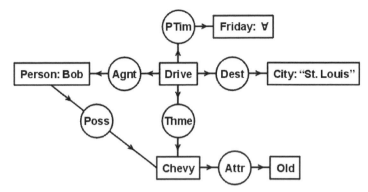

FIGURE 5.4:   Conceptual graph.

an old Chevy, drives it to St. Louis. Figure 5.4 can be translated to the
following formula in a typed version of predicate calculus:

```
(∀x1:Friday)(∃x2:Drive)(∃x3:Chevy)(∃x4:Old)
(Person(Bob) ∧ City("St. Louis") ∧ PTim(x2,x1)
  ∧ Agnt(x2,Bob) ∧ Poss(Bob,x3) ∧ Thme(x2,x3)
  ∧ Attr(x3,x4) ∧ Dest(x2,"St. Louis"))
```

As this translation shows, any concept without a name or a universal quantifier has an implicit existential quantifier. The default assumption for scope
gives the universal quantifier higher precedence than existentials. That leaves
open the question whether Bob drives the same old Chevy or a different one
on each Friday.

A later version of CGs [Sowa, 1984] used Peirce's *existential graphs* (EGs) as
the logical foundation. An important feature of EGs is an explicit enclosure
to delimit the scope of quantifiers and other logical operators. The CG in
Figure 5.5 has a large *context box* as a delimiter; the subgraph for Bob and his
old Chevy is outside that scope. Since the referent for the city of St. Louis is
designated by a proper name, it is a constant, which can be left inside the box
or moved outside the box without any change to the logic. The resulting CG
represents *Bob has an old Chevy, and on Fridays, he drives it to St. Louis.*

The translation of Figure 5.5 to predicate calculus moves the quantifiers
for Bob and his old Chevy in front of the universal. The only existential
quantifier that remains within the scope of the universal is the one for the
concept [Drive]:

```
(∃x1:Chevy)(∃x2:Old)(Person(Bob) ∧ Poss(Bob,x1) ∧ Attr(x1,x2)
    ∧ (∀x3:Friday)(∃x4:Drive)(City("St. Louis") ∧ PTim(x4,X3)
        ∧ Agnt(x4,Bob) ∧ Thme(x4,x1) ∧ Dest(x4,"St. Louis")))
```

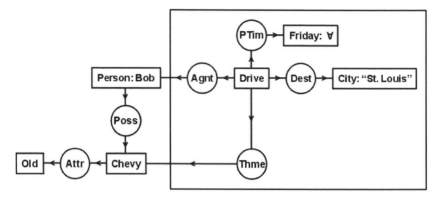

FIGURE 5.5: A context box for delimiting the scope of quantifiers.

The graphical notation illustrated in Figures 5.4 and 5.5 is called the CG *display form*. A linear representation for CGs, called the *Conceptual Graph Interchange Format* (CGIF), is one of the three standard dialects for *Common Logic* (ISO/IEC 24707). CGIF has a one-to-one mapping to and from the nodes of the display form. Following is the CGIF version of Figure 5.5:

```
[Person Bob] [Chevy *x1] [Old *x2] (Poss Bob ?x1) (Attr ?x1 ?x2)
[[Friday @every*x3] [Drive *x4] [City "St. Louis"] (PTim ?x4 ?x3)
   (Agnt ?x4 Bob) (Thme ?x4 ?x1) (Dest ?x2 "St. Louis") ]
```

Square brackets represent the concept nodes, and parentheses represent the relation nodes. Connections between concepts and relations are indicated by names, such as `Bob` and `"St. Louis"` or by *coreference labels* such as `*x1` and `?x1`. Special characters such as ∀ are represented by ASCII strings such as `@every`; the conjunction symbol ∧ is not needed, since the graph implies a conjunction of all the nodes in the same context. The first occurrence of any coreference label, called a *defining label*, is marked by an asterisk (`*x1`); *bound labels*, marked by a question mark (`?x1`), indicate a link to the node that contains the corresponding defining label. Note that `Friday` and `Chevy` represent types, not names of instances. Type labels are placed on the left side of a node; names, quantifiers, and coreference labels are on the right.

Graphs have advantages over linear notations in human factors and computational efficiency. As Figures 5.4 and 5.5 illustrate, the CG display form can show relationships at a glance that are harder to see in the linear notations for logic. Graphs also have a highly regular structure that can simplify many algorithms for searching, pattern matching, and reasoning. Although nobody knows how any information is represented in the brain, graphs minimize extraneous detail: they show connections directly, and they avoid the ordering implicit in strings or trees. The remaining sections of this chapter develop these ideas further: natural logic in Section 2; reasoning methods in Section

3; context and metalanguage in Section 4; research issues in Section 5; and the ISO standard for Common Logic in the appendix, which includes the CGIF grammar.

5.2   Toward a Natural Logic

From Aristotle and Euclid to Boole and Peirce, the twin goals of logic were to understand the reasoning processes in language and thought and to develop a tool that could facilitate reasoning in philosophy, mathematics, science, and practical applications. Various researchers put more emphasis on one goal or the other. Selz, Tesnière, and many linguists and psychologists studied the mechanisms underlying language and thought. Frege and most 20th-century logicians emphasized the applications to mathematics. Peirce and some AI researchers put equal emphasis on both. During his long career, Peirce invented several different notations for logic: an algebra of relations [1870] that is similar to the relational algebra used in database systems; an extension to Boolean algebra [1880, 1885], which became the modern predicate calculus; and existential graphs [1906, 1909], whose inference rules, he claimed, present "a moving picture of thought."

The first complete version of first-order logic was a tree notation by Frege [1879] called *Begriffsschrift* (concept writing). For his trees, Frege used only four operators: assertion (the "turnstile" operator ⊢), negation (a short vertical line), implication (a hook), and the universal quantifier (a cup containing the bound variable). Figure 5.6 shows Frege's notation for the sentence *Bob drives his Chevy to St. Louis.*

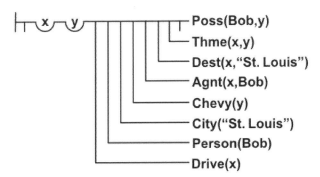

FIGURE 5.6:   Frege's Begriffsschrift for *Bob drives his Chevy to St. Louis.*

Frege had a contempt for language, and set out "to break the domination of the word over the human spirit by laying bare the misconceptions that through the use of language often almost unavoidably arise concerning the relations between concepts." His choice of operators simplified his rules of inference, but they led to cumbersome paraphrases in natural language. A direct translation of Figure 5.6 to predicate calculus would be

$$\sim(\forall x)(\forall y)\ (\text{Drive}(x)\ \supset\ (\text{Person}(\text{Bob})\ \supset\ (\text{City}(\text{"St. Louis"})\ \supset$$
$$(\text{Chevy}(y)\ \supset\ (\text{Agnt}(x,\text{Bob})\ \supset\ (\text{Dest}(x,\text{"St. Louis"})$$
$$\supset\ (\text{Thme}(x,y)\ \supset\ \sim\text{Poss}(\text{Bob},y)))))))))$$

In English, this formula could be read *It is false that for every x and y, if x is an instance of driving then if Bob is a person then if St. Louis is a city then if y is a Chevy then if the agent of x is Bob then if the destination of x is St. Louis then if the theme of x is y then Bob does not possess y.*

Although Peirce had invented the algebraic notation for predicate calculus, he believed that a graph representation would be more cognitively realistic. While he was still developing the algebra of logic, he experimented with a notation for *relational graphs*. Figure 5.7 shows a relational graph for the same sentence as Figure 5.6. In that graph, an existential quantifier is represented by line or by a *ligature* of connected lines, and conjunction is the default Boolean operator. Since those graphs did not represent proper names, monadic predicates isBob and isStLouis are used to represent names.

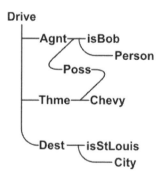

FIGURE 5.7: A relational graph for *Bob drives a Chevy to St. Louis.*

Figure 5.7 contains four ligatures: one for the instance of driving, one for Bob, one for the Chevy, and one for St. Louis. Each ligature maps to an existentially quantified variable in predicate calculus:

$$(\exists x)(\exists y)(\exists z)(\exists w)(\text{Drive}(x)\ \wedge\ \text{Agnt}(x,y)\ \wedge\ \text{Person}(y)\ \wedge\ \text{isBob}(y)$$
$$\wedge\ \text{Poss}(y,z)\wedge\ \text{Thme}(x,z)\ \wedge\ \text{Chevy}(z)\ \wedge\ \text{Dest}(x,w)\ \wedge$$
$$\text{City}(w)\ \wedge\ \text{isStLouis}(w))$$

Peirce experimented with various graphic methods for representing the other operators of his algebraic notation, but like the AI researchers of the 1960s, he couldn't find a good way to express the scope of quantifiers and negation. In 1897, however, he discovered a simple, but brilliant innovation for his new version of *existential graphs* (EGs): an oval enclosure for showing scope. The default operator for an oval with no other marking is negation, but any metalevel relation can be linked to the oval. The graph on the left of Figure 5.8 is an EG for the sentence *If a farmer owns a donkey, then he beats it.* Since an implication $p \supset q$ is equivalent to $\sim(p \wedge \sim q)$, the nest of two ovals expresses **if** p **then** q. To enhance the contrast, Peirce would shade any area enclosed in an odd number of negations.

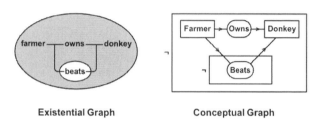

Existential Graph   **Conceptual Graph**

FIGURE 5.8:   EG and CG for *If a farmer owns a donkey, then he beats it.*

The equivalent conceptual graph is on the right of Figure 5.8. Since boxes nest better than ovals, Peirce's ovals are represented by rectangles marked with the symbol $\neg$ for negation. The choice of ovals or boxes, however, is a trivial difference. Three other differences are more significant: first, each rectangle can be interpreted as a concept node to which conceptual relations other than negation may be attached; second, the existential quantifiers, which are represented by EG lines, are represented by CG nodes, which may contain proper names, universal quantifiers, or even *generalized quantifiers*; and third, the type labels on the left side of a concept node restrict the range of quantifiers. Therefore, the EG maps to an untyped formula:

$\sim(\exists x)(\exists y)(\texttt{Farmer}(x) \ \wedge \ \texttt{Donkey}(y) \ \wedge \ \texttt{Owns}(x,y) \ \wedge \ \sim\texttt{Beats}(x,y))$

But the CG maps to the logically equivalent typed formula:

$\sim(\exists x\!:\!\texttt{Farmer})(\exists y\!:\!\texttt{Donkey})(\texttt{Owns}(x,y) \ \wedge \ \sim\texttt{Beats}(x,y))$

In order to preserve the correct scope of quantifiers, the implication operator $\supset$ cannot be used to represent an English *if-then* sentence unless the existential quantifiers are converted to universals and moved to the front:

$(\forall x)(\forall y)((\texttt{Farmer}(x) \ \wedge \ \texttt{Donkey}(y) \ \wedge \ \texttt{Owns}(x,y)) \ \supset \ \texttt{Beats}(x,y))$

In English, this formula could be read *For every x and y, if x is a farmer who owns a donkey y, then x beats y.* The unusual nature of this paraphrase led Kamp [1981] to develop *discourse representation structures* (DRSs) whose logical structure is isomorphic to Peirce's existential graphs (Figure 5.9).

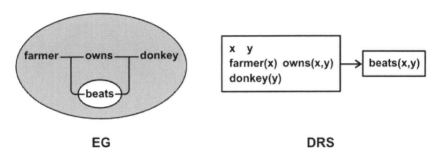

EG   DRS

FIGURE 5.9: EG and DRS for *If a farmer owns a donkey, then he beats it.*

Kamp's primitives are the same as Peirce's: the default quantifier is the existential, and the default Boolean operator is conjunction. Negation is represented by a box, and implication is represented by two boxes. As Figure 5.9 illustrates, the EG contexts allow quantifiers in the *if* clause to include the *then* clause in their scope. Although Kamp connected his boxes with an arrow, he made the same assumption about the scope of quantifiers. Kamp and Reyle [1993] went much further than Peirce in analyzing discourse and formulating the rules for interpreting anaphoric references, but any rule stated in terms of the DRS notation can also be applied to the EG or CG notation.

The CG in Figure 5.8 represents the verbs *owns* and *beats* as dyadic relations. That was the choice of relations selected by Kamp, and it can also be used with the EG or CG notation. Peirce, however, noted that the event or state expressed by a verb is also an entity that can be referenced by a quantified variable. That point was independently rediscovered by linguists, computational linguists, and philosophers such as Davidson [1967]. The CG in Figure 5.10 represents events and states as entities linked to their participants by linguistic relations. The type labels If and Then indicate negated contexts; the two new relation types are Expr for experiencer and Ptnt for patient.

All the notations in this section, including the diagrams and predicate calculus, can express full first-order logic. Since natural languages can also express full FOL, that expressive power is a requirement for a natural logic. But natural languages can express much more or much less, and they can be vague, precise, cryptic, erudite, or naive. Therefore, a natural logic must be flexible enough to vary the level of expression and precision as needed. The choice of logical operators is one consideration. Frege's choice is awkward for the

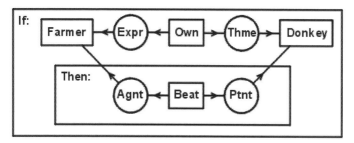

FIGURE 5.10:   CG with case relations shown explicitly.

sample sentence, and Peirce's choice is better. Predicate calculus has five or six common operators, and it can accommodate new ones as needed. For mathematics, many choices of primitives, such as Frege's or Peirce's, can be used to define all the others. For empirical subjects, however, conjunction and the existential quantifier are the only operators that can be observed directly, and the others must be inferred from indirect evidence. Therefore, Peirce's choice of primitives combined with a mechanism for defining other operators seems appropriate.

The structure of logical expressions is another consideration. As Peirce, Tesnière, and the AI researchers have shown, graphs can be directly mapped to natural languages, they can be as precise as any other notation for logic, and they can represent vague sentences as easily as precise ones. Graphs have the minimum syntax needed to show connections, but something more is required to show the scope of the logical operators. The fact that Peirce and Kamp independently discovered isomorphic enclosures for showing scope is an important point in their favor. Those arguments suggest that relational graphs with the Peirce-Kamp enclosures are a good candidate for the structure. The CG extensions show that more features can be added while preserving the overall simplicity. A truly natural logic should support reasoning methods that could also be considered natural.

5.3   A Moving Picture of Thought

Like the graphs themselves, Peirce's rules for graph logics are simple enough to be considered candidates for a natural logic: each rule of inference inserts or erases a single graph or subgraph. Peirce's proof procedure is a generalization and simplification of *natural deduction,* which Gentzen [1935] invented 30 years later. Although Peirce originally stated his rules in terms of the EG syntax, they can be restated in a form that depends only on the semantics. That approach has a striking similarity to recent work on the category-theoretic treatment of proofs [Hughes, 2006, McKinley, 2006]. As a psychological hypothesis, a syntax-independent procedure can be evaluated in terms of external evidence without any assumptions about internal representations.

To implement a proof procedure, semantic rules must be related to syntax. In 1976, the canonical formation rules were presented as a generative grammar for CGs. The 1984 version was also a generative grammar, but the rules were explicitly related to their semantic effects. The latest version [Sowa, 2000a] classifies the rules in three groups according to their semantic effects: *equivalence*, *specialization*, and *generalization*. Each rule transforms a starting graph or graphs u to a resulting graph v:

- **Equivalence**. *Copy* and *simplify* are equivalence rules, which generate a graph v that is logically equivalent to the original: $u \supset v$ and $v \supset u$. Equivalent graphs are true in exactly the same models.

- **Specialization**. *Join* and *restrict* are specialization rules, which generate a graph v that implies the original: $v \supset u$. Specialization rules monotonically decrease the set of models in which the result is true.

- **Generalization**. *Detach* and *unrestrict* are generalization rules, which generate a graph v that is implied by the original: $u \supset v$. Generalization rules monotonically increase the set of models in which the result is true.

Each rule has an inverse rule that undoes any change caused by the other. The inverse of copy is simplify, the inverse of restrict is unrestrict, and the inverse of join is detach. Combinations of these rules, called *projection* and *maximal join*, perform larger semantic operations, such as answering a question or comparing the relevance of different alternatives. The next three diagrams (Figures 5.11, 5.12, and 5.13) illustrate these rules with *simple graphs*, which use only conjunction and existential quantifiers.

The CG at the top of Figure 5.11 represents the sentence *The cat Yojo is chasing a mouse*. The down arrow represents two applications of the copy rule. One application copies the `Agnt` relation, and a second copies the subgraph →(Thme)→[Mouse]. The dotted line connecting the two [Mouse] concepts is a *coreference link*, which indicates that both concepts refer to the same individual; it represents equality in predicate calculus. The up arrow represents two applications of the simplify rule, which performs the inverse operations of erasing redundant copies. Following are the CGIF sentences for both graphs:

```
[Cat: Yojo] [Chase: *x] [Mouse: *y] (Agent ?x Yojo) (Thme ?x ?y)

[Cat: Yojo] [Chase: *x] [Mouse: *y] [Mouse: ?y]
(Agent ?x Yojo) (Agent ?x Yojo) (Thme ?x ?y) (Thme ?x ?y)
```

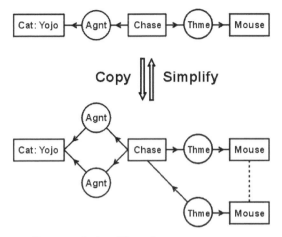

FIGURE 5.11:   Copy and simplify rules.

As the CGIF illustrates, the copy rule makes redundant copies, which are erased by the simplify rule. In effect, the copy rule is $p \supset (p \wedge p)$, and the simplify rule is $(p \wedge p) \supset p$.

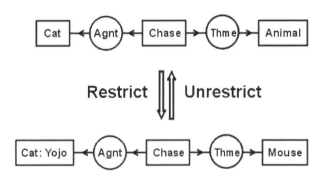

FIGURE 5.12:   Restrict and unrestrict rules.

The CG at the top of Figure 5.12 represents the sentence *A cat is chasing an animal.* By two applications of the restrict rule, it is transformed to the CG for *The cat Yojo is chasing a mouse.* In the first step, the concept [Cat], which says that there exists some cat, is *restricted by referent* to the more specific concept [Cat: Yojo], which says that there exists a cat named Yojo. In the second step, the concept [Animal], which says that there exists an animal, is *restricted by type* to a concept of a subtype [Mouse]. The more specialized graph implies the more general one: if the cat Yojo is chasing a mouse, then a cat is chasing an animal. To show that the bottom graph v implies the top graph u, let c be a concept of u that is being restricted to a more specialized concept d, and let u be $c \wedge w$, where w is the remaining information in u. By hypothesis, $d \supset c$. Therefore, $(d \wedge w) \supset (c \wedge w)$. Hence, $v \supset u$.

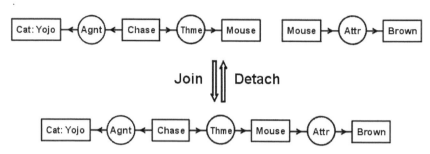

FIGURE 5.13:   Join and detach rules.

At the top of Figure 5.13 are two CGs for the sentences *Yojo is chasing a mouse* and *A mouse is brown*. The join rule overlays the two identical copies of the concept [Mouse] to form a single CG for the sentence *Yojo is chasing a brown mouse*. The detach rule undoes the join to restore the top graphs. Following are the CGIF sentences that represent the top and bottom graphs of Figure 5.13:

```
[Cat: Yojo] [Chase: *x] [Mouse: *y] (Agent ?x Yojo) (Thme ?x ?y)
[Mouse: *z] [Brown: *w] (Attr ?z ?w)
```

```
[Cat: Yojo] [Chase: *x] [Mouse: *y] (Agent ?x Yojo) (Thme ?x ?y)
[Brown: *w] (Attr ?y ?w)
```

As the CGIF illustrates, the bottom graph consists of substituting y for every occurrence of z in the top graph and erasing redundant copies. In general, every join assumes an equality of the form $y = z$ and simplifies the result. If q is the equality and u is the top pair of graphs, then the bottom graph is equivalent to $q \wedge u$, which implies u. Therefore, the result of join implies the original graphs. Together, the generalization and equivalence rules are sufficient for a complete proof procedure for the simple graphs with no negations. The specialization and equivalence rules can be used in a refutation procedure for a proof by contradiction.

To handle full first-order logic, rules for negations must be added to the canonical formation rules. Peirce defined a complete proof procedure for FOL whose rules depend on whether a context is positive (nested in an even number of negations, possibly zero) or negative (nested in an odd number of negations). Those rules are grouped in three pairs: one rule (i) inserts a graph, and the other (e) erases a graph. The only axiom is a blank sheet of paper (an empty graph with no nodes). In effect, the blank is a generalization of all other graphs. Following is a restatement of Peirce's rules in terms of specialization and generalization. These same rules apply to both propositional logic and full first-order logic. In FOL, the operation of inserting a coreference link between two nodes has the effect of identifying them (i.e., inserting an equality); erasing a coreference link has the inverse effect of erasing an equality. In a linear notation, such as CGIF or predicate calculus, the operation of inserting or erasing an equality requires an additional operation of renaming labels. In a pure graph notation, there are no labels and no need for renaming.

1. (i) In a negative context, any graph or subgraph (including the blank) may be replaced by any specialization.

    (e) In a positive context, any graph or subgraph may be replaced by any generalization (including the blank).

2. (i) Any graph or subgraph in any context c may be copied in the same context c or into any context nested in c. (No graph may be copied directly into itself. But it is permissible to copy a graph g in the context c and then make a copy of the copy inside the original g.)

    (e) Any graph or subgraph that could have been derived by rule 2i may be erased. (Whether or not the graph was in fact derived by 2i is irrelevant.)

3. (i) A double negation (nest of two negations with nothing between the inner and outer) may be drawn around any graph, subgraph, or set of graphs in any context.

    (e) Any double negation in any context may be erased.

This version of the rules was adapted from a tutorial on existential graphs by Peirce [1909]. When these rules are applied to CGIF, some adjustments may be needed to rename coreference labels or to convert a bound label to a defining label or vice versa. For example, if a defining node is erased, some bound label ?x may become the new defining label *x. Such adjustments are not needed in the pure graph notation.

All the axioms and rules of inference for classical FOL, including the rules of the *Principia Mathematica*, natural deduction, and resolution, can be proved in terms of Peirce's rules. As an example, Frege's first axiom, written in the algebraic notation, is $a \supset (b \supset a)$. Figure 5.14 shows a proof by Peirce's rules.

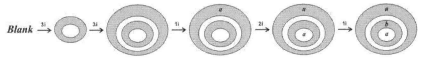

FIGURE 5.14: Proof of Frege's first axiom by Peirce's rules.

In CGIF, the propositions a and b can be represented as relations with zero arguments: (a) and (b). Following are the five steps of Figure 5.14:

1. By rule 3i, Insert a double negation around the blank: ~[~[ ]]

2. By 3i, insert a double negation around the previous one: ~[~[~[~[ ]]]]

3. By 1i, insert (a): ~[(a) ~[~[~[ ]]]]

4. By 2i, copy (a): ~[(a) ~[~[~[(a)]]]]

5. By 1i, insert (b): ~[(a) ~[~[(b) ~[(a)]]]]

The theorem to be proved contains five symbols, and each step of the proof inserts one symbol into its proper place in the final result. All the axioms of any version of FOL could be derived from a blank by similarly short proofs.

Frege's two rules of inference were *modus ponens* and *universal instantiation*. Figure 5.15 is a proof of *modus ponens*, which derives q from a statement p and an implication $p \supset q$:

FIGURE 5.15:   Proof of modus ponens.

Following is the CGIF version of Figure 5.15:

1. Starting graphs: (p) ~[(p) ~[(q)]]

2. By 2e, erase the nested copy of (p): (p) ~[~[(q)]]

3. By 1e, erase (p): ~[~[(q)]]

4. By 3e, erase the double negation: (q)

The rule of *universal instantiation* allows any term t to be substituted for a universally quantified variable in a statement of the form $(\forall x)P(x)$ to derive $P(t)$. In EGs, the term t would be represented by a graph of the form $-t$, which states that something satisfying the condition t exists: $(\exists y)t(y)$. The universal quantifier $\forall$ corresponds to a negated existential $\sim \exists \sim$, represented by a line whose outermost part occurs in a negative area. Since a graph has no variables, there is no notion of substitution. Instead, the proof in Figure 5.16 performs the equivalent operation by connecting the two lines.

The four steps of Figure 5.16 can be written in CGIF, but steps 2 and 3 require some adjustments to the coreference labels:

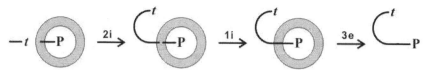

FIGURE 5.16:   Proof of universal instantiation.

1. Starting graphs: [*y] (t ?y) ~[[*x] ~[(P ?x)]]

2. By 2i, copy [*y] and change the defining label *y to a bound label ?y
   in the copy:
   [*y] (t ?y) ~[[?y] [*x] ~[(P ?x)]]

3. By 1i, insert a connection by relabeling *x and ?x to ?y, and erasing
   one copy of [?y]:
   [*y] (t ?y) ~[~[(P ?y)]]

4. By 3e, erase the double negation: [*y] (t ?y) (P ?y)

With the universal quantifier @every, the starting graphs of Figure 5.16
could be written

        [*y] (t ?y) [(P [@every*x])]

The extra brackets around the last node ensure that the existential quantifier [*y] is outside the scope of @every*x. Universal instantiation can be
used as a one-step rule to replace [@every*x] with ?y. Then the brackets
around [(P ?y)] may be erased to derive line 4 above.

In the *Principia Mathematica*, Whitehead and Russell proved the following
theorem, which Leibniz called the *Praeclarum Theorema* (Splendid Theorem).
It is one of the last and most complex theorems in propositional logic in the
Principia, and it required a total of 43 steps:

   $((p{\supset}r) \wedge (q \supset s)) \supset ((p{\wedge}q) \supset (r{\wedge}s))$

With Peirce's rules, this theorem can be proved in just seven steps starting
with a blank sheet of paper (Figure 5.17). Each step inserts or erases one
graph, and the final graph is the statement of the theorem.

After only four steps, the graph looks almost like the desired conclusion,
except for a missing copy of *s* inside the innermost area. Since that area is
positive, it is not permissible to insert *s* directly. Instead, Rule 2i copies the
graph that represents $q \supset s$. By Rule 2e, the next step erases an unwanted
copy of *q*. Finally, Rule 3e erases a double negation to derive the conclusion.

Unlike Gentzen's version of natural deduction, which uses a method of
making and discharging assumptions, Peirce's proofs proceed in a straight

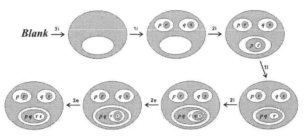

FIGURE 5.17:   Proof in 7 steps instead of 43 in the *Principia*.

line from a blank sheet to the conclusion: every step inserts or erases one subgraph in the immediately preceding graph. As Figure 5.17 illustrates, the first two steps of any proof that starts with a blank must draw a double negation around the blank and insert a graph into the negative area. That graph is usually the entire hypothesis of the theorem to be proved. The remainder of the proof develops the conclusion in the doubly nested blank area. Those two steps are the equivalent of Gentzen's method of making and discharging an assumption, but in Gentzen's approach, the two steps may be separated by arbitrarily many intervening steps, and a system of bookkeeping is necessary to keep track of the assumptions. With Peirce's rules, the second step follows immediately after the first, and no bookkeeping is required.

In summary, generalization and specialization, as performed by the canonical formation rules, are one-step operations that occur frequently in the ordinary use of language. Peirce's rules are also one-step operations that are simpler than the rules that Gentzen called "natural." The canonical formation rules have been implemented in nearly all CG systems, and they have been used in formal logic-based methods, informal case-based reasoning, and various computational methods. A multistep combination, called a *maximal join*, is used to determine the extent of the unifiable overlap between two CGs. In natural language processing, maximal joins help resolve ambiguities and determine the most likely connections of new information to background knowledge and the antecedents of anaphoric references. Stewart [1996] implemented Peirce's rules of inference in a first-order theorem prover for EGs and showed that its performance is comparable to resolution theorem provers. So far, no one has ever proposed a proof procedure that would have a better claim to the title "natural logic."

5.4   Representing Natural Language Semantics

Natural languages are highly expressive systems that can state anything that can be expressed in any formal language or logic. That enormous expressive power makes it difficult or impossible for any formalism to represent every feature of every natural language. To increase the range of expressibility, conceptual graphs constitute an open-ended family of notations with a formally defined core. Each of the following four levels of CGs can be expressed in the graphic display form or the linear CGIF:

- **Core CGs**. A typeless version of logic that expresses the full semantics of Common Logic, as defined by ISO/IEC 24707. This level corresponds to Peirce's existential graphs: its only logical primitives are conjunction, negation, and the existential quantifier. Core CGs permit quantifiers to range over relations, but Peirce also experimented with that option for EGs.

- **Extended CGs**. An upward compatible extension of the core, which adds a universal quantifier; type labels for restricting the range of quantifiers; Boolean contexts with type labels `If`, `Then`, `Either`, `Or`, `Equivalence`, and `Iff`; and the option of importing external CGIF text. Core and extended CGs have exactly the same expressive power, since the semantics of extended CGs is defined by a formal translation to core CGs. Extended CGs are usually more concise than core CGs, and they have a more direct mapping to and from natural languages.

- **CGs with contexts**. An upward compatible extension of core and extended CGs to support *metalanguage*, the option of using language to talk about language. That option requires some way of quoting or delimiting *object level* statements from *metalevel* statements. As a delimiter, CG notations use a type of concept node, called a *context*, which contains a nested conceptual graph (Figure 5.18). CG contexts can be formalized by the IKL semantics, which includes the CL semantics as a subset [Hayes and Menzel, 2006]. Another formalization, which can be adapted to the IKL semantics, is the system of *nested graph models* [Sowa, 2003, 2006]. Either or both of these formalizations can be used in an upward compatible extension of core and extended CGs.

- **Research CGs**. Adaptations of the CG notation for an open-ended variety of purposes. The advantage of a standard is a fixed, reliable platform for developing computer systems and applications. The disadvantage of a standard is the inability to explore useful modifications. But the combination of an ISO standard with the option of open-ended variations gives developers a solid basis for applications without limiting the creativity of researchers. Research extensions that prove to be useful may be incorporated in some future standard.

Peirce's first use for the oval was to negate the graphs nested inside, and that is the only use supported by the ISO standard. But Peirce [1898/1992] generalized the ovals to context enclosures, which allow relations other than negation to be linked to the enclosure. The basic use of a context enclosure is to quote the nested graphs. That syntactic option allows metalevel statements outside the context to specify how the nested (object level) graphs are interpreted. Nested graph models (NGMs) can be used to formalize the semantics of many kinds of modal and intentional logics. A hierarchy of metalevels with the NGM semantics can express the equivalent of a wide range of modal, temporal, and intentional logics. The most useful NGMs can be represented with the IKL semantics, but the many variations and their application to natural languages have not yet been fully explored.

The most common use of language about language is to talk about the beliefs, desires, and intentions of the speaker and other people. As an example, the sentence *Tom believes that Mary wants to marry a sailor*, contains three clauses, whose nesting may be marked by brackets:

```
Tom believes that [Mary wants [to marry a sailor]].
```

The outer clause asserts that Tom has a belief, which is the object of the verb *believe*. Tom's belief is that Mary wants a situation described by the nested infinitive, whose subject is the same person who wants the situation. Each clause makes a comment about the clause or clauses nested in it. References to the individuals mentioned in those clauses may cross context boundaries in various ways, as in the following two interpretations of the original English sentence:

```
Tom believes that [there is a sailor whom Mary wants [to marry]].
There is a sailor whom Tom believes that [Mary wants [to marry]].
```

The two conceptual graphs in Figure 5.18 represent the first and third interpretations. In the CG on the left, the existential quantifier for the concept [Sailor] is nested inside the situation that Mary wants. Whether such a sailor actually exists and whether Tom or Mary knows his identity are undetermined. The CG on the right explicitly states that such a sailor exists; the connections of contexts and relations imply that Tom knows him and that Tom believes that Mary also knows him. Another option (not shown) would place the concept [Sailor] inside the context of type Proposition; it would leave the sailor's existence undetermined, but it would imply that Tom believes he exists and that Tom believes Mary knows him.

The context boxes illustrated in Figures 5.4 and 5.6 express negations or operators such as If and Then, which are defined in terms of negations. But the contexts of type Proposition and Situation in Figure 5.18 raise new issues of logic and ontology. The CL semantics can represent entities of any type, including propositions and situations, but it has no provision for relating such entities to the internal structure of CL sentences. A more expressive language, called IKL [Hayes and Menzel, 2006], was defined as an upward

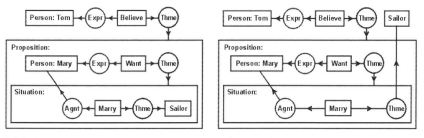

FIGURE 5.18: Two interpretations of *Tom believes that Mary wants to marry a sailor.*

compatible extension of CL. The IKL semantics introduces entities called *propositions* and a special operator, spelled `that`, which relates IKL sentences to the propositions they express. IKL semantics does not have a built-in type for situations, but it is possible in IKL to make statements that state the existence of entities of type `Situation` and relate them to propositions.

The first step toward translating the CGs in Figure 5.18 to IKL is to write them in an extended version of CGIF, which allows CGs to be nested inside concept nodes of type `Proposition` or `Situation`. Following is the CGIF for the CG on the left:

```
[Person: Tom] [Believe: *x1] (Expr ?x1 Tom)
(Thme ?x1 [Proposition:
   [Person: Mary] [Want: *x2] (Expr ?x2 Mary)
   (Thme ?x2 [Situation:
   [Marry: *x3] [Sailor: *x4] (Agnt ?x3 Mary) (Thme ?x3 ?x4)])])
```

This statement uses the option of moving the concept nodes for the types `Proposition` and `Situation` inside the relation nodes of type `Thme`. That option has no semantic significance, but it makes the order of writing the CGIF closer to English word order. A much more important semantic question is the relation between situations and propositions. In the ontology commonly used with CGs, that relation is spelled `Dscr` and called the *description relation*. The last two lines of the CGIF statement above could be rewritten in the following form:

```
(Thme ?x2 [Situation: *s]) (Dscr ?s [Proposition:
   [Marry: *x3] [Sailor: *x4] (Agnt ?x3 Mary) (Thme ?x3 ?x4)])])
```

The last line is unchanged, but the line before it states that the theme of x2 is the situation s and the description of s is the proposition stated on the last line. In effect, every concept of type `Situation` that contains a nested

CG is an abbreviation for a situation that is described by a concept of type
`Proposition` that has the same nested CG. This expanded CGIF statement
can then be translated to IKL (which is based on CLIF syntax with the
addition of the operator `that`).

```
(exists ((x1 Believe)) (and (Person Tom) (Expr x1 Tom)
(Thme x1 (that
   (exists ((x2 Want) (s Situation)) (and (Person Mary) (Expr x2 Mary)
   (Thme x2 s) (Dscr s (that
   (exists ((x3 Marry) (x4 Sailor)) (and (Agnt x3 Mary) (Thme x3 x4)
)))))))))
```

Each line of the IKL statement expresses the equivalent of the corresponding line in CGIF. Note that every occurrence of `Proposition` in CGIF corresponds to `that` in IKL. The syntax of CLIF or IKL requires more parentheses
than CGIF because every occurrence of `exists` or `and` requires an extra closing parenthesis at the end.

As these examples illustrate, the operator `that` adds an enormous amount
of expressive power, but IKL still has a first-order style of semantics. The
proposition nodes in CGs or the `that` operator in IKL introduce abstract
entities of type `Proposition`, but propositions are treated as zero-argument
relations, which are supported by the semantics of Common Logic. Although
language about propositions is a kind of metalanguage, it does not, by itself,
go beyond first-order logic. Tarski [1933], for example, demonstrated how a
stratified series of metalevels, each of which is purely first order, can be used
without creating paradoxes or going beyond the semantics of FOL. In effect,
Tarski avoided paradoxes by declaring that certain kinds of sentences (those
that violate the stratification) do not express propositions in his models. The
IKL model theory has a similar way of avoiding paradoxes: it does not require
every model to include a proposition for every possible sentence. For example,
the following English sentence, which sounds paradoxical, could be expressed
in either IKL or CGIF syntax:

> *There exists a proposition p, p is true,*
> *and p is the proposition that p is false.*

Since IKL does not require every sentence to express a proposition in every
model, there are permissible IKL models in which this sentence is false simply
because no such proposition exists. Therefore, the paradox vanishes because
the sentence has a stable, but false, truth value.

Issues of context and metalanguage require some syntactic and semantic
extensions beyond the CL standard. The syntax for context was proposed by
Sowa [1984] with a semantics that was a subset of the later IKL and NGM
versions. Syntax for the following three extensions has also been available
since 1984, and some CG systems have implemented versions of them. But

they are not yet in ISO standard CGIF because a fully general treatment would involve ongoing research in linguistics.

- **Generalized quantifiers.** In addition to the usual quantifiers of *every* and *some*, natural languages support an open-ended number of quantificational expressions, such as *exactly one*, *at least seven*, or *considerably more*. Some of these quantifiers, such as *exactly one cat*, could be represented as [Cat: @1] and defined in terms of the CL standard. Others, such as *at least seven cats*, could be represented [Cat: @≤7] and defined with a version of set theory added to the base logic. But quantifiers such as *considerably more* would require some method of approximate reasoning, such as fuzzy sets or rough sets.

- **Indexicals.** Peirce observed that every statement in logic requires at least one indexical to fix the referents of its symbols. The basic indexical, which corresponds to the definite article *the*, is represented by the symbol # inside a concept node: [Dog: #] would represent the phrase *the dog*. The pronouns *I*, *you*, and *she* would be represented [Person: #I], [Person: #you], and [Person: #she]. To process indexicals, some linguists propose versions of *dynamic semantics*, in which the model is updated during the discourse. A simpler method is to treat the # symbol as a syntactic marker that indicates a incomplete interpretation of the original sentence. With this approach, the truth value of a CG that contains any occurrences of # is not determined until those markers are replaced by names or coreference labels. This approach supports indexicals in an intermediate representation, but uses a conventional model theory to evaluate the final resolution.

- **Plural nouns.** Plurals have been represented in CGs by set expressions inside the concept boxes. The concept [Cat: *@3] would represent *three cats*, and [Dog: Lucky, Macula] would represent *the dogs Lucky and Macula*. Various methods have been proposed for representing distributed and collective plurals and translating them to versions of set theory and mereology.

Simple versions of these features have been implemented in CG systems. The difficult issues involve generalizing them in a systematic way to cover all the variations that occur in natural languages.

5.5   Ongoing Research

The major difference between natural languages and formal languages is not in the notation. For the world's first formal logic, Aristotle used a subset of Greek. Some computer systems use a *controlled natural language* based on a formally defined subset of a natural language. The defining characteristic of a formal language is that the meaning of any statement is completely determined by its form or syntax. Natural languages, however, are highly context dependent. Ambiguities and indexicals are two kinds of dependencies that have been thoroughly analyzed, but vagueness is more challenging. Peirce [1902, p. 748] defined vagueness in an article for Baldwin's *Dictionary of Philosophy and Psychology*:

> A proposition is vague when there are possible states of things concerning which it is intrinsically uncertain whether, had they been contemplated by the speaker, he would have regarded them as excluded or allowed by the proposition. By intrinsically uncertain we mean not uncertain in consequence of any ignorance of the interpreter, but because the speaker's habits of language were indeterminate.

In other writings, Peirce explained that a vague statement requires some additional information to make it precise, but the sentence itself gives little or no indication of what kind of information is missing. That characteristic distinguishes vagueness from the ambiguity illustrated in Figure 5.18. An ambiguous sentence has two or more interpretations, and the kind of information needed to resolve the ambiguity can usually be determined by analyzing the sentence itself. An indexical is a word or phrase, such as *she* or *the book*, whose referent is not specified, but the kind of thing that would serve as a referent and the means of finding it are usually determined. But the meaning of a vague sentence, as Peirce observed, is "intrinsically uncertain" because the method for finding the missing information is unknown, perhaps even to the speaker.

Peirce's definition is compatible with an approach to vagueness by the logical methods of *underspecification* and *supervaluation*. As an example, the graph on the left of Figure 5.18 is underspecified because it is true in all three of the interpretations of the sentence *Tom believes that Mary wants to marry a sailor*. The graph on the right is the most specific because it is true in only one interpretation. CGs represent indexicals with underspecified concepts such as [Person: #she] for *she* and [Book: #] for *the book*, whose referents could later be resolved by the methods of discourse representation theory. To accommodate the *truth-value gaps* of vague sentences that are neither true nor false, van Fraassen [1966] proposed supervaluation as a model-theoretic method that allows some sentence p to have an indeterminate truth value in a specific model. But other sentences, such as $p \lor \sim p$ would have a supervalue of true in all models, while the sentence $p \land \sim p$ would have the supervalue false in all models.

For language understanding, Hintikka [1973, p. 129] observed that infinite families of models might be theoretically interesting, but "it is doubtful whether we can realistically expect such structures to be somehow actually involved in our understanding of a sentence or in our contemplation of its meaning." As an alternative, he proposed a *surface model* of a sentence S as "a mental anticipation of what can happen in one's step-by-step investigation of a world in which S is true." Instead of a logic of vagueness, Hintikka suggested a method of constructing a model that would use all available information to fill in the details left indeterminate in a given text. The first stage of constructing a surface model begins with the entities occurring in a sentence or story. During the construction, new facts may be asserted that block certain extensions or facilitate others. A conventional model is the limit of a surface model that has been extended infinitely far, but such infinite processes are not required for normal understanding.

After forty years, supervaluations are still widely used by logicians, but Fodor and Lepore [1996] denounced them as useless for linguistics and psychology. Surface models, although largely neglected, can be constructed by constraint satisfaction methods similar to the heuristic techniques in AI and *dynamic semantics* in linguistics. Conceptual graphs are convenient for such methods because they can be used to represent arbitrary statements in logic, to represent formal models of those statements, and to represent each step from an initially vague statement to a complete specification. Any Tarski-style model, for example, can be represented as a potentially infinite simple graph, whose only logical operators are conjunction and existential quantification. Each step in constructing a surface model is a subgraph of a potentially infinite complete model, but for most applications, there is no need to finish the construction. Such techniques have been successfully used for understanding dialogs, in which fragments of information from multiple participants must be assembled to construct a larger view of a subject.

Implementations, in either computers or neurons, can exploit the topological properties of graphs, which include symmetry, duality, connectivity, and cycles. For knowledge representation, the topology often reveals similarities that are masked by different choices of ontology. Figure 5.19 shows two different ways of representing the sentence *Sue gives a child a book*. The CG on the left represents the verb by a triadic relation, while the one on the right represents it by a concept linked to three dyadic relations: agent, theme, and recipient. Different ontologies lead to different numbers of concept and relation nodes with different type labels, but a characteristic invariant of giving is the triadic connectivity of three participants to a central node.

Although the relation type Gives and the concept type Give have similar names, concept nodes, and relation nodes cannot be directly mapped to one another. But when the topology is analyzed, adjacent nodes can be merged to reduce the size of the graphs while preserving invariants such as cycles and connectivity. The invariants can indicate common underlying relationships expressed with different ontologies. Sowa and Majumdar [2003] presented a more complex example that compared one CG derived from a relational database to another CG derived from an English sentence, both of which rep-

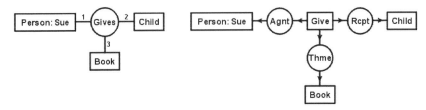

FIGURE 5.19:   Two ways of representing an act of giving.

resented the same physical structure. The database CG had 15 concept nodes
and 9 relation nodes, but the CG from English had 12 concept nodes and
11 relation nodes. Furthermore, no type label on any node of one CG was
identical to any type label on any node of the other. Given only the constraint
that five concept nodes of each CG referred to the same five objects, the algorithms correctly derived the mappings from one ontology to the other. Those
mappings could then be used to compare other representations with the same
two ontologies. Ongoing research on methods for representing graphs has led
to high-performance algorithms for searching, comparing, and transforming
large numbers of very large graphs.

As an example of applied research, one of the largest commercial CG systems is Sonetto [Sarraf and Ellis, 2006], which uses extended versions of earlier
algorithms by Levinson and Ellis [1992]. A key innovation of Sonetto is its
semi-automated methods for extracting ontologies and business rules from un-
structured documents. The users who assist Sonetto in the knowledge extraction process are familiar with the subject matter, but they have no training
in programming or knowledge engineering. CGIF is the knowledge representation language for ontologies, rules, and queries. It is also used to manage
the schemas of documents and other objects in the system and to represent
the rules that translate CGIF to XML and other formats. For the early CG
research, see the collections edited by Nagle et al. [1992], Way [1992], and
Chein [1996]. More recent research on CGs has been published in the annual
proceedings of the International Conferences on Conceptual Structures.

5.6   Appendix: The Common Logic Standard

Common Logic (CL) evolved from two projects to develop parallel ANSI
standards for conceptual graphs and the Knowledge Interchange Format [Gene-
sereth and Fikes, 1992]. Eventually, those projects were merged into a single
ISO project to develop a common abstract syntax and model-theoretic foundation for a family of logic-based notations (ISO/IEC 24707). Hayes and Menzel
[2001] defined a very general model theory for CL, which Hayes and McBride

[2003] used to define the semantics for the languages RDF(S) and OWL. In addition to the abstract syntax and model theory, the CL standard specifies three concrete dialects that are capable of expressing the full CL semantics: the Common Logic Interchange Format (CLIF), the Conceptual Graph Interchange Format (CGIF), and the XML-based notation for CL (XCL). Since the semantics of RDF and OWL is based on a subset of CL semantics, those languages can also be considered dialects of CL: any statement in RDF or OWL can be translated to CLIF, CGIF, or XCL, but only a subset of CL can be translated back to RDF or OWL.

The CL syntax allows quantifiers to range over functions and relations, but CL retains a first-order style of model theory and proof theory. To support a higher-order syntax, but without the computational complexity of higher-order semantics, the CL model theory uses a single domain D that includes individuals, functions, and relations. The option of limiting the domain of quantification to a single set was suggested by Quine [1954] and used in various theorem provers that allow quantifiers to range over relations [Chen et al., 1993].

The CL standard is defined in an abstract syntax that is independent of any concrete notation. It does, however, support the full Unicode character set and the URIs of the Semantic Web. The three dialects defined in the standard (CLIF, CGIF, and XCL) use only the ASCII subset of Unicode for their basic syntax, but they allow any Unicode symbols in names and character strings. Although CGIF and CLIF had different origins, the two notations have many similarities. As an example, following is the core CGIF for the sentence *Bob drives his Chevy to St. Louis*:

```
[*x] [*y]
(Drive ?x) (Person Bob) (City "St. Louis") (Chevy ?y)
(Agnt ?x Bob) (Dest ?x "St. Louis") (Thme ?x ?y) (Poss Bob ?y)
```

In core CGIF, the concept nodes [*x] and [*y] represent the existential quantifiers $\exists x$ and $\exists x$. Following is the CLIF statement, which uses the keyword exists to introduce a list of existentially quantified variables:

```
(exists (x y)
   (and (Drive x) (Person Bob) (City "St. Louis") (Chevy y)
   (Agnt x Bob) (Dest x "St. Louis") (Thme x y) (Poss Bob y) ))
```

Although CGIF and CLIF look similar, there are several fundamental differences:

1. Since CGIF is a serialized representation of a graph, labels such as x or y represent coreference links between nodes, but they represent variables in CLIF or predicate calculus.

2. CGIF distinguishes the labels *x and ?x from a name like Bob by an explicit prefix. CLIF, however, has no special markers on variables; the only distinction is that variables appear in a list after the quantifier.

3. Since the nodes of a graph have no inherent ordering, a CGIF sentence is an unordered list of nodes. Unless grouped by context brackets, the list may be permuted without affecting the semantics.

4. The CLIF operator and does not occur in CGIF because the conjunction of nodes within any context is implicit. Omitting the conjunction operator in CGIF tends to reduce the number of parentheses.

Extended CGIF allows monadic relation names to be used as type labels, and extended CLIF allows them to be used as restrictions on the scope of quantifiers. Following is the extended CGIF for the above sentence:

```
[Drive *x] [Person: Bob] [City: "St. Louis"] [Chevy *y]
(Agnt ?x Bob) (Dest ?x "St. Louis") (Thme ?x ?y) (Poss Bob ?y)
```

And following is the equivalent in extended CLIF:

```
(exists ((x Drive) (y Chevy))
  (and (Person Bob) (City "St. Louis") (Agnt x Bob)
    (Dest x "St. Louis") (Thme x y) (Poss Bob y) ))
```

Since the semantics of any statement in extended CGIF and CLIF is defined by its translation to the core language, neither language makes a semantic distinction between type labels and monadic relations. If a statement in a strongly typed language, such as the Z Specification Language, is translated to CGIF or CLIF, the Z types are mapped to CGIF type labels or CLIF quantifier restrictions. A syntactically correct statement in Z and its translation to CGIF or CLIF have the same truth value. But an expression with a type mismatch would cause a syntax error in Z, but it would merely be false in CGIF or CLIF.

As another example, Figure 5.20 shows a CG for the sentence *If a cat is on a mat, then it is a happy pet.* The dotted line that connects the concept [Cat] to the concept [Pet], which is called a *coreference link*, indicates that they both refer to the same entity. The Attr relation indicates that the cat, also called a pet, has an attribute, which is an instance of happiness.

The dotted line in Figure 5.20, called a *coreference link*, is shown in CGIF by the defining label *x in the concept [Cat: *x] and the bound label ?x in [Pet: ?x]. Following is the extended CGIF:

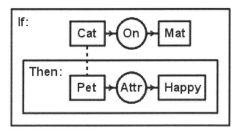

FIGURE 5.20: CG display form for *If a cat is on a mat, then it is a happy pet*.

```
[If: [Cat *x] [Mat *y] (On ?x ?y)
    [Then: [Pet ?x] [Happy *z] (Attr ?x ?z) ]]
```

In CGs, functions are represented by conceptual relations called *actors*. Figure 5.21 is the CG display form for the following equation written in ordinary algebraic notation:

```
y = (x + 7)/sqrt(7)
```

The three functions in this equation would be represented by three actors, which are shown in Figure 5.21 as diamond-shaped nodes with the type labels Add, Sqrt, and Divide. The concept nodes contain the input and output values of the actors. The two empty concept nodes contain the output values of Add and Sqrt.

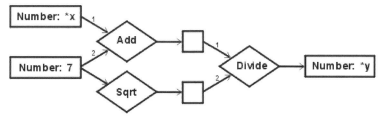

FIGURE 5.21: CL functions represented by actor nodes.

In CGIF, actors are represented as relations with two kinds of arcs: a sequence of *input arcs* separated by a vertical bar from a sequence of *output arcs*.

```
[Number: *x] [Number: *y] [Number: 7]
(Add ?x 7 | [*u]) (Sqrt 7 | [*v]) (Divide ?u ?v | ?y)
```

In the display form, the input arcs of `Add` and `Divide` are numbered 1 and 2 to indicate the order in which the arcs are written in CGIF. Following is the corresponding CLIF:

```
(exists ((x Number) (y Number))
    (and (Number 7) (= y (Divide (Add x 7) (Sqrt 7)))))
```

No CLIF variables are needed to represent the coreference labels *u and *v since the functional notation used in CLIF shows the connections directly.

CLIF only permits functions to have a single output, but extended CGIF allows actors to have multiple outputs. The following actor, which is of type `IntegerDivide`, has two inputs: an integer x and an integer 7. It also has two outputs: a quotient u and a remainder v.

```
(IntegerDivide [Integer: *x] [Integer: 7] | [*u] [*v])
```

When this actor is translated to core CGIF or CLIF, the vertical bar is removed, and the actor becomes an ordinary relation with four arguments; the distinction between inputs and outputs is lost. In order to assert the constraint that the last two arguments are functionally dependent on the first two arguments, the following CGIF sentence asserts that there exist two functions, identified by the coreference labels `Quotient` and `Remainder`, which for every combination of input and output values are logically equivalent to an actor of type `IntegerDivide` with the same input and output values:

```
[Function: *Quotient] [Function: *Remainder]
[[@every*x1] [@every*x2] [@every*x3] [@every*x4]
[Equiv: [Iff: (IntegerDivide ?x1 ?x2 | ?x3 ?x4)]
 [Iff: (#?Quotient ?x1 ?x2 | ?x3) (#?Remainder ?x1 ?x2 | ?x4)]]]
```

Each line of this example illustrates one or more features of CGIF. The first line represents existential quantifiers for two entities of type `Function`. On the second line, the context bracket [encloses the concept nodes with universal quantifiers, marked by @every, to show that the existential quantifiers for `Quotient` and `Remainder` include the universals within their scope. The equivalence on lines three and four shows that an actor of type `IntegerDivide` is logically equivalent to a conjunction of the quotient and remainder functions. Finally, the symbol # on line four shows that the coreference labels

?Quotient and ?Remainder are being used as type labels. Following is the corresponding CLIF:

```
(exists ((Quotient Function) (Remainder Function))
 (forall (x1 x2 x3 x4)
  (iff (IntegerDivide x1 x2 x3 x4)
   (and (= x3 (Quotient x1 x2)) (= x4 (Remainder x1 x2))))))
```

As another example of quantifiers that range over relations, someone might say "Bob and Sue are related," but not say exactly how they are related. The following sentences in CGIF and CLIF state that there exists some familial relation r that relates Bob and Sue:

> [Relation: *r] (Familial ?r) (#?r Bob Sue)

> (exists ((r Relation)) (and (Familial r) (r Bob Sue)))

The concept [Relation: *r] states that there exists a relation r. The next two relations state that r is familial and r relates Bob and Sue.

This brief survey has illustrated nearly every major feature of CGIF and CLIF. One important feature that has not been mentioned is a *sequence marker* to support relations with a variable number of arguments. Another is the use of comments, which can be placed before, after, or inside any concept or relation node in CGIF. The specifications in the CL standard guarantee that any sentence expressed in the dialects CGIF, CLIF, or XCL can be translated to any of the others in a logically equivalent form. Although the translation will preserve the semantics, it is not guaranteed to preserve all syntactic details: a sentence translated from one dialect to another and then back to the first will be logically equivalent to the original, but some subexpressions might be reordered or replaced by semantic equivalents. Following is a summary of the CGIF grammar; see ISO/IEC 24707 for the complete specification of the Common Logic syntax and semantics.

5.6.1   Lexical Grammar Rules

The syntax rules are written in Extended Backus-Naur Form (EBNF) rules, as specified by ISO/IEC 14977. The CGIF syntax rules assume the same four types of names as CLIF:

- namecharsequence for names not enclosed in quotes;

- enclosedname for names enclosed in double quotes;

- numeral for numerals consisting of one or more digits; and

- `quotedstring` for character strings enclosed in single quotes.

But because of syntactic differences between CGIF and CLIF, CGIF must enclose more names in quotes than CLIF in order to avoid ambiguity. Therefore, the only CG names not enclosed in quotes belong to the categories `identifier` and `numeral`.

```
CGname = identifier | '"', (namecharsequence - identifier), '"'
   | numeral | enclosedname | quotedstring;
      identifier = letter, letter | digit | "_";
```

When CGIF is translated to CL, a CGname is translated to a CLIF name by removing any quotes around a name character sequence. CLIF does not make a syntactic distinction between constants and variables, but in CGIF any CGname that is not used as a defining label or a bound label is called a *constant*. The start symbol of CGIF syntax is the category `text` for a complete text or the category `CG` for just a single conceptual graph.

5.6.2   Core CGIF Grammar Rules

An *actor* is a conceptual relation that represents a function in Common Logic. It begins with (, an optional comment, an optional string #?, a CG name, |, an arc, an optional end comment, and). If the CG name is preceded by #?, it represents a bound coreference label; otherwise, it represents a type label. The arc sequence represents the arguments of the CL function and the last arc represents the value of the function.

```
actor = "(", [comment], ["#?"], CGname, arcSequence, "|", arc,
   [endComment], ")";
```

An *arc* is an optional comment followed by a reference. It links an actor or a conceptual relation to a concept that represents one argument of a CL function or relation.

```
arc = [comment], reference;
```

An *arc sequence* is a sequence of zero or more arcs, followed by an option consisting of an optional comment, ?, and a sequence marker.

```
arcSequence = arc, [[comment], "?", seqmark];
```

A *comment* or an *end comment* is a character string that has no effect on the semantics of a conceptual graph or any part of a conceptual graph. A comment begins with "/*", followed by a character string that contains no occurrence of "*/", and ends with "*/". A comment may occur immediately after the opening bracket of any concept, immediately after the opening parenthesis of any actor or conceptual relation, immediately before any arc, or intermixed with the concepts and conceptual relations of a conceptual graph. An end comment begins with ;, followed by a character string that contains no occurrence of] or). An end comment may occur immediately before the closing bracket of any concept or immediately before the closing parenthesis of any actor or conceptual relation.

```
comment = "/*", (character-"*") |
   ["*", (character-"/")], ["*"], "*/";

endComment = ";", character - ("]" | ")");
```

A *concept* is either a context, an existential concept, or a coreference concept. Every concept begins with [and an optional comment; and every concept ends with an optional end comment and]. Between the beginning and end, a context contains a CG; an existential concept contains * and either a CG name or a sequence marker; and a coreference concept contains : and a sequence of one or more references. A context that contains a blank CG is said to be *empty*, even if it contains one or more comments; any comment that occurs immediately after the opening bracket shall be part of the concept, not the following CG.

```
concept = "[", [comment],
      (CG | "*", (CGname | seqmark) | ":", reference- ),
      [endComment], "]";
```

A *conceptual graph* (CG) is an unordered list of concepts, conceptual relations, negations, and comments.

```
CG = concept | conceptualRelation | negation | comment;
```

A *conceptual relation* is either an ordinary relation or an actor. An ordinary relation, which represents a CL relation, begins with (, an optional comment, an optional string #?, a CG name, an optional end comment, and). If the CG name is preceded by #?, it represents a bound coreference label; otherwise, it represents a type label. An ordinary relation has just one sequence of arcs, but an actor has two sequences of arcs.

```
conceptualRelation = ordinaryRelation | actor;

ordinaryRelation = "(", [comment], ["#?"], CGname, arcSequence,
   [endComment], ")";
```

A *negation* is ~ followed by a context.

```
negation = "~", context;
```

A *reference* is an optional ? followed by a CG name. A CG name prefixed with ? is called a *bound coreference label*; without the prefix ?, it is called a *constant*.

```
reference = ["?"], CGname;
```

A *text* is a context, called an *outermost context*, that has an optional name, has an arbitrarily large conceptual graph, and is not nested inside any other context. It consists of [, an optional comment, the type label `Proposition`, :, an optional CG name, a conceptual graph, an optional end comment, and]. Although a text may contain core CGIF, the type label Proposition is outside the syntax of core CGIF.

```
text = "[", [comment], "Proposition", ":", [CGname], CG,
   [endComment], "]";
```

5.6.3   Extended CGIF Grammar Rules

Extended CGIF is a superset of core CGIF, and every syntactically correct sentence of core CGIF is also syntactically correct in extended CGIF. Its most prominent feature is the option of a *type label* or a *type expression* on the left side of any concept. In addition to types, extended CGIF adds the following features to core CGIF:

- More options in concepts, including universal quantifiers.

- Boolean contexts for representing the operators or, if, and iff.

- The option of allowing concept nodes to be placed in the arc sequence of conceptual relations.

- The ability to import text into a text.

These extensions are designed to make sentences more concise, more readable, and more suitable as a target language for translations from natural languages and from other CL dialects, including CLIF. None of them, however, extend the expressive power of CGIF beyond the CG core, since the semantics of every extended feature is defined by its translation to core CGIF, whose semantics is defined by its translation to the abstract syntax of Common Logic.

The following grammar rules of extended CGIF have the same definitions as the core CGIF rules of the same name: `arcSequence`, `conceptualRelation`, `negation`, `ordinaryRelation`, `text`. The following grammar rules of extended CGIF don't occur in core CGIF, or they have more options than the corresponding rules of core CGIF: `actor`, `arc`, `boolean`, `CG`, `concept`, `eitherOr`, `equivalence`, `ifThen`, `typeExpression`.

An *actor* in extended CGIF has the option of zero or more arcs following | instead of just one arc.

```
actor = "(", [comment], ["#?"], CGname,
        arcSequence, "|", arc, [endComment], ")";
```

An *arc* in extended CGIF has the options of a defining coreference label and a concept in addition to a bound coreference label.

```
arc = [comment], (reference | "*", CGname | concept);
```

A *boolean* is either a negation or a combination of negations that represent an either-or construction, an if-then construction, or an equivalence. Instead of being marked with ∼, the additional negations are represented as contexts with the type labels `Either`, `Or`, `If`, `Then`, `Equiv`, `Equivalence`, or `Iff`.

```
boolean = negation | eitherOr | ifThen | equivalence;
```

A *concept* in extended CGIF permits any combination allowed in core CGIF in the same node and it adds two important options: a type field on the left side of the concept node, and a universal quantifier on the right. Four options are permitted in the type field: a type expression, a bound coreference label prefixed with "#", a constant, or the empty string; a colon is required after a type expression, but optional after the other three.

```
concept = "[", [comment],
          ( (typeExpression, ":"
          | ["#?"], CGname, [":"]),
          [["@every"], "*", CGname], {reference}, CG
```

```
     | ["@every"], "*", seqmark
), [endComment], "]";
```

A *conceptual graph* (CG) in extended CGIF adds Boolean combinations of contexts to core CGIF.

```
CG = {concept | conceptualRelation | boolean | comment};
```

An *either-or* is a negation with a type label `Either` that contains zero or more negations with a type label `Or`.

```
eitherOr = "[", [comment], "Either", [":"],
    {"[", [comment], "Or", [":"], CG, [endComment], "]"}
    [endComment], "]";
```

An *equivalence* is a context with a type label `Equivalence` or `Equiv` that contains two contexts with a type label `Iff`. It is defined as a pair of if-then constructions, each with one of the iff-contexts as antecedent and the other as consequent.

```
equivalence = "[", [comment], ("Equivalence" | "Equiv"), [":"],
    "[", [comment], "Iff", [":"], CG, [endComment], "]",
    "[", [comment], "Iff", [":"], CG, [endComment], "]",
    [endComment], "]";
```

An *if-then* is a negation with a type label `If` that contains a negation with a type label `Then`.

```
ifThen = "[", [comment], "If", [":"], CG,
    "[", [comment], "Then", [":"], CG, [endComment], "]",
    [endComment], "]";
```

A *type expression* is a lambda-expression that may be used in the type field of a concept. The symbol @ marks a type expression, since the Greek letter λ is not available in the ASCII subset of Unicode.

```
typeExpression = "@", "*", CGname, CG;
```

Chapter 6

Formal Concept Analysis and Contextual Logic

Rudolf Wille

Technische Universität Darmstadt, Germany

Abstract   *Formal Concept Analysis* and *Contextual Logic* are related mathematical theories about concept hierarchies and concept graphs, respectively, which have been developed to support humans in their thought and knowledge. The aim of this article is to give an introduction to the main themes such as *"The Mathematics of Formal Concept Analysis"*, *"Applications of Formal Concept Analysis"*, *"Contextual Concept Logic"*, and *Contextual Judgment Logic"*. The presentation of the themes begins with a short desription abbout *"The Birth of Formal Concept Analysis."*

6.1   The Birth of Formal Concept Analysis

If one wants to declare the birthday of *Formal Concept Analysis*, then December 7, 1979, would be the most adequate date. At this day, the so-called *"Arbeitsgemeinschaft Mathematisierung"* at the TH Darmstadt continued a preceding discussion about the German standards [Deutsches Institut für Normung, 1979a] "Begriffe und ihre Benennungen. Allgemeine Grundsätze" and [Deutsches Institut für Normung, 1979b] "Begriffssysteme und ihre Darstellungen." The discussion started in repeating an earlier attempt to mathematize the basic notions in those standards. In particular, it was discussed whether a hierarchical system of concepts, as understood in [Deutsches Institut für Normung, 1979b], could be mathematized by a (partially) ordered set. This brought us back to the elementary question: How is a concept in-

ternally structured? According to [Deutsches Institut für Normung, 1979a],
a *concept* consists of a set of objects, called *concept extension* (in German:
Begriffsumfang), and a set of attributes, called *concept intension* (in German:
Begriffsinhalt), where each object of the extension has each attribute of the
intension.

For mathematizing this understanding of a concept structure, my proposal
at that day was that a mathematically abstracted concept should be derived
from given sets G and M combined by a binary relation $I \subseteq G \times M$ where
$(g, m) \in I$ is read: *the object g has the attribute m*; furthermore, for $X \subseteq G$,
there is the *derivation* $X' := \{m \in M \mid (x, m) \in X$ for all $x \in X\}$ and, for
$Y \subseteq M$, there is the *derivation* $Y' := \{g \in G \mid (g, y) \in Y$ for all $y \in Y\}$.
These definitions allow us to mathematically define a *formal concept* of a
binary relational structure (G, M, I), called *formal context*, as a pair (A, B)
with $A \subseteq G$, $B \subseteq M$, $A' = B$, and $B' = A$; A and B are called the *extent* and
the *intent* of the formal concept (A, B). The concept extents of (G, M, I)
are the closures X'' of the closure operator $X \mapsto X''$ on the object set G
and, dually, the concept intents of (G, M, I) are the closures Y'' of the closure
operator $Y \mapsto Y''$ on the attribute set M. In general, the derivations $X \mapsto X'$
$(X \subseteq G)$ and $Y \mapsto Y'$ $(Y \subseteq M)$ can be mathematically characterized by the
following three laws (Z, Z_1, Z_2 are subsets of G or M, respectively):

$$(1) \quad Z_1 \subseteq Z_2 \text{ implies } Z_1' \supseteq Z_2', \quad (2) \quad Z \subseteq Z,'' \quad (3) \quad Z' = Z'''.$$

This mathematization of concepts was first published in an elaborated form
in [Wille, 1982] under the title *"Restructuring lattice theory: an approach based
on hierarchies of concepts"* (the content of this paper was already presented
at the *"Symposium on Ordered Sets"* held at Banff, Canada, Aug 28 to Sept
12, 1981).

The hierarchy of concepts was mathematized by the definition

$$(A_1, B_1) \leq (A_2, B_2) : \iff A_1 \subseteq A_2 \quad (\iff B_1 \supseteq B_2)$$

for formal concepts (A_1, B_1) and (A_2, B_2) of a formal context (G, M, I);
(A_1, B_1) is called a *subconcept* of (A_2, B_2), and (A_2, B_2) is called a *superconcept* of (A_1, B_1). To formulate the so-called *Basic Theorem on Concept Lattices*, which, in particular, states that the set of all formal concepts of a formal context together with the subconcept-superconcept-relation
forms a complete lattice, we need some further definitions: Let $\mathfrak{B}(G, M, I)$
be the set of all formal concepts of the formal context (G, M, I) and let
$\underline{\mathfrak{B}}(G, M, I) := (\mathfrak{B}(G, M, I), \leq)$ be the concept lattice of (G, M, I). A subset
D of a complete lattice $\underline{L} := (L, \leq)$ is called *infimum-dense* (*supremum-dense*)
if $L = \{\bigwedge X \mid X \subseteq D\}$ $(L = \{\bigvee X \mid X \subseteq D\})$.

Basic Theorem on Concept Lattices. *Let $\mathbb{K} := (G, M, I)$ be a formal
context. Then $\underline{\mathfrak{B}}(\mathbb{K})$ is a complete lattice, called the* concept lattice of $\mathbb{K}$,

whose infima and suprema can be described as follows:

$$\bigwedge_{t \in T}(A_t, B_t) = (\bigcap_{t \in T} A_t, (\bigcup_{t \in T} B_t)''), \quad \bigvee_{t \in T}(A_t, B_t) = ((\bigcup_{t \in T} A_t),'' \bigcap_{t \in T} B_t).$$

In general a complete lattice $\underline{L} := (L, \leq)$ is isomorphic to $\mathfrak{B}(\mathbb{K})$ if and only if there exist mappings $\tilde{\gamma} : G \to L$ and $\tilde{\mu} : M \to L$ such that
(1) $\tilde{\gamma}G$ is supremum-dense in $\underline{L}$ (i.e., $L = \{\bigvee X \mid X \subseteq \tilde{\gamma}G\}$),
(2) $\tilde{\mu}M$ is infimum-dense in $\underline{L}$ (i.e., $L = \{\bigwedge X \mid X \subseteq \tilde{\mu}M\}$), and
(3) $gIm \iff \tilde{\gamma}g \leq \tilde{\mu}m$ for $g \in G$ and $m \in M$;
in particular, $\underline{L} \cong \mathfrak{B}(L, L, \leq)$ and, if the set $J(\underline{L})$ of all supremum-irreducible elements is supremum-dense in $\underline{L}$ and the set $M(L)$ of all infimum-irreducible elements is infimum-dense in $\underline{L}$, then $\underline{L} \cong \mathfrak{B}(J(L), M(L), \leq)$.

Since in a *finite lattice $\underline{L}$*, the set of all supremum-irreducible elements is supremum-dense and the set of all infimum-irreducible elements is infimum-dense, the Basic Theorem yields that $\underline{L} \cong \mathfrak{B}(J(L), M(L), \leq)$. In general, the Basic Theorem makes apparent a dual situation that often obtains a concrete meaning by interchanging objects and attributes. Formally, (M, G, I^{-1}) is considered as the *dual context* of the formal context (G, M, I) where the mapping given by $(A, B) \mapsto (B, A)$ is an antiisomorphism from $\mathfrak{B}(G, M, I)$ onto $\mathfrak{B}(M, G, I^{-1})$.

An example may illuminate how a formal context gives rise to a concept lattice. The data table in Figure 6.1 represents a formal context about our planets, the concept lattice of which is shown in Figure 6.2. The crosses in the

	size: small	size: medium	size: large	distance from sun: near	distance from sun: far	moon: yes	moon: no
Mercury	X			X			X
Venus	X			X			X
Earth	X			X		X	
Mars	X			X		X	
Jupiter			X		X	X	
Saturn			X		X	X	
Uranus		X			X	X	
Neptune		X			X	X	
Pluto	X				X	X	

FIGURE 6.1: Formal context about our planets.

data table indicate the context relation between objects and attributes. The

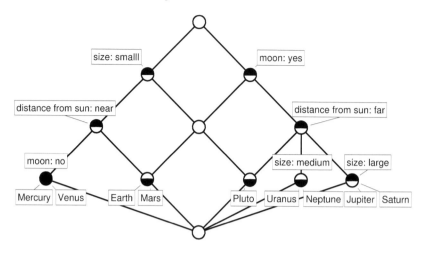

FIGURE 6.2:   Concept lattice of the formal context in Figure 6.1.

labels attached to the circles of the labeled line diagram visualize the mappings $\tilde{\gamma}$ and $\tilde{\mu}$ of the Basic Theorem so that, for each represented concept, its extents (intents) can be determined by collecting all object labels (attribute labels) which can be reached from a circle by a path of line segments downward (upward). For instance, the circle with the label "distance: near" represents the formal concept with the extent {Mercury, Venus, Earth, Mars} and the intent {distance: near, size: small} and the circle in the center with no label represents the formal concept with the extent {Earth, Mars, Pluto} and the intent {size: small, moon: yes} (cf. [Wille, 1982]).

In giving birth to new ideas one has always to be aware that there are, as a rule, *predecessors* who paved the way for those ideas. This is indeed the case for the development of Formal Concept Analysis. In particular, the concept understanding presented in [Deutsches Institut für Normung, 1979a] and [Deutsches Institut für Normung, 1979b] is embeded in a rich history of efforts to understand what concepts are (see [Wille, 1995]).

Already in the philosophy of Greco-Roman times thinking in concepts played a central role. But, according to [Hartmann, 1939], there was no actual *concept of concept* in the ancient world. Only in the Late Middle Ages, the scholastic term *"conceptus"* became the autonomous meaning of a unit of thought formed with regard to the contents. In modern times this understanding of concepts was further developed, first of all, in the school of Port Royal documented by the book *"La Logique ou l'Art de Penser"* written by A. Arnauld and P. Nicole (published in 1662, 1683). In this book the authors particularly discuss (general) *concepts* (named "idée") as the meaning of "general words" and distinguish between the *extension* of a concept (named "étendue") consisting of *objects* (named "sujets") and the *intension* of a concept (named "compréhensions") comprising *attributes* (named "attributs"); furthermore, they explain already formulated notions of *subconcepts* (cf. [Duquenne, 1987]).

With the "Logic of Port Royal," the development of concept-based logics gained increasing significance, above all, by their ordering based on a *general doctrine of elements* comprising doctrines of *concepts, judgments,* and *conclusions,* and extended by a *logic methodology.* This development reached a convincing state in I. Kant's lectures on logic, which were edited by G. B. Jäsche and published in 1800; an excellent English translation was published in [Kant, 1988]), out of which we only cite here §7 of Kant's doctrine of concepts, entitled "Intension and Extension of Concepts":

> "Every concept, as a *partial concept,* is contained *in* the presentation of things; as a *ground of cognition,* i.e., as a *characteristic* [attribute], it has these things contained *under it.* In the former regard, every concept has an *intension* [content]; in the latter, it has an *extension.* Intension and extension of a concept have an inverse relation to each other. The more a concept contains under it, the less it contains in it."

The concept understanding of the "Logic of Port Royal" influenced also mathematicians. A prominent example for this is the mathematician and logician E. Schröder who characterized *concepts* in his "Vorlesungen über die Algebra der Logik," published 1890, as follows (see [Schröder, 1966], p.83):

> "The 'nature' (essentia) of a concept, or as one says, its 'intension' (complexus, intent) forms the common attributes of the things marked by a common name ... . In contrast to this intension, the whole or class of the individuals, combined under the common name, is designated as the 'extension' (ambitus, sphaera, extent) to which the concept belongs.

With his *logical calculus* Schröder even developed order- and lattice-theoretic mathematical structures based on his general understanding of concepts. Fifty years later, the mathematician G. Birkhoff wrote the first book on *lattice theory* [Birkhoff, 1940] in which, in particular, the above discussed derivations $X \mapsto X'$ and $Y \mapsto Y'$ between sets G and M are introduced in lattice theory under the name *polarities,* where the elements of G and M are already called "objects" and "attributes"; Birkhoff understood his polarities as special cases of the so-called *Galois connections* (see DEW04).

Birkhoff's polarities have been used by M. Barbut and B. Monjardet in 1970, Chapter 5, to derive from a binary relation R between two finite sets E_1 and E_2 the lattice of all *product pairs* $A_1 \times A_2 \subseteq E_1 \times E_2$, which are maximal with respect to the ordering $\underline{A}_1 \times \underline{A}_2 \leq A_1 \times A_2 :\iff \underline{A}_1 \subseteq A_1$ and $\underline{A}_2 \supseteq A_2$; they called this lattice *treillis de Galois d'une relation entre deux ensembles.* It is quite instructive how, in [Barbut and Monjardet, 1970], those product pairs and their connections are visualized by shaded rectangles.

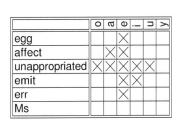

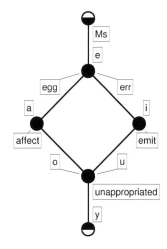

FIGURE 6.3:   A visualization of the infimum and supremum of two maximal product pairs (formal concepts) in a cross-table (formal context).

An example of such a visualization is shown in Figure 6.3 using a small tableau (subcontext) selected from the huge tabbleau (formal context) consisting of all English words (as objects) and the vowels a, e, i, o, u, and y (as attributes), where a word is considered to be related to a vowel if the vowel is a letter within that word. In terms of Formal Concept Analysis, the pairs $(\{affect, unappropriated\}, \{a, e\})$ and $(\{unappropriated, emit\}, \{e, i\})$ are two formal concepts that have as infimum the formal concept $(\{unappropriated\}, \{o, a, e, i, u\})$ and as supremum the concept $(\{egg, affect, unappropriated, emit, err\}, \{e\})$. These four concepts are represented in Figure 6.3 by maximal rectangles full of crosses. This representation particularly shows that, for two concepts, the *union of the two intents* might be smaller than the intent of the infimum and the *union of the two extents* might be smaller than the extent of the supremum, although the intersection of the two extents is always equal to the extent of the infimum and the intersection of the two intents is always equal to the intent of the supremum.

The early theories about *concept lattices* and *treillis de Galois* have been developed independently and also guided by different motivations and aims. The treillis de Galois have been introduced particularly for analyzing data in the field of *Sciences de l'Homme*, while concept lattices have been invented for mathematizing concept hierarchies in general. My first personal contact with the researchers on Galois lattices took place in 1983 when I stayed for a research visit seven weeks at the *Maison de Science de l'Homme* in Paris. During that time we had many discussions about the status of our research in lattice theory and its applications, which also led to research cooperations, most intensely with V. Duquenne.

6.2 The Mathematics of Formal Concept Analysis

For mathematizing concepts and concept hierarchies it is important to understand what mathematics is about, in particular its *semantics*. According to Ch. S. Peirce (see [Peirce, 1992]; p.114ff.), "mathematics ... is the only one of the sciences which does not concern itself to inquire what the actual facts are, but studies hypotheses exclusively"; for this, mathematicians are "gradually uncovering one great Cosmos of Forms, a world of potential being." In the early times just numbers and geometric figures gave rise to the potential being of mathematical structures and forms. Later functions, relations, and finally set structures in general enlarged the basis of the potential semantics of mathematics. For the *Mathematics of Formal Concept Analysis*, the semantics is grounded on basic set structures that have been introduced in Section 6.1 as *formal contexts*, *formal concepts*, and *concept lattices*. The mathematical theory of these basic structures is mainly supported by mathematical order and lattice theory (see e.g., [Davey and Priestley, 2002]).

Mathematical foundations of Formal Concept Analysis have been worked out in the monograph [Ganter and Wille, 1999a] (German: [Ganter and Wille, 1996]). The contents of the monograph are presented in eight sections entitled:

0. Order-Theoretic Foundations;	1. Concept Lattices of Contexts;
2. Determination and Representation;	3. Parts and Quotients;
4. Decompositions of Concept Lattices;	5. Constructions of Concept Lattices;
6. Properties of Concept Lattices;	7. Context Comparison and Conceptual Measurability.

Section 1 (Concept Lattices of Contexts) contains the development of Formal Concept Analysis up to the *Basic Theorem on Concept Lattices*. In addition, *many-valued contexts* (G, M, W, I) are introduced as mathematization of data tables that represent relationships of the form: an object g has an attribute value $m(g)$ of the many-valued attribute m (in mathematical terms: $g \in G$, $m \in M$, and $m(g) \in W$). Here we restrict ourselves to *complete many-valued contexts*, which corresponds to rectangular data tables whose rows are labeled with object names, whose columns are labeled with (many-valued) attribute names, and whose cells are filled with exactly one attribute value, respectively.

To determine a concept lattice of a many-valued context, one has to transform the many-valued context to a formal context. Such a transformation is called *conceptual scaling* because the transformation is formed by using *conceptual scales* that are formal contexts constructed to give meaning to the attribute values of each many-valued attribute. Mathematically, such a transformation of a many-valued context (G, M, W, I) with given conceptual scales (G_m, M_m, I_m) satisfying $m(G) \subseteq G_m$ for each $m \in M$ is performed by defining the derived formal context (G, N, J) with $N := \bigcup_{m \in M} \{m\} \times M_m$ and $gJ(m, n) : \Longleftrightarrow wI_m n$ for $w := m(g)$.

Let us consider an example to support the understanding of conceptual scaling. The many-valued context in Figure 6.4 represents a characterization of the five *platonian bodies* by their quantity of corners, edges, and facets, respectively. For instance, the tetrahedron has 4 corners, 6 edges, 4 facets,

	corners	edges	faces
tetrahedron	4	6	4
hexahedron	8	12	6
octahedron	6	12	8
dodecahedron	20	30	12
isocahedron	12	30	20

FIGURE 6.4:   A many-valued context about the five platonian bodies.

and the dodecahedron has 20 corners, 30 edges, 12 facets. Conceptual scales are used to interpret the columns of the many-valued context. The most simple scales are the *nominal scales*, the attributes of which are just the attribute values in the corresponding column. The derived context in Figure 6.5, combining three conceptual scales, has as attributes 4,6,8,12,20 from the scale "corners," 6,12,30 from the scale "edges," and 4,6,8,12,20 from the scale "facets." The concept lattice of this context is also shown in Figure 6.5.

There is a rich variety of *ordinal scales* that generalize nominal scales. The formal context and its concept lattice in Figure 6.6 are derived from the many-valued context in Figure 6.4 by a one-dimensional ordinal scale. For the many types of conceptual scales the reader is referred to [Ganter and Wille, 1999a, p. 42ff].

Section 2 (Determination and Representation) offers different methods of determining the concept lattice of a formal context. In the case of a small context (G, M, I), it is useful to begin by drawing up a complete list of all formal concepts starting with the largest concept (G, G') and processing in an (arbitrarily choosen) linear order on M. In step $m \in M$, for each concept (A, A') entered in the list in an earlier step, we form the concept $(A \cap \{m\}', (A \cap \{m\}')')$, and include it into the list if it is not already contained in that list.

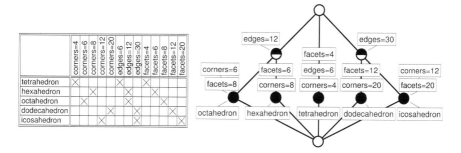

FIGURE 6.5: The formal context and its concept lattice derived from the many-valued context in Figure 6.4 by nominal scaling.

All concepts of (G, M, I) are determined when the last attribute has been worked out.

Often the determination of all concepts becomes easier if the construction of the formal concepts is combined with the task of *drawing a line diagram* of the corresponding concept lattice (cf. [Ganter and Wille, 1999a, p. 64ff]). How this can be performed shall be demonstrated by the formal context about our planets in Figure 6.1. First we draw the little circle representing the greatest concept $\top := (\{Me, V, E, Ma, J, S, U, N, P\}, \emptyset)$ (as shown in Figure 6.7 in the upper left). For this concept, we represent, by further circles, its greatest subconcepts $\mu(ss)$, $\mu(my)$, and their only infimum $\mu(ss) \wedge \mu(my)$, and add four line segments between those circles to represent the subconcept-superconcept-relation. Next we determine the maximal concepts that are not greater or equal to an already represented concept; these are $\mu(dn)$ and $\mu(df)$ giving rise to three further concepts, inclucing the smallest concept $\bot := (\emptyset, \{ss, sm, sl, dn, df, my, mn\})$, and ten new line segments. Finally, the three circles of the concepts $\gamma(Me)$, $\gamma(J)$, $\gamma(U)$ joined by the obvious line segments complete the line diagram of the concept lattice in Figure 6.2.

Since data are often not fully explicit, a method, called *attribute exploration*, has been developed, which allows to infer a large enough explicit context and simultaniously its concept lattice (cf. [Ganter and Wille, 1999a, p. 79ff]). This shall be exemplified by the context $(\mathbb{N}, M_5, I)$ in which $\mathbb{N}$ is the set of all natural numbers, M_5 is the set $\{even, odd, prime, square, square + square\}$ of number properties, and $(n, m) \in I$ means that the number n has the property m. The problem is that we cannot explicitly write down this context because $\mathbb{N}$ is infinite. The subcontext with the first ten natural numbers as objects and its concept lattice is presented in Figure 6.8. The central question is: how can one find a minimal set $\mathbb{N}_{min}$ of natural numbers so that the map

$$\iota : \underline{\mathfrak{B}}(\mathbb{N}, M_5, I) \longrightarrow \underline{\mathfrak{B}}(\mathbb{N}_{min}, M_5, I \cap (\mathbb{N}_{min} \times M_5))$$

is a lattice isomorphism with $\iota(A, B) = (A \cap \mathbb{N}_{min}, B)$ for all $(A, B) \in \underline{\mathfrak{B}}(\mathbb{N}, M_5, I)$? Since the attribute set M_5 is finite, the concept lat-

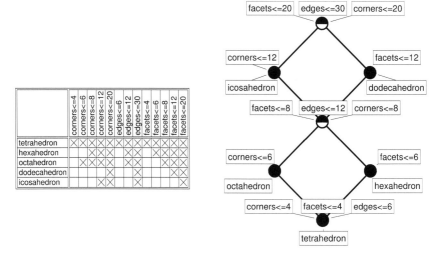

FIGURE 6.6:   The formal context and its concept lattice derived from the many-valued context in Figure 6.4 by ordinal scaling.

tice $\mathfrak{B}(\mathbb{N}, M_5, I)$ must be finite, too, because the context $(\mathbb{N}, M_5, I)$ has only finitely many intents (≤ 32). The *attribute exploration* is a method to algorithmically deduce all intents of a context with finitely many attributes. Figure 6.8 already represents 14 intents of $(\mathbb{N}, M_5, I)$. To complete the list of intents, the exploration algorithm was applied, which asked the following sequence of questions (for the used program "ConImp" see [Burmeister, 2003]):

1. Do "square" and "square+square" imply "even," "odd," "prime"?
   No, because 25 is a counterexample having only the additional attribute "odd"!

2. Do "square" and "square+square" imply "odd"?
   No, because 100 is a counterexample having only the additional attribute "even"!

3. Do "prime" and "square" imply "even," "odd," "square+square"? Yes!

4. Do "even" and "prime" imply "square+square"? Yes!

5. Do "even" and "odd" imply "prime," "square," "square+square"? Yes!

The exploration algorithm stops after answering the last question so that the intents of the formal context in Figure 6.8, extended by the numbers 25 and 100, represents exactly the intents of $(\mathbb{N}, M_5, I)$. According to the Basic Theorem, this context may be reduced, without losing intents, to the numbers that generate a $\bigvee$-irreducible object concept, respectively. Thus, Figure 6.9 represents a reduced context $(\mathbb{N}_{min}, M_5, I)$ and its concept lattice so that there

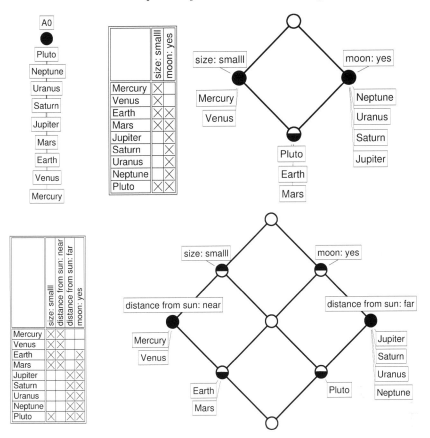

FIGURE 6.7: Constructing a concept lattice by completing its line diagram step by step.

is a lattice isomorphism $\iota : \underline{\mathfrak{B}}(\mathbb{N}, M_5, I) \longrightarrow \underline{\mathfrak{B}}(\mathbb{N}_{min}, M_5, I \cap (\mathbb{N}_{min} \times M_5))$ with $\iota(A, B) = (A \cap \mathbb{N}_{min}, B)$ for all $(A, B) \in \underline{\mathfrak{B}}(\mathbb{N}, M_5, I)$.

Section 3 (Parts and Quotients) discusses mathematical methods concerning *substructures* and *quotient structures*. Here we can only give an idea about the interplay between context structures and concept lattice structures. A powerful notion for structural analysis is the so-called *arrow relation* of finite contexts (G, M, I). For $g, h \in G$ and $m, n \in M$ we define

$g \swarrow m : \iff (g, m) \notin I$ and, if $\{g\}'' \supset \{h\}''$ then $(h, m) \in I$,

$g \nearrow m : \iff (g, m) \notin I$ and, if $\{m\}'' \supset \{n\}''$ then $(g, n) \in I$,

$g \nearrow\!\!\!\!\swarrow m : \iff g \swarrow m$ and $g \nearrow m$.

The significance of the arrow relations for reducing a context is shown by the following equivalences:

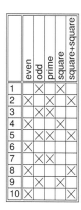 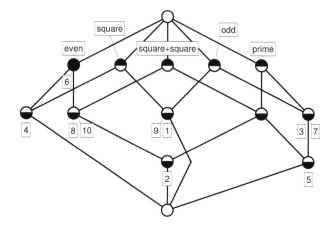

	even	odd	prime	square	square+square
1		X		X	
2	X		X		X
3		X	X		
4	X			X	
5		X	X		X
6	X				
7		X	X		
8	X				X
9		X		X	
10	X				X

FIGURE 6.8: A formal context of the first ten numbers and its concept lattice.

$\gamma(g)$ is $\bigvee$-irreducible	$\Longleftrightarrow$	There is an $m \in M$ with $g \swarrow m$,
$\mu(m)$ is $\bigwedge$-irreducible	$\Longleftrightarrow$	There is a $g \in G$ with $g \nearrow m$.
$\gamma(g)$ is $\bigvee$-irreducible	$\Longleftrightarrow$	There is an $m \in M$ with $g \nearrow m$,
$\mu(m)$ is $\bigwedge$-irreducible	$\Longleftrightarrow$	There is a $g \in G$ with $g \swarrow m$.

A *subcontext* of a formal context (G, M, I) is defined to be a formal context $(H, N, I \cap (H \times N))$ with $H \subseteq G$ and $N \subseteq M$. In general, the concepts of such a subcontext cannot simply be derived from those of (G, M, I). This can be done precisely for compatible subcontexts defined as follows: A subcontext $(H, N, I \cap (H \times N))$ is called *compatible* if the pair $(A \cap H, B \cap N)$ is a concept of the subcontext for every concept $(A, B) \in \mathfrak{B}(G, M, I)$. It can be easily proved that the assignment $(A, B) \mapsto (A \cap H, B \cap N)$ yields a surjective complete homomorphism from $\mathfrak{B}(G, M, I)$ onto the concept lattice of the subcontext $(H, N, I \cap (H \times N))$ (see [Ganter and Wille, 1999a, p. 99ff]).

In the case of finite contexts, the compatible subcontexts can be easily identified by means of the arrow relations. To understand this, it is enough to consider so-called *clarified* finite contexts in which $\gamma(g) = \gamma(h)$ implies $g = h$ and $\mu(m) = \mu(n)$ implies $m = n$. A subcontext $(H, N, I \cap (H \times N))$ of a clarified finite context (G, M, I) is said to be *arrow-closed* if $h \nearrow m$ with $h \in H$ implies $m \in N$, and $g \swarrow n$ with $n \in N$ implies $g \in H$.

Proposition 1 *A subcontext of a clarified finite context (G, M, I) is compatible if and only if it is arrow-closed* (see [Ganter and Wille, 1999a, p. 101]).

An important consequence of the proposition is that the arrow-closed subcontexts of a clarified finite context (G, M, I) are in one-to-one correspondence to the homomorphic images of $\mathfrak{B}(G, M, I)$; in other words, the *arrow-closed subcontexts* correspond one-to-one to the *congruence relations* of $\mathfrak{B}(G, M, I)$.

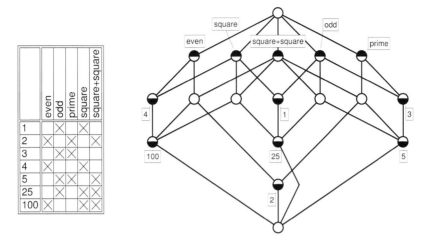

FIGURE 6.9: A reduced context of $(\mathbb{N}, M_5, I)$ and its concept lattice.

How to construct the corresponding congruence relation from an arrow-closed subcontext, this shall be demonstrated by the formal context and its concept lattice presented in Figure 6.10. We consider the arrow-closed subcontext in the lower right corner having the three objects "Special-Liquid-Copy," "Special-Office," "Special-Offset" and the three attributes "For Double-Sided-Copying," "Low-Performance-Copier," "For Type-Writers."

First we construct the three object concepts γ(Special-Liquid-Copy), γ(Special-Office), γ(Special-Offset), and the three attribute concepts μ(For Double-Sided-Copying), μ(Low-Performance-Copier), μ(For Type-Writers) within the whole context. Then we form all minimal intervals, which have as lower bound a supremum of some of the three object concepts and as upper bound an infimum of some ofthe three attribute concepts (we use the convention that $\bigvee \emptyset = \bot$ and $\bigwedge \emptyset = \top$). This construction yields six disjoint intervals covering $\underline{\mathfrak{B}}(G, M, I)$, namely

$[\bot, \mu(\text{For Double-Sided-Copying}) \wedge \mu(\text{For Type-Writers})]$,
$[\gamma(\text{Special-Liquid-Copy}), \mu(\text{For Double-Sided-Copying})]$,
$[\gamma(\text{Special-Office}),$
$\qquad \mu(\text{For Low-Performance-Copier}) \wedge \mu(\text{For Type-Writers})]$,
$[\gamma(\text{Special-Liquid-Copy}) \vee \gamma(\text{Special-Office}),$
$\qquad\qquad\qquad\qquad \mu(\text{For Low-Performance-Copier})]$,
$[\gamma(\text{Special-Offset}), \mu(\text{For Typ-Writers})]$,
$[\gamma(\text{Special-Liquid-Copy}) \vee \gamma(\text{Special-Offset}), \top]$.

These are the congruence classes of the congruence relation that corresponds to the given arrow-closed subcontext. For further correspondencies, in particular the correspondence between *block relations* and *tolerance rela-*

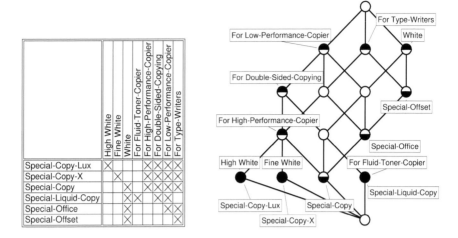

FIGURE 6.10: A formal context and its concept lattice about printing paper.

tions, we refer to the subsections 3.2, 3.3, and 3.4 in [Ganter and Wille, 1999a].

Sections 4 and 5 (Decompositions and Constructions of Concept Lattices) investigate how contexts and concept lattices can be decomposed into smaller components and, conversely, be combined to larger contexts and concept lattices. A basic method of decomposition is the *subdirect decomposition*. Before describing this method in general, we use the context and concept lattice in Figure 6.10 to first give an example of such a decomposition.

The *subdirect decomposition* starts with determining for each double arrow the smallest arrow-closed subcontext containing this arrow. The double arrows in the last three columns give rise to arrow-closed subcontexts with only one object-attribute-pair, namely ("Special-Liquid-Copy," "For Type-Writers"), ("Special Office," "For Double-Side-Copying"), ("Special-Offset," "For Low-Performance-Copier"). Each other double arrow generates the same arrow-closed subcontext that has the four objects "Special-Copy-Lux," "Special-Copy-X," "Special-Copy," "Special-Liquid-Copy" and the four attributes "High White," "Fine White," "White," "For Fluid-Toner-Copier." In general, for each *arrow-closed subcontext*, there exists a unique *congruence relation*. For each of the four arrow-closed subcontexts generated by one double arrow, the congruence classes of the corresponding congruence relation shall be determined:

- (Special-Liquid-Copy, For Type-Writers)
  $\mapsto \{[\gamma(\text{Special-Liquid-Copy}), \top], [(\bot, \mu(\text{For Type-Writers})]\}$

- (Special-Office, For Double-Side-Coppying)
  $\mapsto \{[\gamma(\text{Special-Office}), \top], [(\bot, \mu(\text{For Double-Side-Copying})]\}$

- (Special-Offset, For Low-Performance-Copier)
  $\mapsto \{[\gamma(\text{Special-Offset}), \top], [(\bot, \mu(\text{For Low-Performance-Copier})]\}$

- (Special-Copy-Lux, Fine White)

  $\mapsto$

  $\{(\bot, \bot), [\gamma(\text{Special-Copy-Lux}), \mu(\text{High White})],$
  $[\gamma(\text{Special-Copy-X}), \mu(\text{Fine White})],$
  $[\gamma(\text{Special-Copy}), \mu(\text{White}) \wedge \mu(\text{For Type-Writers})],$
  $[\gamma(\text{Special-Liquid-Copy}), \mu(\text{For Fluid-Toner-Copier})],$
  $[\gamma(\text{Special-Copy}) \vee \gamma(\text{Special-Liquid-Copy}), \mu(\text{White})],$
  $[\gamma(\text{Special-Copy}) \vee \gamma(\text{Special-Copy-X}), \mu(\text{For Typ-Writers})],$
  $[\gamma(\text{Special-Liquid-Copy}) \vee \gamma(\text{Special-Copy-X}), \top]\}.$

The described four congruence relations are not only the $\bigwedge$-irreducible congruence relations of the concept lattice $\mathfrak{L}$ in Figure 6.10, but even the maximal congruence relations. Figure 6.11 exemplifies a subdirect decomposition of $\mathfrak{L}$ by a *nested line diagram*. This diagram represents the direct pro-

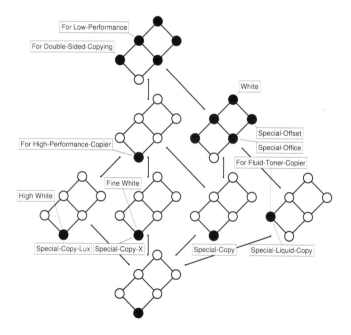

FIGURE 6.11:   A nested line diagram of the concept lattice about printing paper.

thatduct of two quotient lattices of $\mathfrak{L}$ formed by the two congruence relations that correspond to the arrow-closed subcontexts generated by the double arrow (Special-Copy-Lux, Fine White) and the three double arrows (Special-

Liquid-Copy, For Type-Writers), (Special-Office, For Double-Side-Coppying), (Special-Offset, For Low-Performance-Copier), respectively. In Figure 6.11 the elements of the quotient lattice which corresponds to (Special-Copy-Lux, Fine White) are represented by the eight lattice diagrams formed by six circles, respectively; the line segments between those lattice diagrams indicate the lattice order of that quotient lattice. Each of the six-element lattice diagrams represent the second quotient lattice. From the nested line diagram we obtain a non-nested line diagram of the direct product if we replace each line segment of the first quotient lattice by six line segments that link the corresponding circles of the joined six-element lattice diagrams. In the completed line diagram, the black circles represent the concept lattice from Figure 6.10 as sublattice in a direct product of two of its quotient lattices. What we have seen in an example can even be proved in general as the following proposition states:

Proposition 2 *Let (G, M, I) be a finite clarified context with the property that there is at least one double arrow in each of its rows and columns, respectively. Then the subdirect decompositions correspond bijectively to the families of arrow-closed subcontexts $(G_t, M_t, I \cap (G_t \times M_t))$ with $\bigcup_{t \in T} G_t = G$ and $\bigcup_{t \in T} M_t = M$ (see [Ganter and Wille, 1999a, p. 134]).*

Sections 4 and 5 present and discuss further decomposition and construction methods such as *Atlas-Decomposition, Substitution, Tensorial Decompositions*, and *Subdirect Product Constructions, Gluings, Local Doubling*, and *Tensorial Constructions*.

Sections 6 and 7 (Properties, Comparison, and Measurability) analyze how lattice properties, morphisms, and scales might be meaningful for the understanding of formal contexts and concept lattices. Here only the "order dimension" shall be briefly discussed. An ordered set $(P, \leq)$ has the *order dimension* $\dim(P \leq) = n$ if it can be embedded in a direct product of n chains, but not of $n - 1$ chains. A *Ferrers relation* is a relation $F \subseteq G \times M$ for which $(g, m) \in F$, $(h, n) \in F$, $(g, n) \notin F$ imply $(h, m) \in F$. The *Ferrers dimension* $\mathrm{fdim}(G, M, I)$ of a context (G, M, I) is the smallest numner of Ferrers relations $F_t \subseteq G \times M$, $t \in T$, with $I = \bigcap_{t \in T} F_t$ (see [Ganter and Wille, 1999a, p. 237]).

Proposition 3 *The Ferrers dimension of a context (G, M, I) is equal to the order dimension of the concept lattice $\mathfrak{B}(G, M, I)$, i.e., $\dim \mathfrak{B}(G, M, I) = \mathrm{fdim}(G, M, I)$. In particular, $F \subseteq G \times M$ is a Ferrers relation if and only if $\mathfrak{B}(G, M, F)$ is a chain.*

Let us apply the proposition to determine the order dimension of the concept lattice in Figure 6.10. The direct product of two 2-dimensional quotient lattices of the concept lattice about printing paper in Figure 6.11 might suggest that this concept lattice has order dimension 4. But, if we determine the Ferrers dimension of the context in Figure 6.10, we obtain the Ferrers dimension 3 as follows: We begin by dividing the set of all empty cells of the context into three staircase-like subsets where the first consists of the empty cells in the columns headed by "High White," "Fine White" (except the cell in the first row), "For High-Performance-Copier," "For Double-Sided-Copying," "For Low Performance-Copier," the second consists of the empty cells in the column headed by "Fine White" (only the first cell), "White," "For Fluid-Toner-Copier," and the third consists of the single empty cell in the column headed by "For Type-Writers." The complements of the three subsets of cells are three Ferrers relations that have the set of all cross-cells as intersection. Thus, the Ferrers dimension is indeed 3. With the proposition we can conclude that the concept lattice of the context from Figure 6.10 has order dimension 3 (and not 4).

6.3   Applications of Formal Concept Analysis

For applying the *Mathematics of Formal Concept Analysis* it is desirable to basically understand the relationship of mathematics to other sciences and the real world. Such understanding can be supported by *Peirce's classification of sciences*. This classification scales the sciences "in the order of abstractness of their objects, so that each science may largely rest for its principles upon those above it in the scale while drawing its data in part from those below it" ([Peirce, 1992]; p.114). In this treatise we mainly activate the first level of Peirce's classification:

      I. Mathematics   II. Philosophy   III. Special Sciences

where *Mathematics* is viewed as the most abstract science studying hypotheses exclusively and dealing only with potential realities, *Philosophy* is considered as the most abstract science dealing with actual phenomena and realities, while all *other sciences* are more concrete in dealing with special types of actual realities.

Since modern mathematics is essentially based on *set-theoretical semantics*, semantic structures having mathematical meaning can be represented by set structures. Mathematicians are developing those structures in great variety, many of which even in advance. Peirce already wrote that mathematicians are "gradually uncovering a great Cosmos of Forms, a world of potential being" ([Peirce, 1992], p.120). Semantic structures having philosophical meaning are based on *philosophic-logical semantics* that are grounded on networks of philosophical concepts. In traditional philosophical logic, concepts are viewed as the basic units of thought; they and their combinations to judgments and conclusions form "the three essential main functions of thinking" ([Kant, 1988], p. 6), which constitutes the logical semantics of philosophy. Semantic structures having their meaning with respect to special sciences are based on semantics which are grounded on networks of special concepts of those sciences (cf. [Gehring and Wille, 2006], [Eklund and Wille, 2007]).

Experiences have shown that applications of the Mathematics of Formal Concept Analysis need to activate all three semantics: the semantics belonging to the field whose language and understanding are necessary to pose and finally solve a problem, the mathematical semantics of Formal Concept Analysis to structurally represent and formally analyze that problem, and the philosophic-logical semantics to yield the conceptual bridges for making understandable connections between mathematical structures and the actual problem. Already contexts and concepts with their interplay have to be understood within the threefold semantics:

- mathematically, contexts and concepts are set-theoretic structures, named "formal context" and "formal concept," which give rise to "concept lattices";

- philosophically, contexts and concepts are general units of thought, named "(object-attribute-)context" and "(general) concept," which lead to "(general) concept hierarchies";

- concretely, contexts and concepts are special units of thought, named "(elementary) data table" and "(special) concept," from which "(special) concept hierarchies" are derived.

How methods and tools of Formal Concept Analysis may be used in practice, this shall be exemplified by a project of developing an *information system* by which architects may find out which laws, standards, and regulations they have to consider for their building procjects. This system was elaborated over three years by a team of researchers from the Darmstadt University of Technology in cooperation with experts from the "Ministry of Building and Residing" of the German state Nordrhein-Westfalen. The infomation system was realized with the computer program *TOSCANA* that allows to search in larger data sets by touring through prepared line diagrams of concept lattices (cf. [Kollewe et al., 1994]).

As in most application projects, it was important to first establish the basic data contexts to represent the information necessary for the searching procedures. After longer discussions with the experts of the ministry it was decided to take the *paragraphs of the laws, standards, and regulations* as context objects and the *building parts and requirements* as context attributes where a building part or requirement as attribute applies to a paragraph as object if it belongs to the content of the paragraph. Then the second step was to create out of the basic contexts a large number of concept lattices of smaller thematic contexts that could function as *query structures*. These query structures are basic for the searching procedures performed by the TOSCANA program (cf. [Eschenfelder et al., 2000]).

As example, we show in Figure 6.12 the line diagram of the query structure "function room in a hospital." This diagram offers a rich amount of informa-

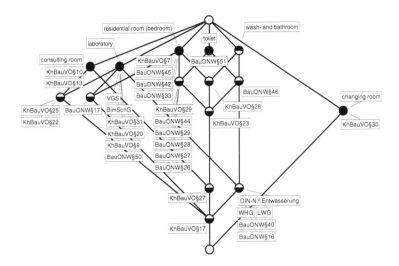

FIGURE 6.12: Query structure "functional rooms in a hospital" of a TOSCANA information system about laws, standards, and regulations concerning building construction.

tion: for instance, an architect gets the information that, for designing the water connections (needed in laboratories, toilets, wash-, and bathrooms), the paragraphs "BauONW§16," "BauONW§40," "DIN-N.f.Entwässerung," "LWG," "WHG," and "KhBauVO§17" have to be observed. In activating and internalizing that information and joining it with the already present knowledge about the building project, an architect will create new knowledge, which can then be utilized in his project. As the example indicates, the logical connections in line diagrams of concept lattices can even *stimulate background knowledge* for improving the knowledge representation; in particular, the example indicates that line diagrams of concept lattices may even stimulate critique and self-correction of the represented knowledge.

In our project we experienced again and again that line diagrams enabled the building experts to discover mistakes in the extensive data context that contributed to a considerable improvement of the data quality. An instructive case of critique and self-correction occurred in a discussion of the line diagram in Figure 6.12: For testing the readability of such a diagram, a secretary of the involved ministry of building and residing, who was not working for our project, was asked to join the discussion. After inspecting the diagram, the secretary expressed her astonishment that, in the diagram, the paragraph

"BauONW§51" of the "Bauordnung Nordrhein-Westfalen" is directly attached to the circle with the label "toilet," which means that only the toilets have to be designed for handicapped people; she could not understand why the wash- and bathrooms do not have to meet requirements for handicapped people, too. Even the experts became surprised when they checked again §51 and saw that only toilets are mentioned in connection with handicapped people. Only after a comprehensive discussion the experts came to the conclusion that, by superior aspects of law, §51 also applies to wash- and bathrooms. Finally, by similar reasons, the consulting rooms and the residential rooms (bedrooms) were also included so that, in the underlying cross-table, three more crosses were added in the row headed by "BauONW§51" so that, in the line diagram of Figure 6.12, the label "BauONW§51" moved down to the circle with the label "KhBauVO§27."

Since 1980 members of the Darmstadt research group on Formal Concept Analysis worked on more than 200 **application projects** in which methods of Formal Concept Analysis have been applied. Unfortunately, not all those projects are well documented. But there are still a large number of projects of which enough documents are available. In particular, there are some publications that present lists of projects chosen under specific views. Some of those lists shall be set out here:

The earliest list can be found in the paper *Liniendiagramme hierarchischer Begriffssyteme* [Wille, 1984a], which covers the following themes (originally written in German):

- An educational film about "living beings and water"

- A semantic field of words about waters

- Examination of mentally and physically handicapped children

- Chinese urns (14th to 9th century B.C.)

- Pattern of a 4-generated experimental plan

- Diatonic harmony forms

- Vowels in German

In the first book publication on Formal Concept Analysis there is an article entitled *Bedeutungen von Begriffsverbänden* [Wille, 1987] in which application projects are subsumed under basic meanings of concept lattices (originally written in German):

- Hierachical classification of objects: Countries of the third world

- System of attribute implications: Driving concepts for motor cars

- Pattern for arranging objects: Evaluation of a physics test

- Schema for determining objects: Symmetry types of planar patterns

- Manifold of concept patterns: Examination of working places for handicaps

- Combination of individual concept systems: Monuments of the Forum Romanum

- Means for returning to basic notions: Computer-ciphers

- Structure for representing and gaining knowledge: Concept exploration

- Dualistic presentation of a lattice structure: Concept lattice of a biordinal scale

How concept lattices may support knowledge communication, has been discussed in the paper *Begriffliche Datensysteme als Werkzeug der Wissenskommunikation* [Wille, 1992]. Examples are classified according to activated cognitive actions (originally written in German):

- **Searching - Recognizing - Identifying**
  Example [Kipke and Wille, 1986]: Concept lattice of symmetry types of planar patterns
  (It supports searching and recognizing symmetries and identifying the symmetry types of the present planar pattern.)

- **Analyzing - Interpreting - Discussing**
  Example [Spangenberg, 1990]: Concept lattice of an interrogation with an anorectic patient
  (It supports analyzing and interpreting the interrogation result and discussing the analysis and interpretation with the patient.)

- **Designing - Planning - Deciding**
  Example [Takàcs, 1984]: Concept lattice for an education film "living being and waters"
  (It supports designing sequences of contents, and planning and deciding the final course of the film.)

- **Ordering - Structuring - Understanding**
  Example [Kohler-Koch, 1989]: Concept lattice of 18 international regimes
  (It supports ordering the attributes, conceptually structuring the regimes, and understanding the connections in its contents.)

- **Learning - Deducing - Impressing**
  Example [Wille, 1989]: Concept lattice of fifteen lattice properties
  (It supports learning lattice-theoretic thinking, and deducing and impressing valid lattice-theoretic propositions.)

How the theoretical foundations of Formal Concept Analysis have been proved in practice; this is the theme of the article *Begriffliche Wissensverarbeitung: Theorie und Praxis* [Wille, 2000a] (see also [Wille, 1997]). Examples are presented to exemplify types of applications that are characterized by a supporting act of thinking (originally written in German):

- **"Exploring"** means looking for something of which one has only a vague idea.
  Example [Rock and Wille, 2000]: Exploring literature with a TOSCANA exploration-system of the library of the Center of Interdisciplinary Technology Research at the Darmstadt University of Technology

- **"Searching"** shall be understood as looking for something which one can more or less specify but not localize.
  Example [Eschenfelder et al., 2000]: Finding paragraphs of laws, standards, and regulations concerning building tasks by a TOSCANA information system established for the Ministry of Building and Residing of the German state Nordrhein-Westfalen

- **"Recognizing"** is understood with the meaning of clearly perceiving circumstances und relationships.
  Example [Hereth, 2000]: Clarifying the purchase behavior of customers in a department store by a TOSCANA-system developed for a data warehouse

- **"Identifying"** shall mean determining the taxonomic position of an object within a given classification.
  Example [Kipke and Wille, 1986]: Identifying the symmetry type of a wallpaper pattern (a computer program for this identification was developed by B. Ganter and J. Richter for the Symmetry Exhibition at the Mathildenhöhe in Darmstadt 1986 [Mazzola, 1986, p. 130])

- **"Investigating"** means studying by close examination and systematic inquiry.
  Example [Vogt et al., 1991]: Investigating international regimes guided by norms and rules (this cooperation project has stimulated the development of the program TOSCANA)

- **"Analyzing"** in the scope of conceptual knowledge processing is understood as examining data in their relationships while guided by theoretical views and declared purposes.
  Example [Großkopf and Harras, 2000]: Using the TOSCANA-methodology for conceptual data analyses in the case of a "TOSCANA-system for speech act verbs"

- **"Making aware"** means bringing something to someone's consciousness.

Example [Spangenberg, 1990]: Evaluating repertory grid tests of anorectic patients by methods of Formal Concept Analysis

- **"Deciding"** shall mean resolving a situation of uncertainty by an order. Example [Strahringer and Wille, 1992]: Deciding about places along the Canadian coast of Lake Ontario, whether swimming may be permitted and taking out water for drinking, respectively

- **"Improving"** has the meaning of enhancement in quality and value. Example [Wolff and Stellwagen, 1993]: Optimizing a chip production by using the results of 81 test runs that have been presented within nested line diagrams of concept lattices

- **"Restructuring"** means reshaping a given structure which, within the scope of our discussion, is conceptual in its nature. Example [Lindig and Snelting, 2000]: Restructuring the X-Window-Tool "xload" by using a concept lattice that represents the configurations in "xload"

- **"Memorizing"** is understood as a process of committing and reproducing what has been learned and retained. Example [Mackensen, 2000]: Memorizing a reconstruction of a "conceptual landscape" in the music esthetics of the 18th century by a TOSCANA-system

- **"Informing"** means comunicating knowledge about something to someone. Example [Eklund et al., 2000]: Informing customers about flights in Austria by an information map based on a conceptual graph in which all flights inside Austria were coded

Applications of Formal Concept Analysis in the field of economics have been viewed in [Wille, 2005b] through the structure of *organizational knowledge management* which is elaborated in the book [Probst et al., 1999]. Examples are given for each of the key processes of knowledge management and their knowledge objectives and knowledge evaluation (for the German version see [Wille, 2002]):

- Key process **"Identifying Knowledge"**
  How can transparency about existing knowledge be obtained internally and externally?
  Example [Kaufmann, 1996]: Conceptual analysis of data on flight movements (see also [Stumme et al., 1998])

- Key process **"Acquiring Knowledge"**
  How is external knowledge acquired?
  Example [Vogel, 1995]: A conceptual system for designing pipelines

- Key Process **"Developing Knowledge"**
  How is new knowledge internally developed?
  Example [Hereth, 2000]: Formal Concept Analysis in data warehousing (cf. [Hereth et al., 2000])

- Key Process **"Distributing and Sharing Knowledge"**
  How can we distribute and share knowledge appropriately?
  Example [Becker et al., 2000]: Information system for IT security management

- Key Process **"Using Knowledge"**
  How is existing knowledge put into active use?
  Example [Eschenfelder et al., 2000]: TOSCANA information system about laws, standards, and regulations concerning building construction (see also [Kollewe et al., 1994])

- Key Process **"Preserving Knowledge"**
  How can we avoid knowledge loss?
  Example [Wille, 2005b], p. 234: Coding systematically comprehensive data about IBM personal computer parts in formal contexts

- System Process **"Establishing Objectives and Evaluations"**
  How can the interrelationship between knowledge objectives and knowledge evaluations be suitably activated?
  Example [Andelfinger, 1997]: Devveloping conceptual models in the field of car manufacture and formally conceptualizing the desired functionalities

Concept lattices of data contexts may also illustrate scientific understandings of concepts as presented in Th. B. Seiler's convincing book, *Begreifen und Verstehen* [Seiler, 2001]. In his book, the psychologist Seiler describes concepts as *cognitive structures* whose development in the human mind is constructive and adaptive. Seiler elaborates his approach within twelve aspects that characterize quite different views concerning concepts. Concept Analysis examples for each of the twelve aspects are discussed in [Wille, 2005a]. Further support of applying Formal Concept Analysis is offered in the article, *Methods of Conceptual Knowledge Processing* [Wille, 2006] in which 38 methods are described and, in many cases, exemplified by concretizing concept lattices. Many methods can be supported by computer programs thatexist in a wide range; in particular, the TOSCANA software has been proven very useful (see [Becker and Hereth Correia, 2005]).

6.4   Contextual Concept Logic

The introduction to *Contextual Concept Logic* shall follow the approach developed in [Wille, 2004a]. There it starts with the statement that *Concepts* are the basic units of thought wherefore a concept-oriented mathematical logic is of great interest. G. Boole has offered the most influential foundation for such a logic, which is based on a general conception of *signs* representing *classes* of objects from a given *universe of discourse* [Boole, 1854]. In the language of Formal Concept Analysis [Ganter and Wille, 1999a], Boole's basic notions can be explicated:

- for a "universe of discourse," by the notion of a *"formal context,"*

- for a "sign," by the notion of an *"attribute"* of a formal context and,

- for a "class," by the notion of an *"extent"* defined in a formal context
  $\mathbb{K} := (G, M, I)$ as a subset $Y' := \{g \in G \mid \forall m \in Y : gIm\}$ for some
  $Y \subseteq M$.

How Boole's logic of signs and classes may be developed as a *Contextual Attribute Logic* by means of Formal Concept Analysis is outlined in [Ganter and Wille, 1999b]. The dual *Contextual Object Logic*, which is, for instance, used to determine conceptual contents of information [Wille, 2003], can be obtained from Contextual Attribute Logic by interchanging the role of objects and attributes so that, in particular, the notion of an "extent" is replaced by

- the notion of an *"intent"* defined in a formal context $\mathbb{K} := (G, M, I)$
  as a subset $X' := \{m \in M \mid \forall g \in X : gIm\}$ for some $X \subseteq G$.

Since a concept, as a unit of thought, combines an extension consisting of objects and an intention consisting of attributes (properties, meanings) (cf. [Schröder, 1966], p. 83ff.), a concept-oriented mathematical logic should be an integrated generalization of a mathematical attribute logic and a mathematical object logic. In our contextual approach based on Formal Concept Analysis, such an integrated generalization can be founded on

- the notion of a *"formal concept"* defined, in a formal context $\mathbb{K} := (G, M, I)$,
  as a pair (A, B) with $A \subseteq G$, $B \subseteq M$, $A = B'$, and $B = A'$ [Wille, 1982],

and its generalizations:

- the notions of a *"⊓-semiconcept"* (A, A') with $A \subseteq G$ and
  a *"⊔-semiconcept"* (B', B) with $B \subseteq M$ [Luksch and Wille, 1991],

- the notion of a *"protoconcept"* (A, B) with $A \subseteq G$, $B \subseteq M$,
  and $A'' = B'$ $(\Leftrightarrow B'' = A')$ [Hereth et al., 2000], and

- the notion of a *"preconcept"* (A, B) with $A \subseteq G$, $B \subseteq M$, and $A \subseteq B'$ $(\Leftrightarrow B = A')$ [Stahl and Wille, 1986].

Clearly, formal concepts are always semiconcepts, semiconcepts are always protoconcepts, and protoconcepts are always preconcepts. Since, for $X \subseteq G$ and $Y \subseteq M$, we always have $X''' = X'$ and $Y''' = Y'$, formal concepts can in general be constructed by forming (X'', X') or (Y', Y''). The basic logical *derivations* $X \mapsto X'$ and $Y \mapsto Y'$ may be naturally generalized to the conceptual level by

- $(X, Y) \mapsto (X, X') \mapsto (X'', X')$ and $(X, Y) \mapsto (Y', Y) \mapsto (Y', Y'')$ for an arbitrary preconcept (X, Y) of $\mathbb{K} := (G, M, I)$.

It is relevant to assume that (X, Y) is a preconcept because otherwise we would obtain $Y \not\subseteq X'$ and $X \not\subseteq Y'$, i.e., (X, X') and (Y', Y) would not be extensions of (X, Y) with respect to the order $\subseteq^2$ defined by

- $(X_1, Y_1) \subseteq^2 (X_2, Y_2) : \iff X_1 \subseteq X_2$ and $Y_1 \subseteq Y_2$.

Notice that, in the ordered set $(\mathfrak{V}(\mathbb{K}), \subseteq^2)$ of all preconcepts of a formal context $\mathbb{K} := (G, M, I)$, the formal concepts of $\mathbb{K}$ are exactly the maximal elements and the protoconcepts of $\mathbb{K}$ are just the elements that are below exactly one maximal element (formal concept).

For contextually developing a *Boolean Concept Logic* as an integrated generalization of the Contextual Object Logic and the Contextual Attribute Logic, *Boolean operations* have to be introduced on the set $\mathfrak{V}(\mathbb{K})$ of all preconcepts of a formal context $\mathbb{K} := (G, M, I)$. That shall be done in the same way as for semiconcepts [Luksch and Wille, 1991] and for protoconcepts [Hereth et al., 2000]:

$$
\begin{aligned}
(A_1, B_1) \sqcap (A_2, B_2) &:= (A_1 \cap A_2, (A_1 \cap A_2)') \\
(A_1, B_1) \sqcup (A_2, B_2) &:= ((B_1 \cap B_2)', B_1 \cap B_2) \\
\neg(A, B) &:= (G \setminus A, (G \setminus A)') \\
\lnot(A, B) &:= ((M \setminus B)', M \setminus B) \\
\bot &:= (\emptyset, M) \\
\top &:= (G, \emptyset)
\end{aligned}
$$

The set $\mathfrak{V}(\mathbb{K})$ together with the operations $\sqcap, \sqcup, \neg, \lnot, \bot$, and $\top$ is called the *preconcept algebra* of $\mathbb{K}$ and is denoted by $\underline{\mathfrak{V}}(\mathbb{K})$; the operations are named *"meet," "join," "negation," "opposition," "nothing",* and *"all".* For the structural analysis of the preconcept algebra $\underline{\mathfrak{V}}(\mathbb{K})$, it is useful to define additional operations on $\mathfrak{V}(\mathbb{K})$:

$$
\mathfrak{a} \sqcup \mathfrak{b} := \neg(\neg\mathfrak{a} \sqcap \neg\mathfrak{b}) \quad \text{and} \quad \mathfrak{a} \sqcap \mathfrak{b} := \lnot(\lnot\mathfrak{a} \sqcup \lnot\mathfrak{b}),
$$
$$
\top := \neg\bot \quad \text{and} \quad \bot := \lnot\top.
$$

	male	female	old	young
father	×		×	
mother		×	×	
son	×			×
daughter		×		×

FIGURE 6.13: A context $\mathbb{K}^f$ of family members.

The semiconcepts resp. protoconcepts of $\mathbb{K}$ form subalgebras $\mathfrak{H}(\mathbb{K})$ resp. $\mathfrak{P}(\mathbb{K})$ of $\mathfrak{V}(\mathbb{K})$, which are called the *semiconcept algebra* resp. *protoconcept algebra* of $\mathbb{K}$. The set $\mathfrak{H}_\sqcap(\mathbb{K})$ of all $\sqcap$-semiconcepts is closed under the operations $\sqcap, \sqcup,$ $\neg, \bot,$ and $\top$; therefore, $\underline{\mathfrak{H}}_\sqcap(\mathbb{K}) := (\mathfrak{H}_\sqcap(\mathbb{K}), \sqcap, \sqcup, \neg, \bot, \top)$ is a Boolean algebra isomorphic to the Boolean algebra of all subsets of G. Dually, the set $\mathfrak{H}_\sqcup(\mathbb{K})$ of all $\sqcup$-semiconcepts is closed under the operations $\sqcap, \sqcup, \lrcorner, \llcorner,$ and $\top$; therefore, $\underline{\mathfrak{H}}_\sqcup(\mathbb{K}) := (\mathfrak{H}_\sqcup(\mathbb{K}), \sqcap, \sqcup, \lrcorner, \llcorner, \top)$ is a Boolean algebra antiisomorphic to the Boolean algebra of all subsets of M. Furthermore, $\mathfrak{B}(\mathbb{K}) = \mathfrak{H}_\sqcap(\mathbb{K}) \cap \mathfrak{H}_\sqcup(\mathbb{K})$, and $(\mathfrak{B}(\mathbb{K}), \wedge, \vee)$ is the *concept lattice* of $\mathbb{K}$ with the operations $\wedge$ and $\vee$ induced by the operations $\sqcap$ and $\sqcup$, respectively. The general order relation $\sqsubseteq$ of $\underline{\mathfrak{P}}(\mathbb{K})$, which coincides on $\mathfrak{B}(\mathbb{K})$ with the subconcept-superconcept-order $\leq$, is defined by

$$(A_1, B_1) \sqsubseteq (A_2, B_2) :\Longleftrightarrow A_1 \subseteq A_2 \text{ and } B_1 \supseteq B_2.$$

The introduced notions found a Boolean Concept Logic in which the Contextual Object Logic and the Contextual Attribute Logic can be integrated by transforming any object sets X to the $\sqcap$-semiconcept (X, X') and any attribute set Y to the corresponding $\sqcup$-semiconcept (Y', Y). In the case of Contextual Attribute Logic [Ganter and Wille, 1999b], this integration comprises a transformation of the Boolean compositions of attributes, which is generated by the following elementary assignments:

$$
\begin{aligned}
m \wedge n &\mapsto (\{m\}', \{m\}) \sqcap (\{n\}', \{n\}), \\
m \vee n &\mapsto (\{m\}', \{m\}) \sqcup (\{n\}', \{n\}), \\
\neg m &\mapsto \neg(\{m\}', \{m\}).
\end{aligned}
$$

In the dual case of Contextual Object Logic, the corresponding transformation uses the operations $\sqcap, \sqcup,$ and $\lrcorner$.

Preconcept algebras can be illustrated by *line diagrams*, which shall be demonstrated by using the small formal context in Figure 6.13. The line diagram of the preconcept algebra of that formal context is shown in Figure 6.14: the formal concepts are represented by the black circles, the proper $\sqcap$-semiconcepts by the circles with only a black lower half, the proper $\sqcup$-semiconcepts by the circles with only a black upper half, and the proper preconcepts (which are even not protoconcepts) by the unblackened circles.

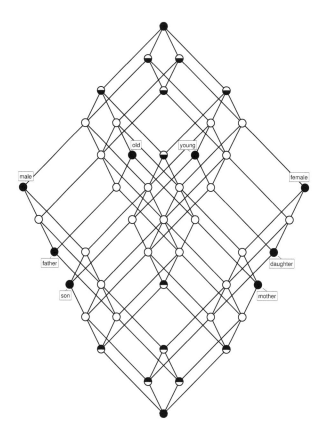

FIGURE 6.14: Line diagram of the preconcept algebra of the formal context $\mathbb{K}^f$ in Figure 6.13.

An object (attribute) belongs to a preconcept if and only if its name is attached to a circle representing a subpreconcept (superpreconcept) of that preconcept. The regularity of the line diagram in Figure 6.14 has a general reason that becomes clear by the following proposition (cf. [Wille, 2004a]):

Proposition 4 *For a formal context* $\mathbb{K} := (G, M, I)$, *the ordered set* $(\mathfrak{V}(\mathbb{K}), \sqsubseteq)$ *is a completely distributive complete lattice, which is isomorphic to the concept lattice of the formal context* $\mathbb{V}(\mathbb{K}) := (G \dot{\cup} M, G \dot{\cup} M, I \cup (\neq \backslash G \times M))$.

A detailed understanding of the *structure of preconcept algebras* is basic for Boolean Concept Logic. To start the necessary structure analysis, we concentrate in this section on determining an equational base of the class all protoconcept algebras and an equational base of the class all preconcept algebras (for the basic theorems on semiconcepts, protoconcepts, and preconcepts see [Vormbrock and Wille, 2005] and [Wille, 2004a], respectively). In the initial paper on *Boolean Concept Logic* [Wille, 2000b] the following proposition is stated:

Proposition 5 *The following equations are valid in all protoconcept algebras:*

1a)	$(x \sqcap x) \sqcap y = x \sqcap y$	*1b)*	$(x \sqcup x) \sqcup y = x \sqcup y$
2a)	$x \sqcap y = y \sqcap x$	*2b)*	$x \sqcup y = y \sqcup x$
3a)	$x \sqcap (y \sqcap z) = (x \sqcap y) \sqcap z$	*3b)*	$x \sqcup (y \sqcup z) = (x \sqcup y) \sqcup z$
4a)	$x \sqcap (x \sqcup y) = x \sqcap x$	*4b)*	$x \sqcup (x \sqcap y) = x \sqcup x$
5a)	$x \sqcap (x \uplus y) = x \sqcap x$	*5b)*	$x \sqcup (x \sqcap\!\!\sqcap y) = x \sqcup x$
6a)	$x \sqcap (y \uplus z) = (x \sqcap y) \uplus (x \sqcap z)$	*6b)*	$x \sqcup (y \sqcap\!\!\sqcap z) = (x \sqcup y) \sqcap\!\!\sqcap (x \sqcup z)$
7a)	$\neg\neg(x \sqcap y) = x \sqcap y$	*7b)*	$\lrcorner\lrcorner(x \sqcup y) = x \sqcup y$
8a)	$\neg(x \sqcap x) = \neg x$	*8b)*	$\lrcorner(x \sqcup x) = \lrcorner x$

9a)	$x \sqcap \neg x = \bot$	*9b)*	$x \sqcup \lrcorner x = \top$
10a)	$\neg\bot = \top \sqcap \top$	*10b)*	$\lrcorner\top = \bot \sqcup \bot$
11a)	$\neg\top = \bot$	*11b)*	$\lrcorner\bot = \top$
12)	$(x \sqcap x) \sqcup (x \sqcap x) = (x \sqcup x) \sqcap (x \sqcup x).$		

The question arises whether the equations of Proposition 5 are enough for determining the equational theory of protoconcept algebras, i.e., whether each equation valid in all protoconcept algebras can be entailed by the equations of Proposition 5. For answering this, a *double Bolean algebra* is defined as an algebraic structure $\underline{D} := (D, \sqcap, \sqcup, \neg, \lrcorner, \bot, \top)$ of type $(2, 2, 1, 1, 0, 0)$ satisfying the equations 1a) to 11a), 1b) to 11b), and 12) of Proposition 5. With this notion it can indeed be proved that the equations of Proposition 5 are enough (see [Wille, 2000b]):

Theorem 1 *The equational class generated by all protoconcept algebras of formal contexts is the equational class of all double Boolean Algebras.*

Proving an analogous result for preconcept algebras needs serious investigations (see [Wille, 2004a], p.9ff.). Finally, the following proposition and theorem can be proved:

Proposition 6 *The equations 1a) to 11a), 1b) to 11b) of Proposition 5 and the equations 12a)* $x_{\sqcap\sqcup\sqcap} = x_{\sqcap\sqcup}$ *and 12b)* $x_{\sqcup\sqcap\sqcup} = x_{\sqcup\sqcap}$ *are valid in all preconcept algebras where* $t_{\sqcap} := t \sqcap t$ *and* $t_{\sqcup} := t \sqcup t$ *is defined for every term t.*

A *generalized double Boolean algebra* algebra $\underline{D} := (D, \sqcap, \sqcup, \neg, \lrcorner, \bot, \top)$ of type $(2, 2, 1, 1, 0, 0)$ is defined as an algebraic structure satisfying the equations 1a) to 12a) 1b) to 12b) of Proposition 6. Now, we obtain:

Theorem 2 *The equational class generated by all preconcept algebras of formal contexts is the equational class of all generalized double Boolean Algebras.*

6.5   Contextual Judgment Logic

This approach to *Contextual Judgment Logic* was strongly stimulated by
J. F. Sowa's theory of conceptual graphs [Sowa, 1984] since those graphs can
be understood as semantic structures that represent logical judgments. The
mathematized conceptual graphs, called *concept graphs*, are mathematical
semantic structures based on formal contexts and their formal concepts (cf.
[Wille, 1997],[Wille, 2002]); those semantic structures are considered as formal
judgments in the underlying Contextual Judgment Logic.

Each step of the presented development of concept graphs shall start with
an example of a judgment represented graphically by a conceptual graph as
standardized by John Sowa (cf. [Sowa, 1992]). Those judgments are deduced
from the following statement written by Charles S. Peirce ([Peirce, 1992], p.
114):

> "Mathematics ... is the only one of the sciences which does not
> concern itself to inquire what the actual facts are, but studies
> hypotheses exclusively."

To obtain an example of a **simple conceptual graph**, we consider the judgment: "The science mathematics studies the hypothesis $2^{\aleph_0} = \aleph_1$" (called
"continuum hypothesis"). This judgment may be represented by the simple
conceptual graph shown in Figure 6.15. In that graph, "science" and "hypothesis" name concepts, while "mathematics" and "$2^{\aleph_0} = \aleph_1$" name objects
that fall under the concepts "science" and "hypothesis," respectively; furthermore, the relational concept "study" links the science "mathematics" with the
hypothesis "$2^{\aleph_0} = \aleph_1$."

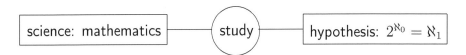

FIGURE 6.15:   Example of a simple conceptual graph.

The example shows that judgments may join plain concepts with relational
concepts so that a mathematization of judgments has to also offer besides
formal concepts "relation concepts." How this has been performed and further
developed shall be explained in the sequel (cf. [Wille, 2004b], pp. 53–55):

A *power context family* is a sequence $\vec{\mathbb{K}} := (\mathbb{K}_0, \mathbb{K}_1, \mathbb{K}_2, \ldots)$ of formal contexts $\mathbb{K}_k := (G_k, M_k, I_k)$ with $G_k \subseteq (G_0)^k$ for $k = 1, 2, \ldots$. The formal
concepts of $\mathbb{K}_k$ with $k = 1, 2, \ldots$ are called *relation concepts*, because they
represent k-ary relations on the object set G_0 by their extents.

A *relational graph* is a structure (V, E, ν) consisting of two disjoint sets V and E together with a map $\nu : E \rightarrow \bigcup_{k=1,2,\dots} V^k$; the elements of V and E are called *vertices* and *edges*, respectively, and $\nu(e) = (v_1, \dots, v_k)$ is read: $v_1, \dots, v_k$ are the *adjacent vertices* of the k-*ary edge* e ($|e| := k$ is the *arity* of e; the arity of a vertex is defined to be 0). Let $E^{(k)}$ be the set of all elements of $V \cup E$ of arity k ($k = 0, 1, 2, \dots$).

A *simple concept graph* of a power context family $\vec{\mathbb{K}} := (\mathbb{K}_0, \mathbb{K}_1, \mathbb{K}_2, \dots)$ with $\mathbb{K}_k := (G_k, M_k, I_k)$ for $k = 0, 1, 2, \dots$ is a structure $\mathfrak{G} := (V, E, \nu, \kappa, \rho)$ for which

- (V, E, ν) is a relational graph,

- $\kappa \colon V \cup E \rightarrow \bigcup_{k=0,1,2,\dots} \underline{\mathfrak{B}}(\mathbb{K}_k)$ is a mapping such that $\kappa(u) \in \underline{\mathfrak{B}}(\mathbb{K}_k)$ for all $u \in E^{(k)}$,

- $\rho \colon V \rightarrow \mathfrak{P}(G_0) \backslash \{\emptyset\}$ is a mapping such that $\rho(v) \subseteq Ext(\kappa(v))$ for all $v \in V$ and, furthermore, $\rho(v_1) \times \cdots \times \rho(v_k) \subseteq Ext(\kappa(e))$ for all $e \in E$ with $\nu(e) = (v_1, \dots, v_k)$;

- in general, $Ext(\mathfrak{c})$ denotes the extent of the formal concept $\mathfrak{c}$.

It is convenient to consider the mapping ρ not only on vertices but also on edges: for all $e \in E$ with $\nu(e) = (v_1, \dots, v_k)$, let $\rho(e) := \rho(v_1) \times \cdots \times \rho(v_k)$.

A *subgraph* of a concept graph $\mathfrak{G} := (V, E, \nu, \kappa, \rho)$ is a concept graph $\mathfrak{G}_s := (V_s, E_s, \nu_s, \kappa_s, \rho_s)$ for which $V_s \subseteq V$, $E_s \subseteq E$, $\nu_s = \nu|_{E_s}$, $\kappa_s = \kappa|_{V_s \cup E_s}$, and $\rho_s = \rho|_{V_s}$. The *union* and *intersection* of subgraphs $\mathfrak{G}_t := (V_t, E_t, \nu_t, \kappa_t, \rho_t)$ ($t \in T$) of a concept graph $\mathfrak{G} := (V, E, \nu, \kappa, \rho)$ are defined by

$$\bigcup_{t \in T} \mathfrak{G}_t := \left(\bigcup_{t \in T} V_t, \bigcup_{t \in T} E_t, \bigcup_{t \in T} \nu_t, \bigcup_{t \in T} \kappa_t, \bigcup_{t \in T} \rho_t \right),$$

$$\bigcap_{t \in T} \mathfrak{G}_t := \left(\bigcap_{t \in T} V_t, \bigcap_{t \in T} E_t, \bigcap_{t \in T} \nu_t, \bigcap_{t \in T} \kappa_t, \bigcap_{t \in T} \rho_t \right).$$

Lemma 1 *The union and intersection of subgraphs of a concept graph $\mathfrak{G}$ is always a subgraph of $\mathfrak{G}$ again.*

From the background knowledge coded in a power context family $\vec{\mathbb{K}}$, two types of material inferences shall be made formally explicit: Let $k = 0, 1, 2, \dots$;

1. *object implications*: for $A, C \subseteq G_k$, $\mathbb{K}_k$ satisfies $A \rightarrow C$ if $A^{I_k} \subseteq C^{I_k}$ and,

2. *concept implications*: for $\mathfrak{B}, \mathfrak{D} \subseteq \mathfrak{B}(\mathbb{K}_k)$, $\mathbb{K}_k$ satisfies $\mathfrak{B} \rightarrow \mathfrak{D}$ if $\bigwedge \mathfrak{B} \leq \bigwedge \mathfrak{D}$.

The *formal implications* $A \to C$ and $\mathfrak{B} \to \mathfrak{D}$ give rise to a closure system $\mathcal{C}(\mathbb{K}_k)$ on $\mathbb{S}^{imp}(\mathbb{K}_k) := \{(g, \mathfrak{b}) \in G_k \times \mathfrak{B}(\mathbb{K}_k) \mid g \in Ext(\mathfrak{b})\}$ consisting of all subsets Y of $\mathbb{S}^{imp}(\mathbb{K}_k)$, which have the following property:

(P_k) If $A \times \mathfrak{B} \subseteq Y$ and if $\mathbb{K}_k$ satifies $A \to C$ and $\mathfrak{B} \to \mathfrak{D}$ then $C \times \mathfrak{D} \subseteq Y$.

For $k = 1, 2, \ldots$, the $\mathbb{K}_k$-*conceptual content* $C_k(\mathfrak{G})$ of a concept graph $\mathfrak{G} := (V, E, \nu, \kappa, \rho)$ of a power context family $\vec{\mathbb{K}}$ is defined as the closure of

$$\{(\vec{g}, \kappa(e)) \mid e \in E^{(k)} \text{ and } \vec{g} \in \rho(e)\}$$

with respect to the closure system $\mathcal{C}(\mathbb{K}_k)$;
the $\mathbb{K}_0$-*conceptual content* $C_0(\mathfrak{G})$ of $\mathfrak{G}$ is defined as the closure of

$$\{(g, \kappa(v)) \mid v \in V \text{ and } g \in \rho(v)\} \cup$$
$$\{(g_i, (G_0, G_0^{I_0})) \mid \exists ((g_1, \ldots, g_k), \mathfrak{c}) \in C_k(\mathfrak{G}) \text{ with } g_i \in \{g_1, \ldots, g_k\}\}$$

with respect to the closure system $\mathcal{C}(\mathbb{K}_0)$. Then

$$C(\mathfrak{G}) := C_0(\mathfrak{G}) \,\dot{\cup}\, C_1(\mathfrak{G}) \,\dot{\cup}\, C_2(\mathfrak{G}) \,\dot{\cup}\, \ldots$$

is called the $(\vec{\mathbb{K}}\text{-})$*conceptual content* of the concept graph $\mathfrak{G}$.

The defined conceptual contents give rise to an *information (quasi-) order* $\lesssim$ on the set of all concept graphs of a power context family: A concept graph $\mathfrak{G}_1 := (V_1, E_1, \nu_1, \kappa_1, \rho_1)$ is said to be *less informative (more general)* than a concept graph $\mathfrak{G}_2 := (V_2, E_2, \nu_2, \kappa_2, \rho_2)$ (in symbols: $\mathfrak{G}_1 \lesssim \mathfrak{G}_2$) if

$$C_k(\mathfrak{G}_1) \subseteq C_k(\mathfrak{G}_2) \text{ for } k = 0, 1, 2, \ldots;$$

$\mathfrak{G}_1$ and $\mathfrak{G}_2$ are called *equivalent* (in symbols: $\mathfrak{G}_1 \sim \mathfrak{G}_2$) if $\mathfrak{G}_1 \lesssim \mathfrak{G}_2$ and $\mathfrak{G}_2 \lesssim \mathfrak{G}_1$ (i.e., $C_k(\mathfrak{G}_1) = C_k(\mathfrak{G}_2)$ for $k = 0, 1, 2, \ldots$). The set of all equivalence classes of concept graphs of a power context family $\vec{\mathbb{K}}$ together with the order induced by the quasi-order $\lesssim$ is a *complete lattice* denoted by $\widetilde{\Gamma}(\vec{\mathbb{K}})$.

To obtain an example of an **existential conceptual graph**, we modify the first judgment as follows: "The science mathematics studies hypotheses." Logically equivalent is the judgment: "There exists at least one hypothesis studied by the science of mathematics." This judgment may be represented by the existential conceptual graph shown in Figure 6.16.

FIGURE 6.16:   Example of an existential conceptual graph.

The example shows that judgments may embody existentially quantified variables that are usually indicated by letters like x, y, z (sometimes they are replaced by a so-called "coreference link"). The mathematization of existential conceptual graphs whose variables are from a variable set X can be based on "free X-extensions" of a power context family. Such mathematization generalizes the approach of section 2 so that it becomes a wider range of applications (cf. [Wille, 2004b], pp. 55–57).

For a set X of variables, an *X-interpretation* into a set G_0 with $G_0 \cap X = \emptyset$ is defined as a mapping $\chi : G_0 \cup X \to G_0$ with $\chi(g) = g$ for all $g \in G_0$; the set of all X-interpretations into G_0 is denoted by $B(X, G_0)$. The *free X-extension* of the power context family $\vec{\mathbb{K}} := (\mathbb{K}_0, \mathbb{K}_1, \mathbb{K}_2, \ldots)$ with $\mathbb{K}_k := (G_k, M_k, I_k)$ for $k = 0, 1, 2, \ldots$ and $G_0 \cap X = \emptyset$ is defined as a power context family $\vec{\mathbb{K}}[X] := (\mathbb{K}_0[X], \mathbb{K}_1[X], \mathbb{K}_2[X], \ldots)$ for which

- $\mathbb{K}_0[X] := (G_0[X], M_0[X], I_0[X])$ with $G_0[X] := G_0 \cup X$,
  $M_0[X] := M_0, \ I_0[X] := I_0 \cup (X \times \{m \in M_0 \mid \{m\}^{I_0} \neq \emptyset\})$,

- $\mathbb{K}_k[X] := (G_k[X], M_k[X], I_k[X])$ $(k = 1, 2, \ldots)$ with
  $G_k[X] := \{(u_1, \ldots, u_k) \in G_0[X]^k \mid \exists \chi \in B(X, G_0) :$
  $$(\chi(u_1), \ldots, \chi(u_k)) \in G_k\},$$
  $M_k[X] := M_k,$ and
  $(u_1, \ldots, u_k) I_k[X] m : \iff \exists \chi \in B(X, G_0) : (\chi(u_1), \ldots, \chi(u_k)) I_k m.$

$\vec{\mathbb{K}}[X]$ is called an *existential power context family*.

For defining existential concept graphs, the surjective $\bigwedge$-homomorphisms $\pi_k^X : \underline{\mathfrak{B}}(\mathbb{K}_k[X]) \to \underline{\mathfrak{B}}(\mathbb{K}_k)$ $(k = 0, 1, 2, \ldots)$ are needed, which are determined by

$$\pi_k^X(A, B) := (A \cap G_k, (A \cap G_k)^{I_k}) \text{ for } (A, B) \in \underline{\mathfrak{B}}(\mathbb{K}_k[X]).$$

An *existential concept graph* of a power context family $\vec{\mathbb{K}}$ is defined as a concept graph $\mathfrak{G} := (V, E, \nu, \kappa, \rho)$ of a free X-extension $\vec{\mathbb{K}}[X]$ for which an X-interpretation χ into G_0 exists such that $\mathfrak{G}^\chi := (V, E, \nu, \kappa^\chi, \rho^\chi)$ with $\kappa^\chi(u) := \pi_k^X(\kappa(u))$ and $\rho^\chi(v) := \chi(\rho(v))$ is a concept graph of $\vec{\mathbb{K}}$; χ is then called an X-interpretation *admissible on* $\mathfrak{G}$. For a fixed variable set X, $\mathfrak{G}$ is more precisely named an existential concept graph of $\vec{\mathbb{K}}$ *over* X.

Lemma 2 *The subgraphs of an existential concept graph over X are existential concept graphs over X, too.*

The *conceptual content* of an existential concept graph $\mathfrak{G}_X$ of a power context family $\vec{\mathbb{K}}$ is defined as the conceptual content of $\mathfrak{G}_X$ understood as a concept graph of the free X-extension $\vec{\mathbb{K}}[X]$. To clarify this, it is helpful to show how variables give rise to object implications of the relational contexts $\mathbb{K}_k[X]$ as indicated in the following lemma:

Lemma 3 *Let $\mathbb{K}_k[X] := (G_k[X], M_k[X], I_k[X])$ with $k \in \{1, 2, \ldots\}$ be a relational context of an existential power context family $\vec{\mathbb{K}}[X]$; furthermore, let α be a map of $G_0 \cup X$ into itself satisfying $\alpha(g) = g$ for all $g \in G_0$. Then*

$\mathbb{K}_k[X]$ *has the object implications* $\{(\alpha(u_1), \ldots, \alpha(u_k))\} \longrightarrow \{(u_1, \ldots, u_k)\}$ *with* $u_1, \ldots, u_k \in G_0 \cup X$.

For a permutation π of the variable set X, let α_π be the map of $G_0 \cup X$ into itself with $\alpha_\pi(g) = g$ for all $g \in G_0$ and $\alpha_\pi(x) = \pi(x)$ for all $x \in X$. Then we obtain the object implication $\{(\alpha_\pi(u_1), \ldots, \alpha_\pi(u_k))\} \longrightarrow \{(u_1, \ldots, u_k)\}$ with $u_1, \ldots, u_k \in G_0 \cup X$. Together with the corresponding object implication for π^{-1}, this yields that changing variables according to a permutation of X in a (k-ary) object of $\mathbb{K}_k[X]$ does not change the intension of that object.

An existential concept graph $\mathfrak{G}_1 := (V_1, E_1, \nu_1, \kappa_1, \rho_1)$ is said to be *less informative (more general)* than $\mathfrak{G}_2 := (V_2, E_2, \nu_2, \kappa_2, \rho_2)$ (in symbols: $\mathfrak{G}_1 \lesssim \mathfrak{G}_2$) if $C_k(\mathfrak{G}_1) \subseteq C_k(\mathfrak{G}_2)$ for $k = 0, 1, 2, \ldots$; $\mathfrak{G}_1$ and $\mathfrak{G}_2$ are called *equivalent* (in symbols: $\mathfrak{G}_1 \sim \mathfrak{G}_2$) if $\mathfrak{G}_1 \lesssim \mathfrak{G}_2$ and $\mathfrak{G}_2 \lesssim \mathfrak{G}_1$ (i.e., $C_k(\mathfrak{G}_1) = C_k(\mathfrak{G}_2)$ for $k = 0, 1, 2, \ldots$). The set of all equivalence classes of existential concept graphs of a power context family $\vec{\mathbb{K}}$ over a fixed set X of variables together with the order induced by the quasi-order $\lesssim$ is an *ordered set* denoted by $\widetilde{\Gamma}(\vec{\mathbb{K}}; X)$.

For representing exactly Peirce's judgment that "mathematics studies hypotheses exclusively," we have to generalize existential conceptual graphs further to **implicational conceptual graphs**. This becomes clear when we consider an equivalent formulation of Peirce's judgment, namely: "If mathematics studies a proposition then mathematics studies a hypothesis." A representation of this judgment by an implicational conceptual graph is pictured in Figure 6.17.

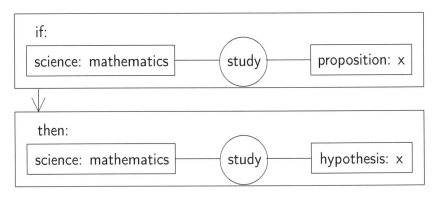

FIGURE 6.17:   Example of an implicational conceptual graph.

The example shows an implicational judgment in which the premise and the conclusion contain the same variable x; this indicates that the proposition x is, more precisely, a hypothesis. The mathematization of implicational conceptual graphs who are composed by two subgraphs representing a premise and a corresponding conclusion, respectively, can be viewed as a generalization of existential concept graphs (cf. [Wille, 2004b], pp. 57–59).

An *implicational concept graph* of a power context family $\vec{\mathbb{K}}$ is defined as an existential concept graph $\mathfrak{G} := (V, E, \nu, \kappa, \rho)$ of $\vec{\mathbb{K}}$ over a variable set X with a designated pair $(p\mathfrak{G}, c\mathfrak{G})$ of subgraphs such that

1. $\mathfrak{G}$ is the union of $p\mathfrak{G}$ and $c\mathfrak{G}$, and

2. each X-interpretation admissible on $p\mathfrak{G}$ is also admissible on $c\mathfrak{G}$ (and hence on $\mathfrak{G}$, too).

$p\mathfrak{G} \to c\mathfrak{G}$ may be written instead of $\mathfrak{G}$; the subgraphs $p\mathfrak{G}$ and $c\mathfrak{G}$ are called the *premise* and the *conclusion*, resp.

For an existential concept graph $\overline{\mathfrak{G}}$ of a power context family $\vec{\mathbb{K}}$ over a variable set X, the formal context $\mathbb{K}(X; \overline{\mathfrak{G}}) := (B(X, G_0), Sub(\overline{\mathfrak{G}}), \rhd)$ is defined where

- the object set $B(X, G_0)$ consists of all X-interpretations into the object set G_0 of the formal context $\mathbb{K}_0$ in $\vec{\mathbb{K}}$,

- the attribute set $Sub(\overline{\mathfrak{G}})$ is the set of all subgraphs of $\overline{\mathfrak{G}}$,

- $\chi \rhd \mathfrak{G}$ means that the X-interpretation χ is admissible on the subgraph $\mathfrak{G}$ of $\overline{\mathfrak{G}}$.

Proposition 7 $\{\mathfrak{G}_s \mid s \in S\} \to \{\mathfrak{G}_t \mid t \in T\}$ *is an attribute implication of* $\mathbb{K}(X; \overline{\mathfrak{G}})$ *if and only if* $\bigcup_{s \in S} \mathfrak{G}_s \to \bigcup_{t \in T} \mathfrak{G}_t$ *is an implicational concept graph of* $\vec{\mathbb{K}}$ *over* X.

Proposition 8 $\mathbb{K}(X; \overline{\mathfrak{G}}) := (B(X, G_0), Sub(\overline{\mathfrak{G}}), \rhd)$ *is always a formal context of which all extents are non-empty attribute extents. Conversely, let* $\mathbb{K} := (G, M, I)$ *be a clarified formal context of which all extents are non-empty attribute extents; then* $\mathbb{K}$ *is isomorphic to the clarified context of the formal context* $\mathbb{K}(\{x\}; \overline{\mathfrak{G}}) := (B(\{x\}, G), Sub(\overline{\mathfrak{G}}), \rhd)$ *where* $\overline{\mathfrak{G}} := (V, E, \nu, \kappa, \rho)$ *is the existential concept graph of the power context family* $\vec{\mathbb{K}} := (\mathbb{K})$ *over* $\{x\}$ *with* $V := M$, $E := \emptyset$, $\nu := \emptyset$, $\kappa(m) := \mu m$, *and* $\rho(m) := \{x\}$.

Corollary 1 *The concept lattices* $\underline{\mathfrak{B}}(\mathbb{K}(X; \overline{\mathfrak{G}}))$ *are up to isomorphism the concept lattices of formal contexts.*

Implicational conceptual graphs can even be generalized to clausal conceptual graphs in which the conclusion consists of a disjunction of propositions (cf. [Wille, 2004b], pp. 59–60). An example of a clausal conceptual graph is shown in Figure 6.18.

A *clausal concept graph* of a power context family $\vec{\mathbb{K}}$ is defined as an existential concept graph $\mathfrak{G} := (V, E, \nu, \kappa, \rho)$ of $\vec{\mathbb{K}}$ over a variable set X with a designated pair $(p\mathfrak{G}, \{c_t\mathfrak{G} \mid t \in T\})$ consisting of a subgraph $p\mathfrak{G}$ of $\mathfrak{G}$ and a set $\{c_t\mathfrak{G} \mid t \in T\}$ of subgraphs of $\mathfrak{G}$ such that

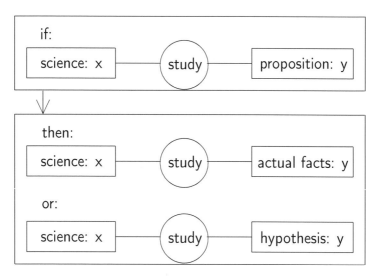

FIGURE 6.18:   Example of a clausal conceptual graph.

1. $\mathfrak{G}$ is the union of $p\mathfrak{G}$ and all the $c_t\mathfrak{G}$ with $t \in T$, and

2. each X-interpretation admissible on $p\mathfrak{G}$ is also admissible on at least one $c_t\mathfrak{G}$ with $t \in T$.

$p\mathfrak{G} \to \bigvee_{t \in T} c_t\mathfrak{G}$ may be written instead of $\mathfrak{G}$; the subgraphs $p\mathfrak{G}$ and $c_t\mathfrak{G}$ $(t \in T)$ are called the *premise* and the *disjunctive conclusions*, resp. For subsets A and B of the attribute set M, $\bigwedge A \to \bigvee B$ is an *attribute clause* of $\mathbb{K}$ if $g \in A^I$ always implies gIm for at least one $m \in B$.

Proposition 9 *Let $\overline{\mathfrak{G}}$ be an existential concept graph of a power context family $\vec{\mathbb{K}}$ over a variable set X and let $\mathfrak{G}_s$ $(s \in S)$ and $\mathfrak{G}_t$ $(t \in T)$ be subgraphs of $\overline{\mathfrak{G}}$. Then $\bigwedge\{\mathfrak{G}_s \mid s \in S\} \to \bigvee\{\mathfrak{G}_t \mid t \in T\}$ is an attribute clause of the formal context $\mathbb{K}(X; \overline{\mathfrak{G}})$ if and only if $\bigcup_{s \in S} \mathfrak{G}_s \to \bigvee_{t \in T} \mathfrak{G}_t$ is a clausal concept graph of $\vec{\mathbb{K}}$ over X.*

Propositions 8 and 9 show that the theory of clausal concept graphs is essentially equivalent to the theory of attribute clauses of formal contexts. The advantage of this equivalence is that many results about attribute clauses can be transferred to clausal concept graphs that substantially enriches the research on Contextual Judgment Logic.

Corollary 2 $\mathfrak{G}_\emptyset \to \bigvee_{t \in T} \mathfrak{G}_t$ *is a clausal concept graph of $\vec{\mathbb{K}}$ over X if and only if for all X-interpretations χ into G_0 there exists a $t_\chi \in T$ such that χ is admissible on $\mathfrak{G}_{t_\chi}$.*

At this point the presentation of the introduction to Formal Concept Analysis and Contextual Logic shall end. For futher information there is quite a number of books and periodicals on the subject of which we recomend, in particular, the *State-of-the-Art Survey* [Ganter et al., 2005b] on Formal Concept Analysis and the Springer Lecture Notes in Computer Science about *Conceptual Structures* and *Formal Concept Analysis*.

Part IV

Text Analysis

Chapter 7

Linguistic Data Exploration

Uta Priss

Napier University, UK

7.1 Introduction

This chapter starts with a caveat: linguists sometimes object to the use of Formal Concept Analysis (FCA) for linguistic applications! The reason for this is that the notion of "concept" in FCA does not correspond exactly to a notion of "concept" as developed by linguistic and philosophical theories. For example, Frege's [1892] notions of "Sinn" (sense) and "Bedeutung" (reference) appear on first sight to contradict FCA concepts because he explains that the words "morning star" and "evening star" have the same reference but a different sense. Thus if "reference" is taken to mean "extension" of a concept and "sense" to mean "intention," then this would contradict the FCA claim that if two concepts have the same extension they also must have the same intension with respect to a given formal context. But this apparent contradiction can in fact be resolved because it is more a misunderstanding than a contradiction.

First, all elements of FCA are mathematical entities, which is why they are usually prefixed with the word "formal," such as "formal concept." Linguistic and philosophical notions of "concept" are not usually defined on a mathematical level. Thus formal concepts can at most be an approximation of concepts as defined in linguistics, but never be exactly the same. Second, FCA concepts are only defined with respect to formal contexts. It is not clear in Frege's example what exactly the formal context is that is modeled. If the formal contexts that underlie this example are explicitly stated (as described by [Priss, 1998a] and elaborated in the next section) then the apparent contradiction is easily explained as a shift between formal concepts or contexts. Third, according to [Eklund and Wille, 2007], FCA applications always function on three different levels: the mathematical, logical, and applied levels. The vocabulary from these three levels should never be mixed. In summary, it is entirely safe to use FCA for linguistic applications as long as it is emphasized that the mathematical vocabulary should be separated from the linguistic vocabulary and, more specifically, that formal concepts are not exactly the same as concepts in linguistics, no matter how similar they may appear in some applications.

The manner in which FCA can be used in linguistics depends to some degree on whether the linguistic analysis is aimed at a "micro"- or "macro"-level. Traditional linguists often conduct research at a "micro"-level by performing extremely detailed analyses of data collected from precise and detailed contexts. Such data often requires careful manual analysis by an expert and is not suitable for quantitative or computerized methods during the initial part of the analysis. If the results of such an analysis are recorded in a somewhat systematic manner, then it may be possible to apply FCA methods in a secondary analysis. For example, FCA methods can be and have been used to visualize word/feature matrices from componential analyses; to explore verb paradigms [Großkopf and Harras, 2000] and to solve a problem about phonological classes in ancient Sanskrit grammar [Petersen, 2005].

FCA is very well suited to "macro"-level analyses of linguistic data. These are usually applications where linguistic data is collected at a large scale from corpora, dictionaries, or lexical databases and analyzed with respect to structures that emerge when hundreds of thousands of entries are taken into consideration or when language is considered at a slightly abstracted, formal level. Such linguistic FCA applications often fall into the areas of computational linguistics or natural language processing and provide an alternative to statistical methods. Examples are the automated construction of thesauri from bilingual dictionaries using Dyvik's [2004] semantic mirrors method (cf. [Priss and Old, 2005] for a modeling of this method with FCA); a formalization of WordNet [Priss, 1998b] and of semantic relations in lexical databases [Priss, 1999], and a formalization of metaphor [Priss and Old, 2001].

Several overviews of linguistic FCA applications have been published elsewhere: a general overview by [Priss, 2005], an overview of how to model lexical databases with FCA by [Priss and Old, 2004], and an overview of FCA applications in information science by [Priss, 2006], which may also be somewhat relevant for linguistics. This chapter does not attempt to repeat these overviews but instead intends to provide an introduction to FCA applications in linguistics using mostly "hands-on" examples. Most of the examples in the text are fairly simple. Several examples are presented as workflows for deriving concept lattices that should be easy to replicate. The goal is to give the reader an idea of what kind of applications are feasible. In order to apply the workflows in a more systematic manner, instead of just for a few manually compiled examples, knowledge of computer programming would be required. The overview papers mentioned above and the references in the text should provide sufficient hints for further reading about more complex applications of these techniques[1].

This chapter has two parts: the first part discusses the difficulties involved in representing linguistic structures as conceptual structures using the example of semantic relations. The second part discusses how to derive concept lattices from corpora and lexical databases, including Roget's Thesaurus and bilingual dictionaries.

7.2 Modeling Semantic Relations

Despite the caveat about the notion of "concept" mentioned in the introduction, the ordering of formal concepts in a concept lattice appears to be similar to the notion of hypernymy/hyponymy in lexical databases - but only for some types of formal contexts! This section introduces the notion of a *(lexical) denotative context* [Priss, 1998a], which is a formal context where the formal objects are words and the formal attributes are features of the denotations of the words chosen in a manner so that the conceptual relations approximately match the semantic relations. Examples of attributes that are not allowed in such formal contexts are meta-level features (for example, "a word with 5 letters") or connotative features ("positive," "neutral," or "pejorative" connotations). The purpose of this section is to highlight the difficulties involved in modeling such contexts and to discuss how other semantic relations can be defined in this framework.

A special notation is used in the diagrams of the concept lattices in this section: the formal attributes are visually distinguished from the formal objects by enclosing the attribute labels in square brackets. Figure 7.1 uses names of concepts in addition to formal attributes and objects, which are indicated by the boxes around the names and by positioning them at the same level as the concepts. The intention of this notation is to highlight the fact that although words occur as formal attributes, objects and names of concepts, the three sets have different functions that need to be carefully distinguished, even though the sets can be overlapping or even be identical.

In the concept lattice in Figure 7.1, the concept named "female" is a hypernym of the concept named "adult female," which in turn is a hyponym of "female." The intension of "adult female," which is the set $\{female, grown\ up\}$ is a superset of the intension of "female," which is the set $\{female\}$. The extension of "adult female," which is the set $\{woman\}$ is a subset of the extension of "female," which is the set $\{female\ person, woman, girl\}$. In this miniature concept lattice, the FCA conditions about the conceptual ordering

---

[1]The overviews are also available at http://www.upriss.org.uk. Links to FCA software, a bibliography and other resources can be found at http://www.fcahome.org.uk.

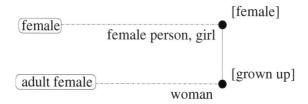

FIGURE 7.1: The conceptual ordering.

match a linguistic intuition about hypernymy/hyponymy. In this sense, the subconcept-superconcept relation of FCA can be considered a formalization of the Aristotelian principle of "genus proximum" and "differentia specifica." In this example, this can be achieved by using a variety of formulas:

- concept name = name of superconcept + differentiating attribute
  ("an adult female is a female who is grown up")

- formal object = name of superconcept + differentiating attribute
  ("a woman is a female who is grown up")

- "formal object = formal object attached to superconcept + differentiating attribute"
  ("a woman is a female person who is grown up")

But these formulas are only correct for "woman" and "female person," not for "girl." The idea for denotative contexts is that the formal objects behave like concept names and the semantic relations between the formal objects correspond to the conceptual relations between their concepts. Thus the formal object "girl" should have some more attributes that distinguish it from "female person."

Unfortunately, lexical databases such as WordNet[2] do not record the data in a format that clearly identifies extensions, intentions and names of concepts. Since words can have many senses, in this chapter the term *disambiguated word* is used for a formal object in a denotative context, i.e., a word in a specific sense. In traditional dictionaries, each entry (or paragraph) represents a word, each sense of the word (indicated by sense numbers in the dictionary) denotes a disambiguated word. In WordNet a concept is defined by a set of synonyms, which is also called a *synset*. Each word in a synset is disambiguated. Using WordNet terminology, *semantic relations* are relations that hold between synsets. They differ from *lexical relations*, such as antonymy, which hold among disambiguated words but not necessarily among their synsets. For example, $\{large, big\}$ is a synset in WordNet where "small"

---

[2]http://wordnet.princeton.edu, [Miller et al., 1990].

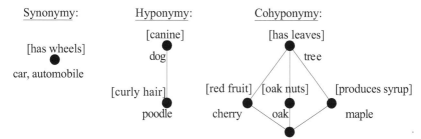

FIGURE 7.2: Semantic relations.

is an antonym of "large" and "little" is an antonym of "big." Not all senses of "large" and "big" have antonyms.

The following linguistic notions can now be explicitly defined for the lattices of denotative contexts (see Figure 7.2):

Synonymy: Disambiguated words are called *synonyms* if they are attached to the same concept. (In the example, "car" and "automobile" are synonyms.)

Hyponymy: A disambiguated word is a *hyponym* of another word if the concept it is attached to is a subconcept of the concept the other word is attached to. (In the example, "poodle" is a hyponym of "dog.")

Cohyponymy: Disambiguated words are called *cohyponyms* if they are attached to immediate subconcepts of the same concept and are not synonymous. (In the example, "cherry," "oak," and "maple" are cohyponyms of "tree.")

If several senses of a word are modeled in the same concept lattice, then they must be distinguished by adding a sense number after the word. For example, "oak 1" could refer to the tree, whereas "oak 2" might refer to the wood as a building material. Homographs (i.e., one word form denotes unrelated notions) and polysemous words (i.e., one word denotes different but related notions) can occur in the same concept lattice, but they would be distinguished by sense numbers. The relationship between "oak 1" and "oak 2" is not a conceptual relation in such a lattice, but would only be apparent by performing a string comparison on the object names.

Relational concept analysis [Priss, 1998a] is an extension of FCA that includes further, not necessarily hierarchical, relations, which are defined on objects and generalized to concepts. This involves quantification similar to the quantification of relations used in relational databases. For example, in a lexical database there can be a *meronymy (part-whole)* relation between the concepts "finger" and "hand." In a prototypical sense, *each* finger is part of *exactly one* hand and *each* hand has *exactly five* fingers. More precisely, for each object that belongs to the concept "finger," there is exactly one object

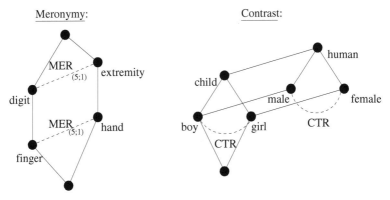

FIGURE 7.3:   Meronymy and contrast.

in the concept "hand" such that the finger belongs to the hand. And for each object in the concept "hand" there are exactly five objects in the concept "finger" such that the fingers belong to the hand. Formally this is denoted by "finger $MER_{(5;1)}$ hand" and graphically this is denoted by dashed lines, such as in Figure 7.3. It should be stressed that the dashed lines are not part of the lattice but indicate additional relations. Other relations such as "contrast," which is a generalization of antonymy, semantically depend on attributes, not objects, as is demonstrated in the definitions below:

Meronymy: A disambiguated word is a *meronym* of another word if the objects in the extension of its concept are parts of the objects of the other concept. There are different types of the meronymy relation depending on the quantifications.

Contrast: Disambiguated words are called *being in contrast* to each other if at least one pair of the attributes of their concepts is in contrast to each other. If the antonymy relation is generalized to hold between concepts it is a contrast relation.

Other meronymy relations use other quantifications. The example in Figure 7.4 is modeled after a small part of WordNet. The hyponymy relation forms a tree hierarchy but can be turned into a lattice by adding a bottom concept. This relation is indicated by solid lines, but only parts of the tree/lattice are shown in the figure. The meronymy relation is shown as dashed lines. In analogy to FCA "nested line diagrams," parallel meronymy relations are abbreviated by drawing boxes around the parts of the lattice and drawing the dashed lines between the boxes. Each concept in the box is in a meronymy relation to each concept in another box between which a bold-faced dashed line is drawn. For example, each general concept of "day" has at least one general time of day and vice versa. More specifically, every "tomorrow" has a "morning" and every "morning" belongs to a "day." This is not true for more

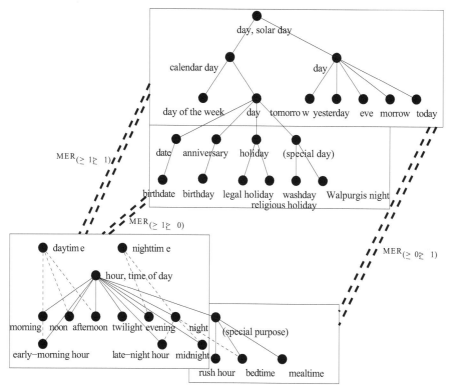

FIGURE 7.4: Meronymy relations in WordNet.

specific days and times of days. For example, not every day has a rush hour. The "≥ 0" quantifier is used in this case to show that a day can have a rush hour, but does not have to. On the other hand, every "rush hour" is part of a "day," but never part of a "holiday." Relational concept analysis facilitates the graphical display and the analysis of such relations. If the quantifiers of a semantic relation are identified, this information can be used to determine a basis of the relation that generates the relation [Priss, 1999].

It can be quite difficult to derive a typology of meronymy relations, distinguishing part-of, substance-of, instance-of, and so on. Many different typologies of meronymy relations have been suggested, for example, six types by [Winston et al., 1987] and eight different types by [Chaffin and Herrmann, 1988]. [Priss, 1996] argues that for meronymy it is much more useful to develop a classification by using quantificational differences (in the sense of relational concept analysis) instead of trying to determine semantic differences because it appears to be impossible to achieve agreement about semantic differences in this case.

As mentioned before, polysemy cannot be modeled directly in a lattice of a denotative context. The following example demonstrates how all polysemous

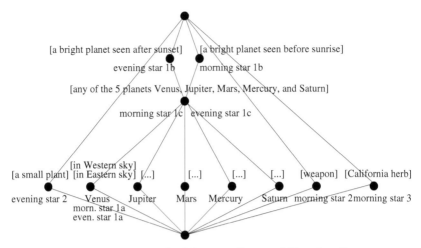

FIGURE 7.5:   The Lattice for Morning Star and Evening Star.

senses of a word in a dictionary can occur in a lattice. The example uses the definitions of "morning star 1" and "evening star 1" in Webster's Third New International Dictionary to demonstrate how the different polysemous senses interrelate. The definitions are

> **Evening star 1a:** a bright planet (as Venus) seen in the western sky after sunset;
>
> **Evening star 1b:** any planet that rises before midnight;
>
> **Evening star 1c:** any of the five planets that may be seen with the naked eye at sunset;
>
> **Evening star 2:** a small bulbous plant of Texas (*Cooperia drummondii*) with grass-like leaves and star-shaped white flowers;
>
> **Morning star 1a:** a bright planet (as Venus) seen in the eastern sky before sunrise;
>
> **Morning star 1b:** a planet that sets after midnight;
>
> **Morning star 1c:** any of the five planets that may be seen with the naked eye if in the sky at sunrise (Venus, Jupiter, Mars, Mercury, and Saturn may be morning stars);
>
> **Morning star 2:** a weapon consisting of a heavy ball set with spikes and either attached to a staff or suspended from one by a chain - also called a "holy water sprinkler;"
>
> **Morning star 3:** an annual California herb with showy yellow flowers;

The lattice diagram in Figure 7.5 shows that the different senses of "morning star" and "evening star" have different degrees of generality and specialization.

In their most specific sense (1a), "morning star" and "evening star" denote only Venus. In a wider sense (1c), they denote any of the five planets Venus, Jupiter, Mars, Mercury, and Saturn. In their widest sense (1b), they denote any planet that can be seen in the morning or evening sky, respectively. In the senses 1a and 1c, morning star and evening star share their denotations and are synonymous. This is because of the denotative features that are invoked by the extensions that are given in the definitions. If "Venus" is mentioned as part of the definition, then the corresponding concept must have the attributes of both evening star and morning star in its intension. If morning and evening star are defined abstractly (as in sense 1b), then they need not be synonymous but are hypernyms of "Venus."

Figure 7.5 is an attempt at solving the apparent discrepancy between Frege's [1892] notion of "reference" and "sense" and the FCA notions of "extension" and "intension." As mentioned in the introduction, in Frege's example of "morning star" and "evening star," both share "Venus" as a reference, but have a different "sense." It all depends on the modeling of the formal context. "Morning star 1a" and "evening star 1a" denote the concept "planet Venus" which has Venus in its extension and attributes of Venus in its intention. "Morning star 1b" and "evening star 1b" belong to different concepts and therefore have different extensions and intensions. The lattice does not specify the exact differences between the denotations of "morning star 1b" and "evening star 1b" because that is external semantic information.

Overall Frege's example and the example from WordNet highlight the difficulties involved in the modeling of semantic relations with FCA. In denotative contexts, semantic relations can be modeled in agreement with conceptual relations. But the attributes need to be carefully selected and there may not be any means for automating the construction of such contexts. In the following section, the conceptual relations in the lattices are not intended to model semantic or other linguistic relations. The special notations in the diagrams (such as square brackets) are not used in the next section.

7.3 Deriving Lattices from Lexical Databases and Corpora

7.3.1 Verb Clusters Derived from a Corpus

If one extracts data from lexical databases or corpora and automatically derives the emerging relationships, it is unlikely that they will ever exactly correspond to hyponymy, synonymy, or any other semantic relations. In this section, words are data retrieved from corpora or lexical databases. Concept lattices are used to explore patterns that emerge within these data, but these patterns can be a mixture of associational, semantic, syntactic, phonological, ontological, coincidental, or any other type of relationship. The underlying idea is that because corpora and lexical databases contain a wealth of information, most of which has been deliberately hand-crafted by people at some stage, the emergent relationships should be meaningful at some level. This section describes how FCA can be used as a tool in the exploration of such emergent relationships.

As explained in the introduction, macro-level linguistic analyses extract data from large sources, for instance, lexical databases or corpora, and analyze patterns among such data. The first example in this section discusses lattices for clustering verbs based on verb subcategorization frames derived from corpora. "Verb subcategorization frame" refers to the syntactic argument structures of the verbs. The assumption is that verbs that have similar syntactic structures often also share semantic features. The idea to model verb subcategorization frames with FCA was first suggested by [Basili et al., 1997]. An even simpler approach of verb clustering using just the prepositions that follow a verb was presented for the verbs of motion by [Priss, 2005]. This chapter shows a similar example for the verbs for "searching." [Cimiano et al., 2003] develop a means for extracting taxonomies from text using FCA, which is basically an extension of Basili's idea but includes detailed analyses of the use of probability measures for noise reduction and discusses the results from experimental comparisons of such methods.

The corpus used in this chapter is the Brown corpus[3] in the version that is part of the NLTK distribution[4]. In the version used here, the words are tagged with syntactic markers. For example, a phrase "searching/vbg for/in" means that the verb "search" is in its -ing form and followed by the preposition "for." The tag "/in" denotes prepositions and the tag "/rp" denotes adverbial particles.

An easy means for processing data from corpora is by using a programming language that supports "regular expressions" for describing linguistic patterns. This could be the NLTK libraries, a scripting language (Perl or Python) or, for small examples, Unix command-line tools. For the example in this chapter as a first step, a list of synonyms of "search" is derived manually from a thesaurus. Then all occurrences of these verbs in the Brown corpus together with the prepositions that immediately follow the verbs are retrieved using a regular expression. The exact workflow is described in Figure 7.6.

The regular expression in Figure 7.6 matches all instances of "search," which are tagged by something starting with "/vb," followed by a space character, followed by any word which is tagged by either "/in" or "/rp." This method is somewhat crude and contains some misleading combinations of verbs and

---

[3]The Brown corpus of Standard American English was compiled by Henry Kucera and W. Nelson Francis at Brown University in the 1960s (cf. http://en.wikipedia.org/wiki/Brown_Corpus).

[4]NLTK - the natural language toolkit is a suite of open source Python modules that support natural language processing. Available at http://nltk.sourceforge.net.

1. Choose a verb. Manually compile a list of synonyms for the verb (e.g., by using a thesaurus).

2. Extract verbs with prepositions (etc.) from a corpus, e.g.,
   grep -h -o ' search[A-z]*/vb[A-z]* [A-z]*/\(in\|rp\)' /brown/*
   to retrieve the verb "search" with its prepositions
   from the Brown corpus files in a Unix directory "/brown."
   This needs to be done for all verbs in the list.

3. Omit those instances that occur less frequently than a threshold.

4. Omit duplicates.

5. Convert to a list of tuples: (verb, preposition).

6. Upload into FCA software and draw the lattice.
   (e.g., using Tupleware, http://tockit.sourceforge.net/tupleware/.
   For Tupleware: the file must have the extension ".tuples,"
   the first line of the file must have tab separated column names,
   e.g., "objects attributes.")

Note: this workflow requires that the Brown corpus in the format of the NLTK distribution is installed.

FIGURE 7.6: Workflow for generating a verb-clustering lattice from the Brown corpus.

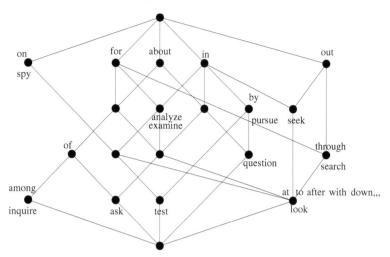

FIGURE 7.7: Verbs for "searching."

prepositions. For example, "examine at" is one of the combinations that occurs in the corpus, but the phrase is actually "examine at least," thus in this case the preposition does not really belong to the verb. Therefore, it is useful to eliminate all phrases that occur only once (or using some other threshold). The concept lattice in Figure 7.7 is then generated by taking the verbs as formal objects and the prepositions or particles as formal attributes.

The lattice shows that "look" is the most versatile word in this lexical field because it uses the most prepositions (the list of formal attributes attached to the concept of "look" is not completely shown in the diagram). The word "look" is more general than "search," "seek," "analyze," "examine" and "spy." "Analyze" and "examine" are very close according to this lattice; "inquire" and "ask" appear to be somewhat close and so are "pursue" and "question." One could argue that different aspects of "look" are expressed: "searching" and "seeking" versus "spying" versus "analyzing/examining." "Test" is more general than "spy," "analyze/examine" and "pursue."

The example described here is very simple because only the prepositions that immediately follow a verb are taken into consideration. To generate a lattice following the workflow in Figure 7.6 should only take about one hour or less. All that is required for more sophisticated analyses is to adapt the regular expression to more complicated patterns. The workflow could be further automated by writing a script that combines steps 2-5 so that the user only needs to enter a word or a list of words and the script produces the resulting list of tuples.

7.3.2 Neighborhood Lattices

Another form of macro-level linguistic analysis employs lexical databases. Lexical databases are usually stored as relational databases or can fairly easily be converted into relational databases. WordNet, for example, is traditionally distributed in form of ASCII files, but there is also a MySQL version[5] available on Sourceforge. The most general means of interacting with such databases is not via a programming language but via SQL. Because the output generated by SQL is in the form of relations (or tables), FCA can be directly applied by choosing any two columns or combination of columns from such tables[6]. The difficulty in extracting information from lexical databases lies in deciding which columns or combinations of columns might be useful and furthermore in coping with the fact that the resulting lattices are often too large and complicated and require some means of reduction or restriction. One such means for selection and restriction is the formation of "neighborhoods" and "neighborhood lattices" [Priss and Old, 2004]. Assuming a relation between formal objects and formal attributes, the idea for neighborhoods is to start with one object, retrieve all attributes that relate to that object, then retrieve all objects that relate to these attributes and so on. A neighborhood lattice is then a lattice which has neighborhoods as sets of object and sets of attributes.

The operation that underlies the selection of elements in a neighborhood is called the "plus operator" as opposed to the "prime operator" used in concept formation. This is because the prime operator applied to a set of objects selects all attributes that are shared among *all* objects in the set whereas the plus operator selects all attributes that belong to *at least one* of the objects in the set:

$$G_1' := \{m \in M \mid \forall_{g \in G_1} : gIm\}$$

$$G_1^+ := \{m \in M \mid \exists_{g \in G_1} : gIm\}$$

Applying the prime operator twice yields a closure operator, which means that further applications of the prime operator do not yield further elements. The plus operator does not usually yield a closure after two applications. In fact the sets generated by the the plus operator continue growing until a "horizontal decomposition" of the lattice is reached [Priss and Old, 2006]. A horizontal decomposition of a lattice refers to the components that a lattice falls into after removing the top and bottom concept. Many small lattices (such as the one in Figure 7.7) consist of only one component. The lattice in Figure 7.5 consists of 4 components: the middle part and the single concepts for "evening star 2," "morning star 2" and "morning star 3." Horizontal decompositions of large linguistic data sets often follow a "power law" distribution, i.e., they consist of one large component that contains most elements, a large number of tiny components and a smaller number of exponentially distributed components

---

[5]http://wnsqlbuilder.sourceforge.net

[6]e.g., by using the Tupleware software available at http://tockit.sourceforge.net

in between. [Priss and Old, 2006] show that a neighborhood decomposition of Roget's Thesaurus follows this distribution.

Ideas similar to the plus operator were re-invented many times. For example, this idea has been suggested for bilingual dictionaries [Wunderlich, 1980]: select all German translations of an English word, then all English translations of those German words, and so on. The Norwegian linguist Helge Dyvik [2004] applies this method to a parallel corpus of English and Norwegian texts and extracts monolingual thesaurus entries. His method has been modeled using FCA by [Priss and Old, 2005].

7.3.3   Roget's Thesaurus

Roget's Thesaurus[7] is an onomasiological dictionary, i.e., a dictionary that is organized by meaning instead of being organized by alphabetical ordering. Modern versions of Roget's Thesaurus usually also contain an alphabetical index at the back at the book, but the main part of the book consists of a tree classification of words into classes of related or nearly synonymous words. In this chapter, RT refers to an on-line searchable database version of Roget's Thesaurus,[8] which has been modeled using FCA methods and researched in detail by [Old, 2004]. In Old's modeling a formal context is constructed by using words as formal attributes and their senses in RT as formal objects. The senses in RT are described by using the class number (out of about 1000 classes), the number of the paragraph in the class and the number of the semicolon group in the paragraph. Figure 7.9 shows a formal context for the neighborhood of "search." The set of attributes consists of the words that share at least one RT sense with "search." The set of formal objects consists of the senses of "search." In Figure 7.9, the objects are described by their sense numbers but also by the names of the classes and the first word of the corresponding paragraph. This context has been retrieved using the on-line interface. The precise workflow is described in Figure 7.8. The on-line generated context does not contain the class names, but these can be retrieved using a different interface on the same website. Figure 7.10 shows the corresponding concept lattice.

It may be of interest to compare the neighborhood lattice from RT in Figure 7.10 with the lattice derived from the Brown corpus in Figure 7.7. In RT, "search" appears to be close to "hunt" in all of its senses (even "seek" in 653:9:9 is under the class "pursuit" in a paragraph with mostly hunting terms). In Figure 7.7, the list of synonyms was manually compiled and "hunt" was omitted. The main distinction in class 484 is between a directed search ("hunting," "quest") and a less directed search ("explore"). In Figure 7.7 this

---

[7]The examples in this chapter are based either on [Roget, 1962] or on the on-line version at http://www.roget.org which is a combination of different editions of the thesaurus.

[8]An interface to RT is available at http://www.roget.org

1. Go to the graphical queries/lattices interface at www.roget.org:
   http://www.roget.org/Anacondacat.htm

2. Enter a word into the box that retrieves the "Burmeister" format.
   If the resulting context is too large, choose the restricted version instead.

3. Copy and paste the result into a text file.
   Save the file with the extension ".cxt."

4. Open the file using FCA software that can read the Burmeister format,
   e.g., Conexp (available at http://sourceforge.net/projects/conexp)
   or Siena (available at http://tockit.sourceforge.net/).

5. Draw the concept lattice.

FIGURE 7.8: Workflow for creating a neighborhood lattice from RT.

	explore	hue and cry	hunt	hunting	look	looksee	quest	search	searching	seek	still hunt
484:25:1 Inquiry:search(verb)	X	.	X	.	X	.	.	X	.	.	.
484:2:1 Inquiry:search(noun)	.	X	X	X	.	X	X	X	X	.	.
653:2:8 Pursuit:hunting(noun)	.	.	.	.	.	.	.	X	.	.	X
653:9:9 Pursuit:hunt(verb)	.	.	.	.	.	.	.	X	.	X	.

FIGURE 7.9: The neighborhood context of "search" in RT.

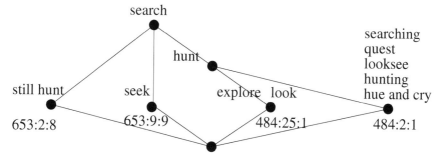

FIGURE 7.10: The neighborhood lattice of "search" in RT.

may correspond to the verbs relating to the preposition "in" as opposed to the verbs under "for" and "about." Unfortunately, again some of the verbs ("quest" and "explore") are missing in Figure 7.7. The comparison of the two diagrams indicates that further analyses might be useful. Figure 7.10 is really not very detailed. The plus operator should probably be applied one more time to retrieve more details about the verbs. On the other hand, Figure 7.7 could be repeated using the verbs from Figure 7.10. The process of deriving neighborhood lattices is often exploratory. Different parts of speech and different semantic areas of a lexical database may display different structures and may require slightly different approaches with respect to data selection and level of detail to be included.

7.3.4   Bilingual Dictionaries

The last example in this section uses a bilingual dictionary. Nowadays bilingual dictionaries are freely available on-line for most major language pairs. Free dictionaries can be of unknown quality. Because there is no absolute right and wrong for dictionaries and no clustering methods are precise, it may not be necessary for a dictionary to be of extremely high quality in order to derive neighborhood lattices, as long as the dictionary contains a sufficient amount of detail. If several different dictionaries are available, the neighborhood lattices for a few selected words can be used to evaluate and compare the dictionaries. If one is only interested in a few words, it is not necessary to download a dictionary. Instead one can follow the instructions given in the workflow in Figure 7.11.

Concept lattices for bilingual neighborhoods as in Figure 7.12 usually display three layers: the atoms and co-atoms correspond to the words of either language; the middle layer represents the degree of similarity between the two languages. If both languages have exactly the same words with the same meanings, then the lattice would be symmetric with respect to the middle layer. In Figure 7.12 a symmetry appears to occur between the polysemies of quest/search and forschen/suchen, which are mutual translations of each other. An extreme *lexical gap* would be a word that does not have a translation at all apart from a phrasal description. Since phrases were omitted in the construction of the neighborhoods, an extreme lexical gap would be easily visible in the lattice because the word would be attached to the top or bottom node and would not have any translation.

Any lack of symmetry corresponds to at least a mild lexical gap in either language, which means that the words can be translated into the other language as disambiguated words but not simultaneously in all their polysemous senses. This means that there can be slight connotational differences between the translated words. For example, there is a concept between "search/scan" and "suchen/durchsuchen" which is labeled by the German word "absuchen," but not by an English word. Thus the overlap between "search" and "scan" is lexicalized in German but not in English. On the other hand, "readout"

1. Find an on-line bilingual dictionary that prints the result as a table:
   word1 | first translation
   word1 | second translation

   ...

   word2 | first translation
   For example, on http://www.leo.org
   choose your preferred language, e.g., German/English.

2. Start with a word, for example "to search."
   Copy and paste the list for this word and its translations into a textfile.
   (In order to simplify, restrict yourself to one part of speech;
   only choose words/translations that are not phrases or compounds.)

3. Repeat Step 2 but with the translations.
   e.g., find the translations of "absuchen," "auslesen," "durchforschen,"
   "durchsuchen," "fahnden," "forschen," "suchen."

4. Repeat Step 3 but with the translations of the translations.

5. Compile a tab delimited file that has the words of one language in the
   first column and the translations in the second column.
   This may require some manual deletion of articles, inflected forms,
   etc.
   Remove duplicate entries (in Unix this can be done using sort/uniq).

6. Open the file using FCA software that can read a tab delimited list
   of objects and attributes.
   (e.g., using Tupleware, http://tockit.sourceforge.net/tupleware/.
   For Tupleware: the file must have the extension ".tuples,"
   the first line of the file must have tab separated column names,
   e.g., "objects attributes.")

7. Draw the concept lattice.

FIGURE 7.11: Workflow for creating a neighborhood lattice from a bilingual dictionary.

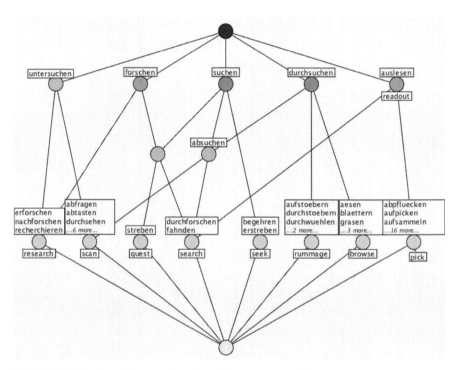

FIGURE 7.12: The neighborhood lattice for a bilingual data set.

appears to be a very specific English term that does not have an exact translation and can only be translated into a German word of higher polysemy.

It should be noted that because the plus operator is not a closure operator, a neighborhood lattice is always somewhat incomplete. In this case, some of the German words were added in the last step in the workflow in Figure 7.11. (All of these are attached to the concepts adjacent to the bottom concept, apart from "untersuchen.") The information about these words is not complete because all of their translations have not yet been included. Thus no conclusions should be made about these words.

A bilingual neighborhood lattice is also the first step in Dyvik's [2004] semantic mirrors method that makes use of the symmetry in such lattices. As [Priss and Old, 2005] have discussed, there is a difference between using bilingual dictionaries and bilingual corpora for this method. Dyvik uses an aligned bilingual English-Norwegian corpus, which means that the English words in each sentence have been manually or semi-automatically aligned to their Norwegian counterparts. The neighborhood lattices generated from such a corpus tend to be "richer" than the ones generated from bilingual dictionaries because in a corpus of sufficient size, more subtle shades of meaning may be displayed for each word than in dictionaries. Unfortunately, bilingual corpora with aligned sentences are expensive to produce and usually not freely available. Lattices such as in Figure 7.12 are still useful, but they do not usually display a strong degree of symmetry (or "semantic mirrors").

One application for such lattices is the exploration of lexical gaps. The lattice in Figure 7.13 is generated using the workflow from Figure 7.11 but in this case the starting point was the German word "Ohrwurm" and the components of its English translations, i.e., the words "catchy," "song" and "tune." "Ohrwurm" is a lexical gap in English. It means "catchy tune," but because in German it is a metaphoric use of "earwig," it has the connotation of literally being stuck in one's ear, which the phrase "catchy tune" does not have. The lattice in Figure 7.13 explores the semantic surroundings of "Ohrwurm." The biological component ("earwig") is represented by only a single concept because it is not mirrored in English. A horizontal decomposition of the lattice produces three components: "Ohrwurm" itself, the translations of "catchy" and the translations of "song," "tune," "melody." "Song" and "melody" are semantic mirrors of "Gesang" and "Lied" in the lattice. Both "catchy" and "earwig" have some negative connotations, in the case of "catchy" this is represented in the concept translated as "schwierig," in the case of "earwig" this can only be deduced from background knowledge, which is not actually represented in the lattice. Thus an "Ohrwurm" is pleasurable (because of its musical content), but also somewhat annoying.

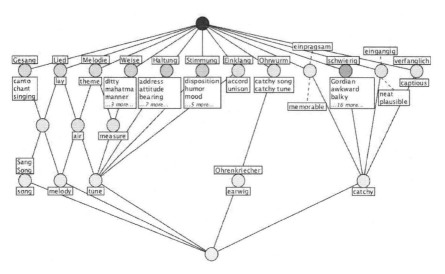

FIGURE 7.13:   The lexical gap "Ohrwurm."

7.4   Student Projects

After some of the projects information in square brackets indicates whether the project requires: [P] programming skills (scripting language or NLTK library use), [R] knowledge of regular expressions or whether a project is rather complex and difficult [!].

7.4.1   Semantic Relations

1. In analogy to the example in Figure 7.5 using a dictionary that contains sufficient amount of detail, try to model the entries of a word with several senses or a lexical field of related words as a concept lattice. What are the difficulties involved in this modeling? Is it easy or difficult to determine suitable formal attributes? Is the resulting concept lattice useful? Would it ever be possible to automate the process of generating denotative concept lattices from dictionaries? If yes, can you describe how this could be done?

2. In analogy to the example in Figure 7.4, can you model the relations for a word or set of words in WordNet (http://wordnet.princeton.edu/) in a similar manner using FCA? How can you derive a formal context? If you start with a word and collect all related words, where do you stop collecting? How do you determine the formal attributes? Can you see a way of how the process of generating concept lattices from WordNet

could be automated?

7.4.2 Neighborhood Lattices as Described in the Workflow in Figure 7.8 or 7.11.

1. Follow the instructions given in the workflow in Figure 7.8 for a word or set of words of your choice. Experiment with different words. Try the restricted neighborhood for some of the words.

2. Follow the instructions given in the workflow in Figure 7.11 for a word or set of words of your choice. Experiment with different words. Can you find any words where Steps 3 or 4 do not add any more words?

3. Choose a lexical field and analyze the words in the field using both *RT* neighborhood lattices and bilingual lattices.

4. Download a bilingual dictionary from the Web. Such dictionaries should be available for most major language pairs. You want to look for one that uses a format that is easy to parse (e.g., a tab delimited file that has one pair of words per line). Automate steps 1-5 in the workflow in Figure 7.11. [P,!]

5. Neighborhood closure: implement an algorithm that calculates the neighborhood closure for any object or attribute of a formal context, i.e., the set of objects and attributes that belong to the same component of the lattice (after horizontal decomposition) as the original object or attribute. It is easy to determine the sets visually from the line diagram of the lattice. But it is not so easy to write an efficient algorithm that calculates these sets for formal contexts of the size of *RT* (i.e., which have hundreds of thousands objects and attributes). For further information: [Priss and Old, 2006] provide more background on neighborhood closure lattices. [P,!]

7.4.3 Lattices for Verb Clusters Derived from a Corpus

1. Install the version of the Brown corpus from the NLTK distribution. Then follow the instructions in the workflow in Figure 7.6.

2. Choose a lexical field of a verb and analyze the verbs in the field using RT neighborhood lattices, bilingual lattices and lattices for verb clusters.

3. The last paragraph in the section on *Roget's Thesaurus* discusses ways in which both Figure 7.7 and Figure 7.10 could be improved by adding more words, etc. Perform these improvements. With respect to the plus operator in RT: the interface will not actually allow you to modify

the number of times the plus operator is applied. But you can conduct separate searches for all of the words in Figure 7.10 and then manually combine all of the data into one formal context to achieve one further application of the plus operator.

4. Modify the regular expression in Figure 7.6 so that it looks for a particular sentence pattern, for example, verbs that are followed by two prepositional phrases. [R]

5. Can you identify other parts of speech that could be clustered using this method and determine suitable regular expressions for these part of speech? [R]

6. Write software that automates steps 2-5 from the workflow. (This could be done using the NLTK libraries or a scripting language.) [P,R,!]

Chapter 8

Ontology Learning Using Corpus-Derived Formal Contexts

Philipp Cimiano

AIFB, Universität Karlsruhe (TH), Germany

8.1 Introduction

Ontologies can be defined as logical theories describing some aspect of reality. Typically, such logical theories describe a specific domain, i.e., some part of reality within a certain field, topic, area, branch, situation, etc. For example, we could define an ontology for the domain of university organizations (compare the SWRC ontology described in [Sure et al., 2005]). We could also formalize the aspects of reality in the domains of medicine [Grenon et al., 2004] or biochemistry [The Gene Ontology Consortium, 2000]. Ontologies, as they are declarative in nature (they represent a logical theory) can be used for many applications in which a description of the domain is useful, for example, for inferring knowledge which is implicit in the current state of an information system, for integrating different data sources, etc.

Popular definitions of ontologies in computer science are the following ones:

- "An ontology is an explicit specification of a conceptualization" (Gruber [Gruber, 1993]),

- "Ontology is the term used to refer to the shared understanding of some domain of interest [...] An ontology necessarily entails or embodies some sort of world view with respect to a given domain. The world view is often conceived as a set of concepts (e.g., entities, attributes, processes), their definitions and their interrelationships; this is referred to as conceptualization." (Uschold and Grüninger [Uschold and Grüninger, 1996]),

- "An ontology is a formal, explicit specification of a shared conceptualization." (Studer et al. [Studer et al., 1998]).

Common to all these definitions is the term *conceptualization*, which refers to an abstraction of reality captured with respect to the expressiveness of the inventory at hand. Gruber as well as Studer et al. require this conceptualization to be *explicit* in the sense that it is not just in someone's head but written down somewhere, e.g., on a sheet of paper, an electronic file, etc. Studer et al. further require an ontology to be *formal* in the sense that it should have some formal semantics and thus be machine-interpretable. Further, some researchers such as Uschold and Grüninger as well as Studer et al. require an ontology to be *shared* among different people, but this is not an aspect of ontologies we will emphasize in this chapter. In this chapter we adhere to ontologies as (formal) logical theories used to conceptualize a certain domain, i.e., to formalize an abstraction of some part of reality relevant to this domain.

While ontologies have currently the status of resources in computer science, originally *ontology* was a discipline concerned with the study of existence and with structuring reality in a formal way. In fact, according to Smith, "Ontology as a branch of philosophy is the science of what is, of the kinds and structures of objects, properties, events, processes, and relations in every area of reality." [Smith, 2003]. In this view ontology is a universal discipline and deals with existence independent of any concrete domains.

Platon's student Aristotle (384 BC - 322 BC) may certainly be called the first ontologist. However, Aristotle never talked about *ontology* as such but talked about *metaphysics*. The term *ontology* seems to have been actually introduced by the German scholar Jacob Lorhard (1561 - 1609) in his *Ogdoas scholastica* [Lorhard, 1606] (compare [Øhrstrøm et al., 2005][1]). The term was later taken up by Rudolf Göckel (1547-1628) in his *Lexicon philosophicum* [Göckel, 1613] and later established in philosophy by Christian Wolff (1679 - 1754) (see again [Øhrstrøm et al., 2005]).

Nevertheless, Aristotle fundamentally shaped the nature of ontologies as we understand them today (compare [Sowa, 2000a]). In his books *Metaphysics* and *Categories*, he introduced the important notion of a *taxonomy*, in which objects existing in reality are hierarchically organized according to their properties. Further, Aristotle introduced the notions of *genus* and *species*, which in modern computer science terminology would be called *classes* or *concepts* as well as *subclasses* or *subconcepts*, respectively. Most importantly, he introduced the notion of *differentiae* as those properties that distinguish different subspecies from each other and thus allow us to express the rationale for a specific taxonomic organization of objects.

Aristotle already recognized an important duality, which is that the more properties are ascribed to some object, the more special it is, while the less

---

[1]See also http://www.formalontology.it/jacob_lorhard.htm.

properties are ascribed to it, the more general it is. This duality actually formsthe philosophical foundation of Formal Concept Analysis.

An important part of an ontology is in fact the organization of the objects that exist in a hierarchy or taxonomy according to their properties. While Formal Concept Analysis seems definitely an appealing tool to do this, the question arises: which attributes should we use to describe our objects of interest? While manually entering attributes might be an option, this represents neither a principled nor a scalable solution. In fact, what are the criteria we would apply to decide whether one attribute should describe an object or not?

Deriving attributes from (textual) data might offer a principled solution to both problems. As textual data is massively available, an approach to deriving attributes from text might give us as many attributes as we desire (as we can continue adding more and more data). Further, the choice of attributes is a principled one as we only consider an attribute as relevant if it was derived from a collection of texts. As lattices can be seen as logical theories, the net result is thus that we have at the very end learned ontologies from textual data. We might ask: is this a meaningful way to proceed? In fact, there is many evidence in the philosophical literature as well as in the linguistic literature that meaning can be approximated by analyzing linguistic behavior. The German philosopher Wittgenstein, for example, stated: *Man kann für eine grosse Klasse von Fällen der Benützung des Wortes "Bedeutung" [...] dieses Wort so erklären: Die Bedeutung eines Wortes ist sein Gebrauch in der Sprache.* ("For a large class of cases – though not all – in which we employ the word "meaning" it can be defined thus: the meaning of a word is its use in the language.") (compare [Wittgenstein, 1953]).

Structural linguists such as Firth [Firth, 1957] claimed that *"You shall know a word by the company it keeps"*. In these lines, it makes very much sense to describe meaning in terms of attributes derived from the way words are used in their linguistic context. And this is what we aim for in this article. In particular, we intend to highlight different ways in which attributes can be extracted from textual data, discussing the different types of lattices that can be derived.

As a final note to this introduction, we should still clarify the nature of the ontologies that can be learned from text. Smith [Smith, 2003] distinguishes between *internal metaphysics* and *external metaphysics*. External metaphysics deals with reality beyond language, belief, or other cognitive biases, while internal metaphysics is concerned with the ontological commitments shared by the adherents to a certain theory or belief system or by the speakers of a certain community or the scientists working in a certain area. External vs. internal metaphysics is thus essentially the difference between the things that exist as such in reality beyond any other biases on the one hand and those things some community perceives and talks about on the other. For internal metaphysics, truth, or grounding in reality are thus not important. When learning ontologies from textual data we are essentially analyzing how a certain community is using its vocabulary. In this sense, when applying ontology learning methods to textual data we are in essence playing the role of an internal metaphysicist, analyzing the way a vocabulary is used rather than analytically examining reality beyond the way it is described in texts.

The remainder of this article is structured as follows: in Section 8.2 we discuss the general idea of using formal concept analysis for ontology learning form text. We discuss different methods for deriving attributes from textual data, i.e., by extracting grammatical relations (Section 8.2.1), adjectives (Section 8.2.2) as well as qualia structures (Section 8.2.3). We describe related work in Section 8.3, conclude in Section 8.4 and present ideas for exercises and further research in Section 8.5.1.

8.2   FCA for Ontology Learning from Text

Several approaches have been suggested to learn taxonomies from textual data. The most prominent class of approaches presented is the class of *similarity-based* clustering algorithms. Similarity-based algorithms rely on a representation of a word's context as a vector as well as on a measure of similarity defined over this vector space. Similar words are then grouped to clusters that somehow capture the shared meaning of their members. In contrast, the advantage of using a set-theoretic approach such as formal concept analysis is that objects are organized into concepts according to their properties (differentiae) as well as taxonomically ordered according to the inclusion relations between their extent or intent (see Chapter 1). This allows to formally interpret the resulting concept hierarchy, a feature that most clustering approaches lack. This is due to the fact that clusters are typically only described by some aggregated or representative vector such as their centroid, i.e., an average vector. We refer the reader at this point to the more detailed discussion of related approaches in Section 8.3.

In order to construct a formal context on the basis of textual data, we need to derive meaningful attributes. Without loss of generality, let us assume that the objects we are interested in have already been fixed. However, they could also have been extracted from text performing a process known as *term extraction* and applying term weighting measures known from information retrieval, such as tf.idf (compare [Cimiano et al., 2007]). In general, for the task of extracting attributes from text, some preprocessing of the text collection will be necessary. Some of the components typically applied in extracting attributes from text are:

- a **tokenizer**: detecting sentence and word boundaries;

- a **morphology component**: for some languages, a lemmatizer reducing words to their base form might even suffice, but for languages with a richer morphology (e.g., German) a component for structuring a word into its components (lemma, prefix, affix, etc.) might be necessary. For most machine learning-based algorithms a simple stemming of the word might be enough (compare [Porter, 1980]);

- a **part-of-speech tagger**: assigning every word to its syntactic category in context, thus determining whether it is a noun, a verb, an adjective, etc. An example for a part-of-speech-tagger is TreeTagger [Schmid, 1994];

- a **regular expression matching engine**: this is, for example, provided by the JAPE engine implemented in GATE [Cunningham et al., 1997];

- a **chunker**: identifying larger syntactic constituents in a sentence. Chunkers are also called *partial parsers*. An example of a publicly available chunker is Steven Abney's CASS [Abney, 1996];

- a **parser**: a full parser determining the full syntactic structure of a sentence might be needed for some ontology learning algorithms (compare [Cimiano et al., 2005]).

However, when using texts, we can yield an arbitrary large number of attributes that will not all be relevant for discriminating different concepts. Thus, it is crucial to select the most relevant attributes. The question of which attributes are relevant is of course very difficult to answer. In practice, we typically apply some measure of relevance computed on the basis of the text collection. We will discuss several alternatives for such measures below. In general, the use of a measure for determining relevance allows to select those attributes that are relevant beyond some chosen threshold and thus to control the number of attributes and thus the size of the derived formal context. This is important from several perspectives. On the one hand, it is important to keep the context size tractable from the point of view of algorithmic performance. Indeed, some researchers have empirically shown that for sparsely populated formal contexts, the lattice computation is quadratic in the size of the formal context (compare [Carpineto and Romano, 1996, Lindig, 2000]). On the other hand, the produced lattice might also need to be inspected manually, such that their size and granularity should be controllable. Possibly, some mechanism for *zooming in* might be desired, in case of which the threshold could be lowered to yield more attributes.

The overall process of automatically deriving concept hierarchies from text by FCA is depicted in Figure 8.1. The process starts with a text collection which is linguistically analyzed applying some of the techniques described above. From the linguistically processed corpus, we extract attributes, weight and prune them, thus resulting in a formal context from which a lattice can be constructed.

FIGURE 8.1: Process for ontology jearning from text with FCA.

In what follows we discuss in more detail different possibilities for extracting a formal context from textual data, i.e., relying on (i) grammatical relations, (ii) adjectives, as well as (iii) lexico syntactic patterns extracting the qualia structure for a word.

8.2.1 Grammatical Relations

The constituents of a sentence typically play a specific role and have a very specific grammatical function. Some constituents play the role of the *subject* or *object*[2] of the sentence, the role of a *prepositional complement* etc. In the sentence: *The boy eats a lolly.*, the constituent *the boy* plays the role of subject, while *a lolly* plays the role of the object of the sentence. Grammatical roles are indeed essential to grasp the meaning of a sentence. We typically say that *the boy* is the subject and *a lolly* is the object of the verb *eats* (in a certain sentence). Hereby, it is interesting to see that verbs typically impose some constraints on the words that can play the role of subjects or objects. While the sentence *The boy ate the moon.* might be acceptable in certain circumstances, it is for sure not as intuitive (in a null context) as the above example. Given some appropriate context, almost every sentence is acceptable. Nevertheless, it seems to be true that verbs (and other linguistic predicates) impose certain defaults on the words they feature as subject or object. These defaults are typically known as *selectional restrictions* or *preferences* in the linguistics literature. For us here it is important to conclude that verbs tell us something about the nature of the words playing the role of subject or objects. The verb *eat*, for example, would per default have an object that is *eat*-able (edible) as well as some physical agent able to perform the act of eating, i.e., an *eat*-er. Thus, the fact that something occurs as subject or object of a certain verb can tell us something about its (default) meaning. In our example, we might well assume that *lollies* are *eat*-able (edible) and *boys* can play the role of *eat*-ers. As a first attempt, we might thus use grammatical relations as a way to extract attributes from text describing a certain word. For example, we

---

[2]In this article, we will use the term object both to denote the *Gegenstände* in Formal Concept Analysis as well as the object as grammatical function in a sentence. The meaning should be clear from the context.

	book-able	rent-able	drive-able	ride-able	join-able
hotel	x				
apartment	x	x			
car	x	x	x		
bike	x	x	x	x	
excursion	x				x
trip	x				x

TABLE 8.1:   Example Formal Context for the Tourism Domain.

could create a formal context on the basis of words (as *formal objects*) as well as the verbs for which they play the role of (grammatical) object as *formal attributes*. So far so good, but how can we extract the words standing in a certain grammatical relation for all verbs? For this purpose, one can rely on parse trees produced by most of the available linguistics parsers. A parse tree essentially encodes the syntactic structure of the words in a certain sentence. For our example sentence, we would have a (simplified) parse tree as follows:

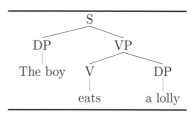

We could then define rules that assume that the subject is always the DP-constituent directly dominated by the top S-node dominating the verb, while the object is the DP-constituent directly dominated by the VP-node dominating the main verb. The definition of such rules is possible with tools such as tgrep, developed at the University of Pennsylvania in the context of the Penn Treebank Project[3], which allows to define path expressions over tree structures.

Now let us assume we are interested in creating an ontology for the domain of tourism and we perform the above described extraction for the words *hotel, apartment, car, bike, excursion,* and *trip*. We could, for example, find that *hotel* appears as the object of *book*, while an apartment can be *booked* and *rented* as well. Further, whereas a car can be *booked, rented* and *driven*, a bike can be, in addition, *ridden*.

In an approach we described previously (see [Cimiano et al., 2005]), we used the LoPar parser [Schmid, 2000] as well as tgrep to extract paths corresponding to grammatical relations. The corresponding verbs can then be used as attributes to automatically derive a formal context as depicted in Table 8.1.

---

[3]http://www.ldc.upenn.edu/ldc/online/treebank/

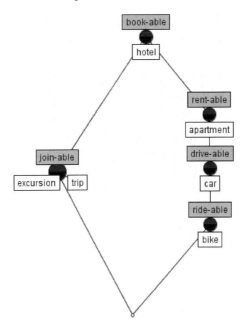

FIGURE 8.2:   Example concept lattice for the tourism example.

In general, when extracting attributes in this way, there are two important issues to consider:

1. the output of the parser can be erroneous, i.e., not all derived verb–argument dependencies are correct,

2. not all the derived dependencies are interesting in the sense that they will help to discriminate between the different objects

One solution to deal with these issues is to introduce some measure that statistically weighs the importance of a certain object-attribute pair. A threshold can then be used to only consider the most relevant pairs. In previous work (compare [Cimiano et al., 2003] and [Cimiano et al., 2004]), we have examined more closely the *conditional probability*, the *pointwise mutual information* (PMI) as well as a measure based on Resnik's *selectional preference strength* (SPS) (compare [Resnik, 1997]). The formulas are given in the following:

$$Conditional(n, v_{arg}) = P(n|v_{arg}) = \frac{f(n, v_{arg})}{f(v_{arg})} \qquad (8.1)$$

$$PMI(n, v_{arg}) = log_2 \frac{P(n|v_{arg})}{P(n)} \qquad (8.2)$$

$$SPS(n, v_{arg}) = S_R(v_{arg}) P(n|v_{arg}) \qquad (8.3)$$

where $S_R(v_{arg}) = \sum_{n'} P(n'|v_{arg}) \, log \frac{P(n'|v_{arg})}{P(n')}$.

In the above formulas, $f(n, v_{arg})$ is the total number of occurrences of a term n as argument arg of a verb v, $f(v_{arg})$ is the number of occurrences of verb v with a corresponding argument and $P(n)$ is the relative frequency of a term n compared to all other terms. The first information measure is simply the conditional probability of the term n given the argument arg of a verb v. The second measure $PMI(n, v_{arg})$ is the pointwise mutual information and was used by Hindle [Hindle, 1990] for discovering groups of similar terms. The third measure is inspired by the work of Resnik [Resnik, 1997] and introduces an additional factor $S_R(v_{arg})$, which takes into account all the terms appearing in the argument position arg of the verb v in question. In particular, the factor measures the relative entropy of the prior and posterior (considering the verb it appears with) distributions of n and thus the *selectional strength* of the verb at a given argument position. It is important to mention that in our approach the values of all the above measures are normalized into the interval $[0,1]$ by the function $f(x) = \frac{x}{max-min}$. For illustration purposes, let us discuss the above measures on the basis of the example verb–object–occurrence matrix given in Table 8.2.

	book$_{obj}$	rent$_{obj}$	drive$_{obj}$	ride$_{obj}$	join$_{obj}$
hotel	10				
apartment	6	5			
car	3	4	5		
bike	2	3	2	2	
excursion	1				3
trip	2				2

TABLE 8.2: Example Occurrences of Nouns as Objects of Verbs.

First of all, we have a total number of 50 occurrences of nouns as objects of some verbs. Therefore, we get the following prior probabilities for the nouns: P(hotel)=$\frac{10}{50}$ = 0.2, P(apartment)=$\frac{11}{50}$ = 0.22 , P(car)=$\frac{12}{50}$ = 0.24, P(bike)=$\frac{9}{50}$ = 0.18, P(excursion)=$\frac{4}{50}$ = 0.08, P(trip)=$\frac{4}{50}$ = 0.08. The posterior probabilities, i.e., the probability of a noun given the object position of a certain verb, are given in Table 8.3.

Now, the PMI of hotel and book$_{obj}$ is, for example, $PMI(hotel, book_{obj}) = log_2 \frac{P(hotel|book_{obj})}{P(hotel)} \approx log_2(\frac{0.42}{0.2}) \approx 1.07$. The selectional strength of book$_{obj}$ should be much lower than the one of *drive$_{obj}$* according to our intuitions. In fact, we get: $S_R(book_{obj}) \approx 0.42 \cdot log_2(\frac{0.42}{0.2}) + 0.25 \cdot log_2(\frac{0.25}{0.22}) + 0.13 \cdot log_2(\frac{0.13}{0.24}) + 0.08 \cdot log_2(\frac{0.08}{0.18}) + 0.04 \cdot log_2(\frac{0.04}{0.08}) + 0.08 \cdot log_2(\frac{0.08}{0.08}) \approx 0.25$, and $S_R(drive_{obj}) \approx 0.71 \cdot log_2(\frac{0.71}{0.24}) + 0.29 \cdot log_2(\frac{0.29}{0.18}) \approx 1.31$. The above results certainly correspond to our intuitions about the selectional strength of *book*

	$P(n\vert book_{obj})$	$P(n\vert rent_{obj})$	$P(n\vert drive_{obj})$	$P(n\vert ride_{obj})$	$P(n\vert join_{obj})$
hotel	$\frac{10}{24} \approx 0.42$				
apartment	$\frac{6}{24} = 0.25$	$\frac{5}{12} \approx 0.42$			
car	$\frac{3}{24} \approx 0.13$	$\frac{4}{12} \approx 0.33$	$\frac{5}{7} \approx 0.71$		
bike	$\frac{2}{24} \approx 0.08$	$\frac{3}{12} = 0.25$	$\frac{2}{7} \approx 0.29$	$\frac{2}{2} = 1$	
excursion	$\frac{1}{24} \approx 0.04$				$\frac{3}{5} = 0.6$
trip	$\frac{2}{24} \approx 0.08$				$\frac{2}{5} = 0.4$

TABLE 8.3:   Posterior Probabilities.

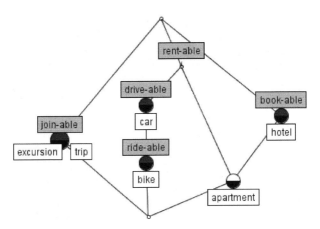

FIGURE 8.3: Lattice for the tourism formal context pruned on the basis of the conditional probability (threshold 0.2).

and *drive*.

By normalizing the values of the different measures, say, into the interval [0..1], we would then introduce a certain threshold and then only consider those pairs over the threshold as relevant. Thus, depending on the relevance measure and threshold selected, we would get different formal contexts and resulting lattices. Considering all the attributes above a value of 0.2 for the posterior probabilities given in Table 8.3, we would yield the lattice depicted in Figure 8.3.

A few comments on this notion of relevance seem appropriate here. On the one hand, the notion of relevance is here thus a statistical one and has no formal logical background. Nevertheless, it is an objective and deterministic measure of relevance that can be computed on the basis of a given corpus. In general, the notion of relevance can be used to rank attributes according to relevance and thus to directly control the size of the resulting formal contexts, which is a very important issue from the point of view of computational tractability as well as visualization.

Having discussed how verbs can be used as attributes, we now turn to the discussion how adjectives can be used to describe objects.

8.2.2 Adjectives

Adjectives are typically formalized as monadic predicates in formal semantics (compare [Portner, 2006]). The noun phrase *a sweet lolly* is, for example, typically formalized using the logical expression $\exists x \; sweet(x) \wedge lolly(x)$. In fact, the above formula corresponds to the interpretation of so-called *intersective* adjectives. However, it is well-known that many adjectives cannot be interpreted in this way. Nevertheless, as a rough simplification, we can regard an adjective as denoting the value of a property of the modified noun. For example, in the expression *a red car red* can be seen as the value of a property *colour*. In the expression *the sweet lolly*, *sweet* might refer to the value of a property *taste* or *sweetness*.

An important question is how pairs of adjectives and modified nouns can be extracted from a text collection. This is in fact relatively simple. In order to extract pairs of adjectives and modified nouns from a text collection, we would first tokenize and part-of-speech tag the corpus using TreeTagger [Schmid, 1994], for example. Then we could apply suitable regular expressions to match the following two patterns and extract adjective/noun pairs:

- {DET}? {NN}+ is{VBZ} {JJ}

- {DET}? {JJ} {NN}+

The first pattern extracts sequences consisting of an optional determiner (DET), a sequence of nouns (NN), the verb "is" as well as an adjective (JJ). This pattern would for example match the sequence *The car is red*. The second regular expression matches sequences consisting of an optional determiner (DET), an adjective (JJ) and a sequence of nouns (NN). The second pattern would match the expression *"a sweet lolly"*, for example.

For the same reasons as described in Section 8.2.1, these pairs are weighted using the conditional probability, i.e.,

$$Cond(n, a) := \frac{f(n, a)}{f(n)} \tag{8.4}$$

where $f(n, a)$ is the joint frequency of adjective a and noun n and $f(n)$ is the frequency of noun n. On the basis of this measure, we can then extract all the adjective-noun pairs over a certain threshold. At a second step, for each of the adjectives we look up the corresponding attribute in WordNet[4] [Fellbaum, 1998], further considering only adjectives that actually have an

---

[4]WordNet is a lexical database in which words are organized into sets of word sharing some meaning (synsets). Further, lexical relations such as hyponym/hypernym, meronym/holonym, adjective/attribute, etc. are defined on the basis of these synsets. The relation between an adjective and its attribute has been used in the approach described here.

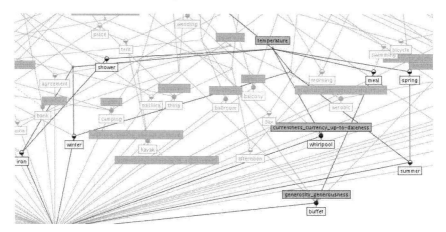

FIGURE 8.4:   Adjective-derived concept lattice: objects with temperature.

attribute. For example, the adjectives *hot, warm, or cold* are all associated with the attribute *temperature* in WordNet.

When extracting attributes in this way from our tourism corpus and organizing the nouns as objects in the form of a lattice, interesting structures do indeed emerge. Figure 8.4, for example, shows all objects that appear with adjectives belonging to the attribute *temperature* in WordNet. Figure 8.5 shows objects that appear with adjectives belonging to the attribute *duration/length*. Objects which have a *temperature* are for example: *iron, shower, whirlpool, buffet, meal* as well as seasons such as *winter, summer,* and *spring*. Objects with a *duration/length* are, for example: *promenade, terrace, iron, distance* (these are things primarily with a length) as well as *time, trip, ball, journey, period, holiday, walk, winter* (all with a temporal duration).

8.2.3   Qualia Structures

According to Aristotle, there are four basic factors or causes by which the nature of an object can be described (cf. [Kronlid, 2003]): i) the *material cause*, i.e., the material an object is made of, ii) the *agentive cause*, i.e., the source of movement, creation or change, iii) the *formal cause*, i.e., its form or type, and iv) the *final cause*, i.e., its purpose, function or intention.

In his Generative Lexicon (GL) framework, Pustejovsky [Pustejovsky, 1991] reused Aristotle's basic factors for the description of the meaning of lexical elements.[5] In fact he introduced so-called *qualia structures* by which the

---

[5]See the work of Walter [Walter, 2001] for a critical discussion of the compatibility of Aristotle's basic causes or factors and Pustejovsky's theory of the generative lexicon.

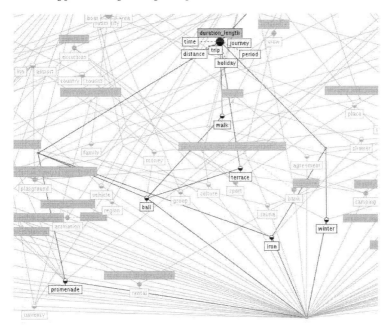

FIGURE 8.5:   Adjective-derived concept lattice: objects with duration or length.

meaning of a lexical element is described in terms of four roles:

- *Constitutive*: describing physical properties of an object, i.e., its weight, material as well as parts and components

- *Agentive*: describing factors involved in the *bringing about* of an object, i.e., its creator or the causal chain leading to its creation

- *Formal*: describing that properties that distinguish an object within a larger domain, i.e., orientation, magnitude, shape, and dimensionality

- *Telic*: describing the purpose or function of an object

Most of the qualia structures described in the literature (see e.g., [Pustejovsky, 1991]), however, seem to have a more restricted interpretation. In fact, in most examples the *Constitutive* role seems to describe the parts or components of an object, while the *Agentive* role is typically described by a verb denoting an action that typically brings the object in question into existence. The *Formal* role normally consists of typing information about the object, i.e., its hypernym. Finally, the *Telic* role describes the purpose or function of an object either by a verb or nominal phrase. The qualia structure for *knife*, for example, could look as follows (cf. [Johnston and Busa, 1996]):

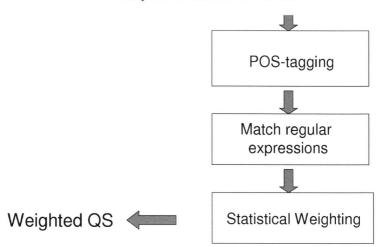

FIGURE 8.6:　General approach.

Formal:	artifact_tool
Constitutive:	blade,handle,...
Telic:	cut
Agentive:	make

Recently, we have developed an approach able to automatically derive the qualia structure for a word from the Web (see [Cimiano and Wenderoth, 2007]). In essence, this approach relies on patterns that are matched on the Web to derive fillers of the above-described roles for a given word. The approach consists of 5 phases and is depicted in Figure 8.6. For each *qualia term* (the word we want to find the qualia structure for) we:

1. generate for each qualia role a set of so-called *clues*, i.e., search engine queries indicating the relation of interest,

2. download the snippets (abstracts) of the 50 first Web search engine results matching the generated clues,

3. part-of-speech tag the downloaded snippets,

4. match patterns in the form of regular expressions conveying the qualia role of interest, and

5. weight and rank the returned qualia elements according to some measure.

The patterns defined in our pattern library are actually tuples $\langle c, r \rangle$ where c is a clue that is sent to the search engine (e.g., Google in our case) to download a number of promising pages where the corresponding regular expression r is likely to be matched. In what follows we briefly present the clues and patterns for the different roles. For ranking the different fillers of a qualia role, we rely on a measure based on the conditional probability (actually the relative frequency) as it was found to perform best in an earlier study (compare [Cimiano and Wenderoth, 2007]).

8.2.3.1 Patterns for the Formal Role

The patterns for the formal role are defined as follows:

Clue	Pattern
Singular	
"a(x) x is a kind of "	NP_{QT} is a kind of NP_F
"a(x) x is"	NP_{QT} is a kind of NP_F
"a(x) x and other"	NP_{QT} (,)? and other NP_F
"a(x) x or other"	NP_{QT} (,)? or other NP_F
Plural	
"such as p(x)"	NP_F such as NP_{QT}
"p(x) and other"	NP_{QT} (,)? and other NP_F
"p(x) or other"	NP_{QT} (,)? or other NP_F
"especially p(x)"	NP_F (,)? especially NP_{QT}
"including p(x)"	NP_F (,)? including NP_{QT}

The patterns have been designed to return a noun as filler for the formal qualia role. In particular, we use similar patterns as already defined by Marti Hearst [Hearst, 1992], who defined a set of patterns to extract hyponym relations. In the above patterns, we use the following notation: while $p(x)$ denotes the plural of the term x we want to acquire the qualia structure for, $a(x)$ stands for the appropriate indefinite article to be used, i.e., "a," "an" or no article at all. The plural $p(x)$ is determined by look-up in a lexicon with inflected forms, while $a(x)$ is determined relying on some heuristics. In particular, the heuristic is on calculating the ratio of the Web counts of occurrences of the word in question with the articles "a" or "an" compared to the overall occurrences of the word in the Web. A threshold value is then used to decide whether the article "a," "an" or no article at all should be used. Further, NP stands for a noun phrase and NP_F for an unknown noun phrase with the filler of the formal role as (nominal) head. NP_{QT} stands for a noun phrase with the qualia term as head. Overall, the patterns are relatively straightforward and need no further discussion. In general, all the patterns described in this section have been created by hand in an iterative process, starting from a few examples to yield initial patterns. The patterns were then refined on the basis of examples corresponding to different categories, i.e., words denoting artifacts, naturally occurring objects, abstract things, etc.

8.2.3.2 Patterns for the Constitutive Role

The patterns for acquiring fillers of the constitutive role are given in the following:

Clue	Pattern
Singular	
"a(x) x is made up of "	NP$_{QT}$ is made up of NP'$_C$
"a(x) x is made of"	NP$_{QT}$ is made of NP'$_C$
"a(x) x comprises"	NP$_{QT}$ comprises (of)? NP'$_C$
"a(x) x consists of"	NP$_{QT}$ consists of NP'$_C$
Plural	
"p(x) are made up of "	NP$_{QT}$ is made up of NP'$_C$
"p(x) are made of"	NP$_{QT}$ are made of NP'$_C$
"p(x) comprise"	NP$_{QT}$ comprise (of)? NP'$_C$
"p(x) consist of"	NP$_{QT}$ consist of NP'$_C$

They are in principle similar to the ones of the formal role and need no further motivation. In the case of the constitutive role, as an additional heuristic, we test if the lemmatized head of NP$_C$ is an element of the following list containing nouns denoting an indication of amount: *{variety, bundle, majority, thousands, million, millions, hundreds, number, numbers, set, sets, series, range}* and furthermore this NP$_C$ is followed by the preposition "of." In that case we would take the head of the noun phrase after the preposition 'of' as a potential candidate for the *Constitutive* role. For example, when considering the expression *a conversation is made up of a series of observable interpersonal exchanges*, we would take *observable interpersonal exchange* as a potential qualia element candidate instead of *series*. Further, while we do not explain in detail the structure of NP$_C$, which is in turn defined as a regular expression, it is important to mention that it matches enumerations. Thus the patterns can match expressions like: *"a knife is made of a handle, a blade, steel and metal."*. Overall, the fillers of the constitutive role are also nouns.

8.2.3.3   Patterns for the Telic Role

The *telic* role is in principle acquired in the same way as the *formal* and *constitutive* roles with the exception that the filler of the qualia role is not only the head of a noun phrase, but also a verb or a verb followed by a noun phrase. The patterns used to discover fillers of the telic role are given below:

Clue	Pattern
Singular	
"purpose of a(x) x is"	purpose of (a|an) x is (to)? PURP
"a(x) x is used to"	(a|an) x is used to PURP
Plural	
"purpose of p(x) is"	purpose of p(x) is (to)? PURP
"p(x) are used to"	p(x) are used to PURP

In particular, the returned candidate qualia elements are the lemmatized underlined expressions in PURP:=\w+[VB] NP | <u>NP</u> | be[VB] \w+[VBD]).

The purpose (telic role) can thus be either filled with a noun phrase as expressed in the expression *"the purpose of a movie is entertainment"*, a verb followed by a noun phrase as in *"the purpose of a computer is to perform calculations"* or simply a verb in a passive construction: *the purpose of a book is to be read*.

8.2.3.4 Patterns for the Agentive Role

The most difficult role to acquire is certainly the agentive role (compare [Cimiano and Wenderoth, 2005]). We have used the following patterns in our approach:

Clue	Pattern
Singular	
"to * a(x) new x"	to [RB]? [VB] a? new x
"to * a(x) complete x"	to [RB]? [VB] a? complete x
"a(x) new x has been *"	a? new x has been [VBD]
"a(x) complete x has been *"	a? complete has been [VBD]
Plural	
"to * new p(x)"	to [RB]? [VB] new p(x)
"to * complete p(x)"	to [RB]? [VB] complete p(x)

Interestingly, the patterns make use of specific adjectives such as *new* and *complete*. In fact, we found that using these specific adjectives, it was more likely to find an actual filler of the agentive role vs. finding a general verb related to the object in question. The creation verb is then found at the position of the '*' wildcard in the clues. In order to reduce the number of spurious fillers, we filter out verbs such as *buy, purchase*, and *win*.

8.2.3.5 Example Qualia Structures and Lattice

Let us assume that with the above-mentioned patterns we have extracted the following qualia structures for *cutlery, spoon*, and *fork*. While the qualia structures are not the actual structures produced by our approach, they are actually inspired in the results produced by our online demo of the approach.[6] Thus, while the structures have been manipulated for the purposes of presentation, we can reasonably assume that they have been extracted automatically via our approach.

---

[6] Available at http://www.cimiano.de/qualia

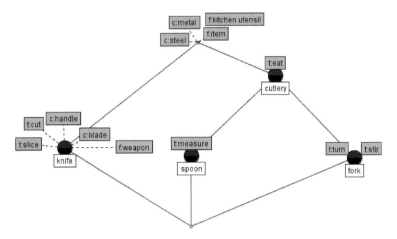

FIGURE 8.7:   Cutlery lattice.

	Formal	Constitutive	Telic
cutlery	item	steel	eat
	kitchen utensil	metal	
spoon	item	steel	eat
	kitchen utensil	metal	stir
			measure
fork	item	steel	eat
	kitchen utensil	metal	turn
knife	item	steel	cut
	kitchen utensil	blade	slice
	weapon	handle	

The lattice derived from the above qualia structures is shown in Figure 8.7. In the figure, the attributes corresponding to the formal role are prefixed with an "f:," while the attributes derived from the telic and constitutive role are prefixed with a "t:," and "c:," respectively. We have omitted the attributes for the *agentive* role in the above qualia structures as well as in the lattice. The top formal concept of the lattice (the one containing all objects) consists of *items* and *kitchen utensils* (formal role) as well as of those things that are made of *steel* and *metal*. Further, we see that the formal concept containing *cutlery* subsumes both *spoon* and *fork*. A spoon is more special as it can be used to *measure* (telic role), while a fork is more special as it can be used to *stir* and *turn* (telic role). Interestingly, *knife* is not subsumed by the concept containing *cutlery* as *eat* does not appear in the telic role of *knife*. Further, *knife* is the only *kitchen utensil* that is also a *weapon* (formal role). Overall, the qualia structures seem to yield quite interesting structures.

8.2.4 Discussion

We have discussed different ways of extracting attributes from text, relying on i) grammatical relations, ii) adjectives as well as iii) qualia structures derived from the Web. All these types of attributes yield substantially different formal contexts as well as corresponding lattices. Certainly, it would be important to characterize the properties of the different lattices more formally, but this is beyond the scope of this article and an issue for future work. Overall, we have seen that very interesting structures do indeed emerge in all the lattices considered, so that we conclude that it seems definitely worthwhile to further explore approaches for deriving lattices from text and to examine their properties more closely. We have further argued in this article that a crucial question is to define some notion of relevance that allows us to control the number of attributes and thus the size of the lattice. We have presented different such measures and suggested how they can be used to select only the most relevant attributes for building the formal context. In the future, it would be desirable to explore techniques which allow a user to directly interact and influence the notion of relevance and thus the attributes selected.

8.3 Related Work

Various methods have been proposed in the literature of ontology learning from text to address the problem of (semi-) automatically deriving a concept hierarchy from text (see [Buitelaar and Cimiano, 2008] for a recent overview of the state-of-the-art in ontology learning from text). In essence, these methods can be grouped in two classes: the *similarity*-based methods on the one hand, and the *set-theoretical* approaches on the other hand.

Similarity-based approaches represent an object – or word in our case – as a vector based on the vector-space model. The dimensions of the vectors correspond to features extracted from the text and describe the nature of the word in question. In information retrieval, for example, a so-called bag-of-words model is used in which the dimensions of the vector used to describe a document corresponds to all the words found in the collection of documents. The vectors used are in most cases real-valued and denote the importance, relevance, or frequency of the feature in question. In information retrieval, the well-known tf.idf value (compare [Baeza-Yates and Ribeiro-Neto, 1999]) is, for example, used to measure the relative importance of a word in a document. We have presented more elaborated measures in this article to weight the importance of a certain text-derived attribute. Given such a vector-based representation, similarity-based methods exploit some measure for calculating similarity in vector space to guide a clustering process that groups objects together into clusters. Here we mainly distinguish between top-down (partitional) or bottom-up (agglomerative) approaches. The former ones are called in this way because they start with the cluster containing all the elements and iteratively partition this cluster into clusters of smaller size. The agglomerative clustering techniques start from a configuration in which every word belongs to exactly one cluster and then the most similar clusters are iteratively merged. Examples of agglomerative methods are [Faure and Nedellec, 1998, Caraballo, 1999, Cimiano and Staab, 2005]. Examples of partitional techniques can be found in [Pereira et al., 1993] or [Cimiano et al., 2005], where Bi-Section-KMeans is used as a partitional clustering algorithm, always partitioning a cluster into two subclusters. One problem of such similarity-based clustering approaches is that the content of a cluster cannot be described intensionally. The most typical representation used for a cluster is its so-called *centroid*, i.e., an average vector of all the data points contained in the cluster. However, such a representation is in general quite unsatisfactory as it does not tell us why these elements have been grouped exactly in this way (and not in another way).

Set-theoretical methods describe a concept by a set of attributes. As in FCA, objects are ordered hierarchically on the basis of the relations between their attribute sets. In particular, FCA also gives intensional descriptions of the (formal) concepts obtained. In this sense it can be called a *conceptual clustering* method as it actually explains the rationale for grouping the objects in a certain way. In this sense, set-theoretical methods and in particular FCA are superior to the above-mentioned similarity-based methods. In the computational linguistics community, such approaches have been applied to lexicon representation [Petersen, 2002, Sporleder, 2002]. The advantage of such an approach is that linguistic properties of words can be represented concisely relying on inheritance. The application of FCA for the induction of word hierarchies has been already suggested by Haav [Haav, 2003].

As described in this article, the features used to describe words greatly differ between the different approaches examined in the literature. Verb-argument dependencies have been widely used (compare [Hindle, 1990, Pereira et al., 1993, Grefenstette, 1994, Faure and Nedellec, 1998]). Caraballo [Caraballo, 1999] relied on appositions, while some researchers have in general suggested the use of modifiers, e.g., adjectives modifiers, nominal modifiers, etc. (compare [Gamallo et al., 2005]). In general, there is little work on comparing different types of attributes quantitatively. Gasperin et al. [Gasperin et al., 2001] comment that for their purposes a more fine-grained representation of modifiers delivered better results than the representation used by Grefenstette [Grefenstette, 1992]. This is also in line with the work of ourselves [Cimiano et al., 2004] where we showed that using different types or verb-argument dependencies yields better results, in comparison to using verb-object dependencies alone. Recently, Poesio and Almuhareb have examined the role of attributes vs. attribute values with respect to cluster quality (see [Almuhareb and Poesio, 2004]). While *attribute values* denote concrete values that a certain property can take, *attributes* stand for the properties themselves. Both attribute values and attributes are derived using similar patterns as the ones we have discussed in Sections 8.2.2 and 8.2.3.

8.4   Conclusion

We have discussed in this chapter how formal concept analysis can be used for learning ontologies from text. When learning ontologies from text using formal concept analysis, one crucial question is what the objects and what their attributes are. In this article, without loss of generality, we have assumed that the objects are words appearing in the text and given *a priori*. In fact, this is a reasonable assumption given the fact that there is a plethora of techniques available for extracting the relevant terminology from a text corpus. Concerning which attributes to use, we have presented three different approaches to derive attributes from text data and given examples for the lattices that can be induced from these. Further work should indeed clarify which approach is more suitable for which purpose. In this sense, we have only scratched at the surface of the problem in the present article. Thus, we propose a number of issues for future work in the following section. Overall, it is important to mention that the object-attribute representation and tools (FCA in particular) described in this article are not expressive enough to yield contradictions in the derived logical theories. This will obviously change the more we move to a more expressive logical language such as, for example, the Web Ontology Language OWL [Bechhofer et al., 2004], which is based on description logics. Moving to a more expressive logical representation with modeling primitives such as negation or disjointness such as inluded in OWL means that the logical theory learned can be inconsistent (in the proper logical sense). Thus, a proper treatment of inconsistency should form part of any ontology learning process. This issue is discussed in more detail in Chapter 9.

8.4.1   Further Research

8.4.1.1   Incrementality and Dynamicity

On the one hand, it would be interesting to examine methods for explicitly calculating changes to the created lattice on the basis of corpus updates. This follows the idea of an incremental ontology learning process as suggested in [Cimiano and Völker, 2005]. One challenging issue is to avoid the complete recalculation of the whole lattice when the corpus and consequently the formal context changes. This is an issue of incremental construction of the lattice. However, it would also be useful to account for the dynamicity of ontology learning and be able to explicitly track changes in the lattice to changes in the underlying corpus, thus explaining why the earlier structure has changed.

8.4.1.2   Tractability

Formal Concept Analysis can in principle yield lattice structures with a number of concepts exponential in the size of the objects or attributes. Such a situation is undesirable in the sense that it imposes high computational costs and furthermore the resulting lattices cannot be anymore understood by users. Therefore, it makes sense to control the number of attributes (which are typically much higher than the number of objects in the context of ontology learning) in order to yield lattices, which are tractable computationally and from a visualization and browsing point of view. Indeed, some researchers have shown empirically that the size of the lattices remains approximately quadratic in the size of objects/attributes in case the formal contexts are sparsely populated, where sparsely populated means a fill ratio under 1% [Carpineto and Romano, 1996, Lindig, 2000]. Thus, we need to investigate methods that produce sparsely populated formal concepts, which nevertheless would contain the most relevant attributes capturing the essential properties of a certain object. A method to control the size of the formal context could, for example, be the Repertory Grid Technique reported in [Spangenberg and Wolff, 1993]. At the user interface level such a method would allow users to specify a level of granularity by directly selecting the number of attributes considered.

8.4.1.3   Attribute Granularity

A crucial issue directly coupled with the granularity of the formal contexts is the degree of abstraction of the attributes used in the formal context. While we can definitely resort to words as appearing in text, we could also move one level of abstraction up and consider word classes rather than words themselves, thus leading to more abstract attributes and thus less fine–granular formal contexts and corresponding lattices. Moving a level of abstraction up can be achieved in several ways. On the one hand, we can directly use more abstract linguistic categories such as the top levels of the WordNet lexical database [Fellbaum, 1998] or linguistic classes of verbs such as those suggested by Beth Levin [Levin, 1993].

8.4.1.4   Choice of Attributes

In general, while we have presented different ways of extracting attributes from text in this article, further investigations are needed to clarify which attributes are suitable for which purposes and to determine the properties of each type of lattice depending on the type of attributes chosen.

8.5 Student Projects

In what follows, we give some suggestions for research topics that could be pursued in the context of ontology learning from text with formal concept analysis. The following topics can provide the basis for the formulation of a student project or even a PhD thesis.

8.5.1 Student Exercises

8.5.1.1 Construct a Lattice from Text

In order to construct a lattice from textual data, consider a Wikipedia article of your choice. Write down all the adjectives that appear in the article together with the terms they modify. Introduce the adjectives in Word-Net online (`http://wordnet.princeton.edu/perl/webwn`) and note the corresponding attribute for each adjective. Download the Concept Explorer (`http://conexp.sourceforge.net/`) software, for example, and create a formal context with the terms as objects and their corresponding attributes derived from the adjectives. Explore the corresponding lattice. Repeat the same process with all the verbs and their objects appearing in the Wikipedia article. What are the differences between the lattices?

8.5.1.2 Automatic Extraction of Adjectives

Instead of manually extracting the attributes, try to use some off-the-shelf natural language processing software to extract the attributes. For example, try GATE (General Architecture for Text Engineering) (`http://gate.ac.uk/`). Use the Wikipedia article from the exercise above as a (tiny) corpus. Apply a part-of-speech-tagger to assign the appropriate syntactic category to every word. Write a JAPE regular expression to extract all adjectives and the nouns they modify. Are these the same adjective / noun pairs you extracted manually?

8.5.1.3 Extracting Qualia Structures

Download a larger number of Wikipedia articles and create a GATE corpus. Implement some of the qualia patterns mentioned in this article as JAPE patterns and apply them to the corpus. Create a formal context using for example Concept Explorer and inspect the lattice. Is the resulting lattice reasonable?

8.5.1.4 Formalizing Views as Query-Based Multicontexts

Given our notion of attribute relevance and an appropriate threshold, the filtered formal contexts can be seen as a view on the original context. Try to use the theory of *Query-Based Multicontexts* as defined by Tane et al. [Tane et al., 2006] to formalize such views.

Part V

Web Semantics

Chapter 9

A Lexico-Logical Approach to Ontology Engineering

Sebastian Rudolph

AIFB, Universität Karlsruhe (TH), Germany

Johanna Völker

AIFB, Universität Karlsruhe (TH), Germany

9.1 Introduction

Building ontologies is a difficult and time-consuming task, requiring to combine the knowledge of domain experts with the skill and experience of ontology engineers resulting in a high demand on scarce expert resources. Moreover, the size of knowledge bases needed in real-world applications easily exceeds the modeling capabilities of any human expert. On the other hand, both quality and expressivity of the ontologies generated automatically by the state-of-the-art ontology learning systems fail to meet the expectations of people who argue in favor of powerful, knowledge-intensive applications based on ontological reasoning.

In order to overcome this bottleneck, it is necessary to thoroughly assist the modeling process by providing hybrid semi-automatic methods that (i) intelligently suggest the extraction of potential knowledge elements (complex domain axioms or facts) from resources such as domain relevant text corpora and (ii) provide guidance during the knowledge specification process by asking decisive questions in order to clarify still undefined parts of the knowledge base.

Obviously, those two requirements complement each other. The first one

clearly falls into the area of natural language processing. By using existing methods for knowledge extraction from texts, passages can be identified
that indicate the validity of certain pieces of knowledge. For the second requirement, strictly logic-based exploration techniques are needed that yield
logically crisp propositions. We believe that integrating these two directions
of knowledge acquisition in one scenario will help to overcome disadvantages
of either approach. The framework proposed in this chapter realizes this integration and shows its potential for practical applications.

In Section 9.2, we briefly introduce the description logic $\mathcal{SHOIN}$. Section 9.3 sketches the field of ontology learning before presenting LExO as one
method for acquiring DL axioms from texts. Section 9.4 gives the necessary
background for Relational Exploration (RE), a technique used for interactive
knowledge specification based on Formal Concept Analysis. In Section 9.5,
we describe in detail how LExO and RE (possibly assisted by other ontology learning components) can be synergetically combined into an integrated
framework. Implementation details as well as an example are given in Section 9.6. Finally, Section 9.7 concludes and gives an outlook to future research.

9.2   Preliminaries

Here, we will very briefly introduce the description logic $\mathcal{SHOIN}$ that
serves as the theoretical basis for the Web Ontology Language OWL as defined
in [McGuinness and van Harmelen, 2004].

A $\mathcal{SHOIN}$ knowledge base (KB, also: ontology) is based on sets N_R (*role
names*) **C** (*atomic concepts*) and **I** (*individuals*). The set of $\mathcal{SHOIN}$ *roles* is
$\mathbf{R} = \mathsf{N}_R \cup \{R^- \mid R \in \mathsf{N}_R\}$. In the following, we leave this vocabulary implicit
and assume that A, B are atomic concepts, a, b, i are individuals, and R,
S are roles. Those can be used to define concept descriptions employing
the constructors from the upper part of Table 9.1. We use C, D to denote
concept descriptions. Moreover, a $\mathcal{SHOIN}$ KB consists of two finite sets of
axioms that are referred to as *TBox* and *ABox*. The possible axiom types
for each are displayed in the lower part of Table 9.1.[1] Note that we do not
explicitly consider concept or role equivalence $\equiv$, since it can be modeled via
mutual concept or role inclusions. We adhere to the common model-theoretic
semantics for $\mathcal{SHOIN}$ with general concept inclusion axioms (GCIs): an
interpretation $\mathcal{I}$ consists of a set Δ called *domain* together with a function $\cdot^{\mathcal{I}}$
mapping individual names to elements of Δ, class names to subsets of Δ, and
role names to subsets of $\Delta \times \Delta$. This function is inductively extended to roles

---

[1] As usual, we require to restrict number restrictions to simple roles, being (roughly speaking
and omitting further technical details) roles without transitive subroles.

Name	Syntax	Semantics	
inverse role	R^-	$\{(x,y) \mid (y,x) \in R^{\mathcal{I}}\}$	
top	$\top$	Δ	
bottom	$\bot$	$\emptyset$	
nominal	$\{i\}$	$\{i^{\mathcal{I}}\}$	
negation	$\neg C$	$\Delta \setminus C^{\mathcal{I}}$	
conjunction	$C \sqcap D$	$C^{\mathcal{I}} \cap D^{\mathcal{I}}$	
disjunction	$C \sqcup D$	$C^{\mathcal{I}} \cup D^{\mathcal{I}}$	
universal restriction	$\forall R.C$	$\{x \mid (x,y) \in R^{\mathcal{I}} \text{ implies } y \in C^{\mathcal{I}}\}$	
existential restriction	$\exists R.C$	$\{x \mid \exists y \in \Delta,\ (x,y) \in R^{\mathcal{I}}, y \in C^{\mathcal{I}}\}$	
(unqualified) number	$\leq n\,R$	$\{x \mid \#\{y \in \Delta \mid (x,y) \in R^{\mathcal{I}}\} \leq n\}$	
restriction	$\geq n\,R$	$\{x \mid \#\{y \in \Delta \mid (x,y) \in R^{\mathcal{I}}\} \geq n\}$	
role inclusion	$S \sqsubseteq R$	$S^{\mathcal{I}} \subseteq R^{\mathcal{I}}$	TBox
transitivity	$\mathsf{Trans}(S)$	$S^{\mathcal{I}}$ is transitive	TBox
general concept inclusion	$C \sqsubseteq D$	$C^{\mathcal{I}} \subseteq D^{\mathcal{I}}$	TBox
concept assertion	$C(a)$	$a^{\mathcal{I}} \in C^{\mathcal{I}}$	ABox
role assertion	$R(a,b)$	$(a^{\mathcal{I}}, b^{\mathcal{I}}) \in R^{\mathcal{I}}$	ABox

FIGURE 9.1: Role/concept constructors and axiom types in $\mathcal{SHOIN}$. Semantics refers to an interpretation $\mathcal{I}$ with domain Δ.

and concept descriptions and finally used to decide whether the interpretation satisfies given axioms (according to Figure 9.1).

9.3 Lexical and Logical Knowledge Acquisition

Ontology generation from natural language text, or lexical resources – most commonly referred to as "ontology learning" – is a relatively new field of research that aims to support the tedious task of knowledge acquisition by automatic means.

However, many of today's ontology learning approaches build upon methods and ideas that were developed by (computational) linguists long before ontologies became a popular means of knowledge representation. Ontology learning techniques based, e.g., on lexico-syntactic patterns [Hearst, 1992], or Harris' distributional hypothesis [Harris, 1954] draw from previous advances in lexical acquisition, and terminology research, which have been to a major extent focusing on the extraction of lexical relations. However, there is a tacit agreement in the ontology learning community that there exists a certain correspondence between lexical relations (e.g., hyponymy, synonymy), and ontological axioms (e.g., subsumption, equivalence). This assumption, which is not only prevalent in ontology learning, but also influences manual ontology engineering[2] led to a kind of "lexical" (i.e., lexically inspired) ontology generation–implemented in frameworks such as Text2Onto [Cimiano and Völker, 2005]. One may argue that due to the differences between lexical semantics, and the model-theoretic semantics of description logics (see also [Völker et al., 2007a]), this type of approach will always yield at best semi-formal ontologies without precisely defined semantics, being grounded in natural language more than in logics.

On the other hand, lexical approaches to ontology generation offer a lot of advantages: They can benefit from large amounts of lexical resources such as machine-readable dictionaries, encyclopedias, and all kinds of Web documents that are available in abundance on the Web. The resulting ontologies are grounded in natural language, and usually close to the human way of modeling, which makes them easily comprehensible and reusable. And finally, most of these approaches are very flexible with respect to the degree of user interaction, and relatively easy to combine with other, complementary or supporting ontology learning methods.

Besides lexical ontology learning, i.e., linguistically inspired ontology generation, another direction of ontology learning has received more and more attention during the last couple of years. Approaches based on Inductive Logic Programming (ILP) and Formal Concept Analysis (FCA) have been developed in the logics community [Fanizzi et al., 2004, Rudolph, 2006] for some reason widely unappreciated by (lexical) ontology learning research. Although there are a few approaches aiming to reconcile the two worlds by using either FCA [Stumme and Maedche, 2001, Cimiano et al., 2005] or ILP [Nedellec, 1999] for lexical ontology acquisition, none of them has been designed specifically for the refinement of OWL DL ontologies or knowledge bases. Common to all those approaches is their idea to acquire knowledge based on presented domain entities and their properties. However, this type of logical ontology generation is often less efficient than lexical approaches, and requires a relatively large amount of manually acquired knowledge (e.g., ABox statements for taxonomy induction). The resulting ontologies often lack the traceability of a natural language grounding, and meaningful labels for complex class descriptions. Their expressivity is typically restricted to some variant of $\mathcal{ALC}$. On the other hand, those approaches have several advantages. Since they are based on already structured, formal data, they naturally come with a precisely defined, formal set-theoretic semantics. Thus being on "safe logical grounds," it is guaranteed that the acquired knowledge is also logically consistent.

Despite their respective advantages, both lexical and logical approaches to automatic (or semi-automatic) ontology engineering have failed to meet all the expectations of people arguing in favor of knowledge-intensive, reasoning-based applications, e.g., in domains such as bio-informatics or medicine. In

---

[2]In fact, if one tries to explain the semantics of subsumption to a non-logician, one often resorts to "clue phrases" similar to lexico-syntactic patterns, which themselves reflect lexical relations.

particular, expressivity and quality of the resulting axiomatizations are often insufficient for practical use. In order to meet these fundamental require-ments, a few lexical approaches toward learning more expressive ontologies, i.e., ontologies featuring the expressiveness of OWL DL, have been proposed recently [Völker et al., 2007a,b]. But these approaches have to face a lot of challenges that need to be overcome in order to make them useable in practice. Obviously, the more expressive learned (or manually engineered) ontologies become, the more important it will be to provide automatic support for qual-ity assurance, because the difficulty of a purely manual revision obviously rises with the growing complexity of the ontology. On the other hand, applications relying on reasoning over complex ontologies make it necessary to consider a larger variety of qualitative aspects that must be taken into account as an ontology is being learned or constructed, including logical consistency, and completeness. Notwithstanding, there exist only very few frameworks aim-ing at a tight integration of methods for ontology learning and evaluation. Whereas, e.g., Haase et al. [Haase and Völker, 2005] propose a way to deal with logical inconsistencies in lexically generated ontologies the problem of modeling completeness has been largely neglected up to now.

In this chapter, we therefore present an approach to ontology acquisition that effectively combines the strengths of the two complementary directions of research while at the same time compensating for many of their regarding disadvantages. It relies upon Relational Exploration, an FCA-based approach to systematic, logical refinement (cf. Section 9.4), and the automatic gener-ation of formal class descriptions by means of natural language processing techniques which is described in the remainder of this Section.

LExO (Learning EXpressive Ontologies) [Völker et al., 2007a] is an ap-proach toward the automatic generation of ontologies featuring the expressive-ness of OWL DL. The core of LExO is a syntactic transformation of definitory natural language sentences into description logic axioms. Given a natural lan-guage definition of a class, LExO starts by analyzing the syntactic structure of the input sentence. The resulting dependency tree is then transformed into a set of OWL axioms by means of manually engineered transformation rules. Possible input resources for LExO include all kinds of definitory sentences, i.e. universal statements about concepts, that can be found in online glossaries such as Wikipedia,[3] comments in the ontology, or simply given by a domain expert.

In order to exemplify the approach, we assume that we would like to refine the description of the class *Reviewer* the semantics of which could be infor-mally described as follows: *A reviewer is a person who reviews a paper that has been submitted to a conference or workshop.*[4] We will come back to this

---

[3]http://en.wikipedia.org/wiki/Main_Page

[4]Depending on the intended meaning of *Reviewer* other, broader definitions (e.g., covering reviews of journal articles, or research projects) might be more adequate, but we wanted to

example in Section 9.6.

Initially, LExO applies the Minipar dependency parser to produce a structured output as shown in Figure 9.2. Every node in the dependency tree contains information about the token such as its lemma (base form), its syntactic category (e.g., *N* (noun)) and role (e.g., *subj*), as well as its surface position. Indentation in this notation visualizes direct dependency, i.e., each child node is syntactically dominated by its parent.

```
( E2 () (fin) C
  ( 3 is (be) VBE i
    ( 2 reviewer (~) N s
      ( 1 A (~) Det det )
    )
    ( 5 person (~) N pred
      ( E4 () (reviewer) N subj 2 )
      ( 4 a (~) Det det )
      ( E1 () (fin) C rel
        ( 6 who (~) N whn 5 )
        ( 7 reviews (review) V i
          ( E5 () (who) N subj 5 )
          ( 9 paper (~) N obj
            ( 8 a (~) Det det )
            ( E0 () (fin) C rel
              ( 10 that (~) THAT whn 9 )
              ( 13 submitted (submit) V i
                ( 11 has (have) have have )
                ( 12 been (be) be be )
                ( E6 () (that) THAT obj 9 )
                ( 14 to (~) Prep mod
                  ( 16 conference (~) N pcomp-n
                    ( 15 a (~) Det det )
                    ( 17 or (~) U punc )
                    ( 18 workshop (~) N conj )
) ) ) ) ) ) ) ) ) )
```

FIGURE 9.2: Dependency tree (minipar).

This dependency structure is then transformed into an XML-based format in order to facilitate the subsequent transformation process, and to make LExO more independent of the particular parsing component.

The set of rules which are then applied to the XML-based parse tree make use of XPath expressions for transforming the dependency structure into one or more OWL DL axioms. Figure 9.3 shows a few examples of such transfor-

---

keep the example as simple as possible.

mation rules in compact syntax.[5] Each of them consists of several arguments (e.g., *arg_1:...*), the values of which are defined by an optional prefix, i.e., a reference to a previously matched argument (*arg_0*), plus an XPath expression such as */C[@role='rel']* being evaluated relative to that prefix. The last lines of each transformation rule define one or more templates for OWL axioms in KAON2-internal syntax, with variables to be replaced by the values of the arguments. Complex expressions such as *0-1* allow for "subtracting" individual subtrees from the overall tree structure.

```
rule: relative clause {
    arg_0:   //N
    arg_1:   arg_0 /C[@role='rel']
    arg_2:   arg_1 /V
    result:  [equivalent 0 [and 0-1 2]]
}
rule: verb with prepositional complement {
    arg_0:   //V
    arg_1:   arg_0 /Prep
    arg_2:   arg_1 /N[@role='pcomp-n']
    result:  [equivalent 0 [some 0-2 2]]
}
```

FIGURE 9.3:   Transformation rules.

A minimum set of rules for translating the "Reviewer" example into a DL class description is given by Figure 9.4 (for a more complete listing of possible transformation rules, and further explanations see [Völker et al., 2007a]). It is important to emphasize that this set of rules does not define the only possible way to perform the transformation. In fact, there are many different modeling possibilities, and the choice of appropriate rules very much depends on the particular application or individual preferences of the user.

Rule	Natural Language Syntax	OWL Axioms
Disjunction	X: NP_0 or NP_1	$X \equiv (NP_0 \sqcup NP_1)$
Copula	X: NP_0 VBE NP_1	$NP_0 \equiv NP_1$
Relative Clause	X: NP_0 C(*rel*) VP_0	$X \equiv (NP_0 \sqcap VP_0)$
Verb with Prep. Compl.	X: V_0 $Prep_0$ NP(*pcomp-n*)$_0$	$X \equiv \exists V_0\_Prep_0.NP_0$

FIGURE 9.4:   Sample transformation rules for *Reviewer* example.

---

[5]LExO also supports an XSLT-based rule representation, which is much more expressive and allows for standardized serialization and exchange of transformation rules.

Depending on the concrete set of translation rules and modeling preferences of the user, a translation of the sentence into OWL DL could then yield the following axioms:

reviewer
$\equiv$ *a_person_who_reviews_a_paper_that_has_been_submitted_to_a_\ conference_or_workshop*

a_person_who_reviews_a_paper_that_has_been_submitted_to_a_\ conference_or_workshop
$\equiv$ *a_person* $\sqcap$ *reviews_a_paper_that_has_been_submitted_to_a_conference_\ or_workshop*

reviews_a_paper_that_has_been_submitted_to_a_conference_or_\ workshop
$\equiv$ $\exists$*reviews.a_paper_that_has_been_submitted_to_a_conference_or_workshop*

a_paper_that_has_been_submitted_to_a_conference_or_workshop
$\equiv$ *a_paper* $\sqcap$ *has_been_submitted_to_a_conference_or_workshop*

has_been_submitted_to_a_conference_or_workshop
$\equiv$ $\exists$*has_been_submitted_to.a_conference_or_workshop*

a_conference_or_workshop $\equiv$ (*a_conference* $\sqcup$ *workshop*)

Obviously, the above set of axioms can be normalized, and turned into a semantically equivalent, unfolded, representation:

Reviewer $\equiv$ *Person* $\sqcap$ $\exists$*review.*(*Paper* $\sqcap$ $\exists$*submitted_to.*(*Conference* $\sqcup$ *Workshop*))

While such a compact class description might be easier to grasp at first glance (at least for ontology engineers being familiar with logics), the first axiomatization obviously conveys a lot of additional information to the human reader. The fact that each part of the overall class description (e.g., *Conference* $\sqcup$ *Workshop*) is associated with an equivalent atomic class (e.g., *a_conference_or_workshop*) makes completely transparent how this axiomatization was constructed, and at the same time provides the user with an intuitive explanation of the semantics of each class description. Further advantages of the extended axiomatization are discussed in Section 9.5.

9.4   Relational Exploration

The technique of Relational Exploration (short: RE, introduced in [Rudolph, 2004] and thoroughly treated in [Rudolph, 2006]) extends the attribute exploration algorithm presented in Chapter 1 of this book to a DL setting: Given an interpretation $\mathcal{I}$ on a domain Δ and a set M of $\mathcal{SHOIN}$ concept descriptions, the corresponding $\mathcal{I}$-*context* is defined by $\mathbb{K}_\mathcal{I}(M) := (\Delta, M, I)$ with $\delta I C :\Leftrightarrow$

$\delta \in C^{\mathcal{I}}$. Then it can easily be shown, that implications in $\mathbb{K}_{\mathcal{I}}$ coincide with certain axioms w.r.t. their validity in $\mathcal{I}$: for $\mathcal{C}, \mathcal{D} \subseteq M$, the implication $\mathcal{C} \to \mathcal{D}$ holds in $\mathbb{K}_{\mathcal{I}}$ if and only if $\mathcal{I}$ models the DL axiom $\bigsqcap \mathcal{C} \sqsubseteq \bigsqcap \mathcal{D}$. Hence it is possible to explore DL axioms (more precisely: general concept inclusion axioms, short: GCIs) with this techniques, i.e., in an interview-like process, a domain expert has to judge, whether a proposed GCI is valid in the domain (formally: the interpretation $\mathcal{I}$) he is describing and in the negative case provide a counterexample.[6] Since OWL DL [McGuinness and van Harmelen, 2004] – the standard language for representing ontologies – is based on description logics, the RE method easily carries over to any kind of ontologies specified in that language.

The advantage of RE is that the obtained results are logically crisp and naturally consistent. Moreover, the acquired information is complete with respect to certain well-defined logic fragments of OWL DL.[7] Yet, one major shortcoming of RE is the following: due to the aimed-at completeness, the number of asked questions (and therefore, the runtime and the workload for the expert) grows rapidly with the number of involved concepts and roles which threatens to exceed the ontology designers resources.

We combine two strategies to counter this: firstly, we use an OWL DL reasoner to determine whether the answer to a question posed by the exploration algorithm can be deduced from a previously given background knowledge ontology. Secondly we use lexical ontology learning to determine a relatively small number of relevant classes to focus on. Both points will be elaborated in the next section.

9.5 An Integrated Approach to Ontology Refinement

In the sequel, we will describe how LExO and RE can be synergetically combined by giving a comprehensive description of the integrated algorithm. En route, we will briefly mention how other lexical ontology learning techniques could be beneficially used within that process. In addition to the LExO and RE component, an OWL DL reasoner will be applied in order to draw conclusions that are already implicitly present, i.e., entailed by the actual knowledge base making an intervention of the user obsolete.

Creation of New Definitions and Mappings. We start with an OWL DL ontology $\mathcal{KB}$ to be refined with respect to a (new or already contained) class C, for which a natural language definition is provided by some textual resource. This textual definition is then analyzed by LExO yielding a set $\mathcal{KB}'$

---

[6]This will be further elaborated and demonstrated in the subsequent sections.

[7]depending on which variant of RE is used

of OWL DL axioms as described in Section 9.3. Most likely, some (or even most) of the named classes those axioms refer to will not be present in $\mathcal{KB}$. Therefore, at least the primitive classes among those – i.e., those classes not stated to be equivalent to a complex class description[8] – should be linked to $\mathcal{KB}$. There are several ways for doing that. If textual definitions are available, LExO could be employed "recursively," i.e., it might be applied to the definitions of the classes in question in order to obtain other classes that can be linked to $\mathcal{KB}$ more easily. In any case, ontology *mappings* between $\mathcal{KB}$ and $\mathcal{KB}$' could be either added manually or established by one or several of the well-known mapping tools like FOAM [9][Ehrig and Sure, 2005]. So let Map be a (possibly empty) set of respective mapping axioms.

Selection of Relevant Classes. In the next step, we stipulate the focus of the subsequent exploration, by selecting the named classes from $\mathcal{KB} \cup \mathcal{KB}'$ whose logical dependencies shall be further clarified. A natural default choice for this would be the set of all named classes from $\mathcal{KB}$', as we might suppose the (remaining) classes from $\mathcal{KB}$ to be modeled in a sufficiently precise way – an assumption that might be disproved later on. However, it might be reasonable to include some of the classes from $\mathcal{KB}$ as well. Knowledge extraction methods that determine the relevance of terms (like those offered by Text2Onto [Cimiano and Völker, 2005]) could be employed for an automatic selection or to generate reasonable suggestions. In any case, let $\mathbf{C}$ denote the set of selected attributes.

After this selection of relevant named classes, a basic fact from FCA allows to further restrict $\mathbf{C}$: put into DL notation, it assures the dispensability of a class $C \in \mathbf{C}$ whenever there is a set $\mathbf{D} = \{D_1, \ldots, D_n\} \subseteq \mathbf{C} \setminus \{C\}$ such that $C \equiv D_1 \sqcap \ldots \sqcap D_n$ follows from all knowledge $\mathcal{KB}_\Sigma := \mathcal{KB} \cup \mathcal{KB}' \cup Map$ stated so far. It takes just a little consideration that this is the case iff

$$\mathcal{KB}_\Sigma \models \prod \left\{ D \mid D \in \mathbf{C} \setminus \{C\}, \ \mathcal{KB}_\Sigma \models C \sqsubseteq D \right\} \sqsubseteq C,$$

such that the elimination of redundant classes from $\mathbf{C}$ requires just $O(|\mathbf{C}|^2)$ reasoner calls in the worst case. Let $\mathbf{C}'$ denote the result of this reduction process.

Exploration. Now we start RE as described in Section 9.4 on the concept set $\mathbf{C}'$. A work flow diagram of the procedure is displayed in Figure 9.5. For every hypothetical DL axiom $C_1 \sqcap \ldots \sqcap C_n \sqsubseteq D_1 \sqcap \ldots \sqcap D_m$ brought up by the exploration algorithm:

- Employ the reasoner to check whether this GCI is a consequence of $\mathcal{KB}_\Sigma$. If so, confirm the implication and continue the exploration with the next hypothesis.

---

[8]which are those occurring explicitly in the normal form from Section 9.3

[9]http://www.aifb.uni-karlsruhe.de/WBS/meh/foam/

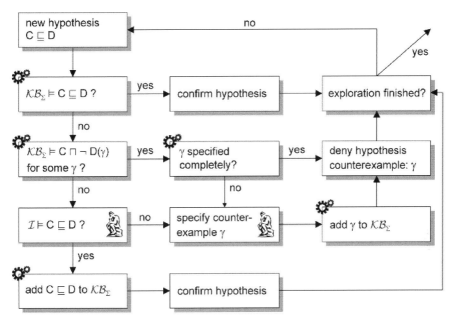

FIGURE 9.5: Relational exploration process (the gear wheels indicate ontology management activities including reasoning and updates, whereas the thinker icon marks user involvement).

- Employ the reasoner to query for all individuals γ with $C_1 \sqcap \ldots \sqcap C_n \sqcap \neg(D_1 \sqcap \ldots \sqcap D_m)(\gamma)$, i.e., for instances of the class that characterizes exactly the property for being a counterexample for the hypothetical GCI. Let Γ be the set of individuals retrieved this way. If $\Gamma \neq \emptyset$, check for each $\gamma \in \Gamma$ and every $C \in \mathbf{C}$ whether $C(\gamma)$ or $\neg C(\gamma)$. If for *one* γ, all the class membership for *each* $C \in \mathbf{C'}$ can be proved or disproved, this very γ will be passed to the exploration algorithm and the exploration continues without consulting the expert. Otherwise the human expert will be asked to complete the assertions for one γ accordingly – possibly assisted by lexical knowledge retrieval tools – before passing it as a counterexample to the exploration algorithm. In any case, after providing γ the exploration will proceed with the next hypothesis.

- If the DL axiom in question can be neither automatically proved nor declined (the latter meaning $\Gamma = \emptyset$), the human will be asked for the ultimate decision whether the axiom is satisfied in the described domain $\mathcal{I}$ or not. Again, ontology learning tools could support him by suggesting answers endowed with a probability, or simply scanning a corpus for potential hints and presenting selected passages.

The exploration terminates after finitely many steps, yet it may also be stopped by the user beforehand. In this case, the internal order of the classes from the set $\mathbf{C}'$ is relevant since it determines the order of the posed questions. Hence, it is beneficial to sort those classes w.r.t. their relevance, possibly based on textual information. After the exploration cycle being finished, we have obtained a refined knowledge base $\mathcal{KB}_\Sigma$ containing the (possibly new) class C endowed with its definition (as extracted from the textual definition) and its interrelationships with concepts from the original knowledge base. Additionally, the "ontological context" of C has been made logically explicit by interactive exploration. In fact, any subsumption between conjunctions of classes from $\mathbf{C}$ can be decided (i.e., proven or disproven) based on the refined knowledge base. This also shows the advantage of introducing atomic classes for the complex concept descriptions occurring in the LExO output as demonstrated in Section 9.3: although RE as applied in this case[10] deals only with conjunctions on atomic classes, we introduce more expressivity "through the back-door" by having complex definitions for those named classes in our ontological background ready to be used by the reasoner.

The synergies provided by the presented combination are manifold: Firstly, the classes contained in the definitions provided by LExO provide a reasonable small to medium size "exploration scope" being crucial for a reasonable application of the RE technique. Secondly, we can use textual information for generating ontological information (a source not accessible to purely logical approaches) yet being able to interactively clarify logical dependencies that have been left open by the text. The latter is done in a guided way ensuring completeness.

9.6   Implementation and Example

In order to prove the feasibility of a synthesis of ontology learning and RE as described in Section 9.5, we implemented a prototypical application named *RELExO*.[11] It relies upon KAON2[12] as an ontology management back-end and features a simple graphical user interface. The architecture is depicted by Figure 9.6.

LExO, possibly complemented by other ontology learning components, generates or extends the initial set of axioms $\mathcal{KB}$ (mappings can be added by

---

[10] Actually, RE provides means for exploring GCIs in whole $\mathcal{ALC}$, however we restrict to conjunctions on atomic classes in this example.

[11] Both sources and binaries of RELExO are available for public use and can be downloaded from http://relexo.ontoware.org

[12] http://kaon2.semanticweb.org

FOAM, if necessary), and initializes the context $\mathbb{K}$ by suggesting a set of attributes **C** to the user. The actual refinement process is handled by a RE component that manages the context $\mathbb{K}$ and the implication set $\mathfrak{I}$. Both are updated based on answers obtained from the "expert team" constituted by the KAON2 reasoner, an optional ontology learning component as well as the human knowledge engineer.

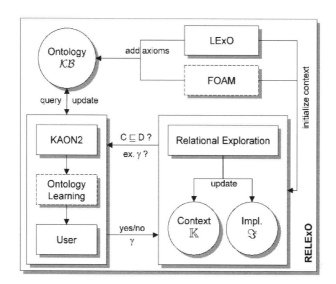

FIGURE 9.6:   RELExO architecture.

We now illustrate the integrated ontology refinement process that has been elaborated on in Section 9.5 by means of a real-world example. The complete material necessary for reproducing this example, i.e., ontologies and screenshots, is contained in the RELExO distribution.

The **SWRC** (Semantic Web for Research Communities)[13] [Sure et al., 2005] ontology is a well-known ontology modeling the domain of Semantic Web research. Version 0.7 contains 71 classes, e.g., for different types of persons, publication, and events, 48 object properties, 46 datatype properties, and an overall number of 672 axioms. Its expressiveness is slightly beyond OWL DLP featuring subsumption, properties, and a few disjointness axioms. The ontology serves as a basis for semantic annotation in the AIFB Web portal,[14] which manages information about more than 2.000 persons, projects, and publications. For the purpose of our experiment, we exported all in-

---

[13]http://ontoware.org/projects/swrc/

[14]http://www.aifb.uni-karlsruhe.de

stance data stored in the AIFB portal into one single OWL file (more than 3 Megabytes in RDF syntax), and merged it with the corresponding TBox, i.e., the latest version of SWRC. After minor syntactic corrections (removing non-XML-compliant characters), we obtained a considerably large ontology. Debugging with Radon[15] revealed two inconsistencies due to conflicting range specifications of data properties that could be fixed without difficulty.

Subsequently (in order to keep the example simple and rule out a few trivial questions that would otherwise come up in the exploration phase), we added axioms stating the disjointness of the SWRC top-level concepts *Person*, *Event*, and *Publication*, obviously true axioms yet not present in the current version of this ontology. Those axioms could also have been generated automatically by techniques for learning disjointness from [Völker et al., 2007b]. However, adding these axioms turned the ontology inconsistent again as some individuals were inferred to instantiate both *Person* and *Publication*. The reason for this inconsistency was an incorrect use of the *editor* relationship in SWRC. Although its domain was restricted to *Person* ("*editor_ of*"), the property was apparently conceived to have "*has_ editor*" semantics by most of the annotators. We fixed this inconsistency by changing the definition of *editor* accordingly. Another problem became apparent after we had already started the exploration of the resulting ontology with RELExO. An individual (in our opinion) belonging to the class *ResearchPaper* was proposed as a counterexample, but could not be classified as such. A closer look at both individual and ontology showed that it was assigned to the class *InProceedings*, which was declared disjoint from *ResearchPaper*, the latter actually being empty. Since we found that this modeling decision is not justified by the associated comments in the ontology, we simply removed the disjointness axiom.

To demonstrate the RELExO approach, we assume that we would like to add a new class *Reviewer* to the SWRC ontology. Part of a change request could be a natural language description of this class such as *a reviewer is a person who reviews a paper that has been submitted to a conference or workshop* (cf. Section 9.3). Given this definitory sentence, LExO automatically suggests an axiomatization of *Reviewer* to the user who can correct or remove some of the generated axioms before they are added to the ontology. Applying FOAM for suggesting mappings between the newly introduced class names and those already present in SWRC, we find *Paper* to be equivalent to *ResearchPaper* and add a corresponding equivalence axiom to the extended ontology. Likewise, we find *Person*, *Conference*, and *Workshop* already present in the original ontology.

In the next step, the set of "relevant" classes has to be selected. As mentioned in Section 9.5, it is reasonable to choose those atomic classes present in the definition of *Reviewer*. We decided to add two more classes denoting undergraduate and PhD students and (introducing abbreviations for over-length

---

[15]http://radon.ontoware.org

concept names from $\mathcal{KB}$') we set:

$$\mathbf{C}' := \{\perp, CoW, Conference, SubCoW, Person, PhDStudent,$$
$$ResearchPaper, RevPSubCoW, Undergraduate, Workshop\}.$$

Based on this set of classes, the RE algorithm is started. The first hypothetical DL axiom, the exploration comes up with is $\top \sqsubseteq \perp$. Naturally, this hypothesis cannot be deduced from the ontology. Hence, following the description in Section 9.5, KAON2 will query the knowledge base for instances of $\top \sqcap \neg \perp$, which is equivalent to $\top$. Hence *all* ABox individuals are retrieved. Unfortunately, for none of those, the class membership w.r.t. *SubCoW* can be decided, i.e. none is completely specified. Hence, the expert is asked to complete the specification of one of the retrieved individuals, in our case *id1289instance*, being an instance of *ResearchPaper* as well as *SubCoW* but of none of the other classes in question.

In a similar way, the next hypothesis posed – $\top \sqsubseteq ResearchPaper \sqcap SubCoW$ – is handled. Clearly, not every ABox individual is a research paper witnessed by the counterexample *id1478instance* being a book chapter and hence neither a research paper nor submitted to a conference or workshop.

However, the subsequent hypothesis $CoW \sqsubseteq \perp$ can neither be proved nor disproved by KAON2 using the information actually present in the ontology – since it does not contain any individuals being a conference or workshop. Therefore, the human expert will be asked for the final decision. Obviously, this hypothesis has to be denied and a counterexample for it is just any conference, so we enter *ISWC_2007* and specify it as instance of *CoW* and *Conference*.

Consequently, the next question $CoW \sqsubseteq Conference$ comes up and has to be denied as well by entering the workshop instance *IWOD_2007*.

Equally, the hypothesis $SubCoW \sqsubseteq ResearchPaper$ cannot be decided based on the present knowledge and is thus passed to the expert. In fact, this is the first "design decision" to make depending on the intended scope of the ontology. A look into the SWRC taxonomy reveals that there is a class *Poster* to denote posters presented at conferences. Indeed, any submitted poster would be a counterexample for the presented hypothesis, so we add *iMapping_Poster_SWUI_2006* to the knowledge base.

The next hypothesis brought up is $CoW \sqcap SubCoW \sqsubseteq \perp$ being an integrity constraint saying that nothing being a conference or workshop can be submitted (to a conference or workshop). Figure 9.7 shows how it is presented to the user.

Here we encounter another design decision. Although it might be reasonable to say that a workshop (actually: a workshop proposal) has been submitted to a conference, we decide to restrict the domain of the role *submitted_to* to publications and hence confirm the validity of the presented hypothesis.

The hypothesis $Person \sqsubseteq \perp$, coming up next, is refuted by the reasoner retrieving an individual being a PhD student at the AIFB institute. Figure 9.8

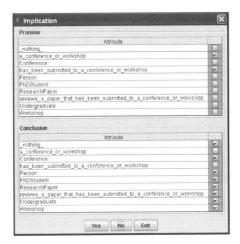

FIGURE 9.7: Dialog for displaying the hypothetical axiom $CoW \sqcap SubCoW \sqsubseteq \bot$.

shows the dialog where the user is asked to add the missing information about this individual.

In this way, the exploration continues. During the process, some individuals are added and the following additional axioms are confirmed:

- $SubCoW \sqcap Person \sqsubseteq \bot$ – a person cannot be submitted;

- $ResearchPaper \sqsubseteq SubCoW$ – every research paper has been submitted (to a conference or workshop)[16];

- $RevPSubCoW \sqsubseteq Person$ – everybody reviewing a submitted paper is a person;

- $Person \sqcap PhDStudent \sqcap Undergraduate \sqsubseteq \bot$ – PhD students are disjoint with undergraduates[17];

- $RevPSubCoW \sqcap Undergraduate \sqcap Person \sqsubseteq \bot$ – actually a "policy decision": undergraduates are not allowed to review papers;

The formal context with the examples acquired during the exploration is displayed in Figure 9.9.

We end up with a refined SWRC ontology containing the new class *Reviewer* fully integrated into the existing ontology. Any subsumption between conjunctions of the specified interesting classes can be directly decided based on this refined SWRC ontology. This can be nicely demonstrated by starting

---

[16]justified by the existence of a class *Unpublished* disjoint to *ResearchPaper*

[17]Another modeling flaw: this axiom should have been present in SWRC.

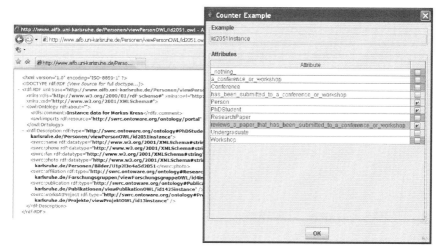

FIGURE 9.8:   Specifying a counterexample. Every(non-)class-membership deducible from the knowledge base is automatically entered leaving just the open questions to the expert.

RELExO again with the refined ontology: it terminates without ever asking the human expert for a decision, showing that all upcoming questions can be answered by the reasoner alone.[18]

9.7   Conclusion and Outlook

In this chapter, we have sketched a way to combine complementary approaches to ontological knowledge acquisition: the more intentional approach of distilling conceptual information from textual resources, and the extensional method of extracting hypothetical domain axioms based on given entities. We have instantiated this approach by designing and implementing a prototypical framework that integrates the LExO ontology learning application, a Relational Exploration component, and the KAON2 reasoner. In an example – using the well-known SWRC ontology – we have demonstrated the feasibility of our approach, and its applicability to real-world ontology engineering tasks.

Altogether, we are confident that the proposed framework will considerably

---

[18]In fact, this only works reliably, if the algorithm tries to find among all automatically retrieved counterexamples a completely specified one. As this feature can significantly slow down the exploration process in case of large ABoxes, it can be disabled by the user.

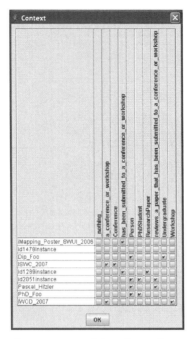

FIGURE 9.9:   Formal context resulting from the exploration.

alleviate the task of designing comprehensive and complex, yet logically consistent ontologies for knowledge-intensive applications. The number of design decisions to be made by the human user is minimized by the usage of textual resources and the employment of a reasoning back-end. Relational exploration provides guidance, ensuring that neither redundant information will be asked for nor important information is simply forgotten in the modeling process (in fact, we were well-nigh inevitably confronted with several modeling flaws in SWRC while preparing our example).

Without any doubt, human intervention will always remain indispensable, especially for complex knowledge modeling tasks. Notwithstanding, the workload to ontology engineers and domain experts can be drastically decreased by intelligently integrated lexico-logical components. By reducing the costs for the acquisition of expressive OWL ontologies, we hope to foster the development of more sophisticated, reasoning-based applications, and help to put semantic technologies into practice.

Pursuing this promising goal, we identify several central issues for future research. Firstly, we are planning to extend the approach of Relational Exploration, and RELExO to the situation with partially specified objects as described in [Baader et al., 2007] as well as to incorporate the just recently proposed technique of role exploration from [Rudolph, 2008]. The implementation of RELExO could be further extended by an additional (automatic) expert which uses ontology learning techniques, and online resources for confirming hypotheses, or suggesting counterexamples. Additional ontology learning components could be used to complement the LExO-generated axiomatizations by other modeling primitives (e.g., disjointness axioms), or to sort the attributes, i.e., class descriptions, with respect to the current domain or the user's interests. Finally, we will integrate RELExO into an ontology engineering environment such as the NeOn Toolkit,[19] and improve its usability by adding a natural language generation component for translating hypotheses, i.e., logical implications, into natural language questions. In the end, we are confident that further extensive evaluations in real-world application scenarios will demonstrate the advantages of our combined, lexico-logical approach.

9.8   Student Projects

9.8.1   Project 1

In this exercise, we try to convey a feeling about advantages and drawbacks of exploration techniques in the ontology engineering and design area. Depending on the comprehensiveness of the chosen domain, the task might take up to several days. The task requires basic knowledge about modelling ontologies in OWL and using OWL reasoners.

1. Study the literature in order to get familiar with the technical details of the attribute exploration algorithm. Implement a prototype if possible.

2. Select an arbitrary domain and specify a set of "interesting" OWL class descriptions.

3. Model the obvious features of your chosen domain in an OWL DL ontology (e.g., using Protégé).

4. Carry out an exploration with respect to those class descriptions, either manually (feasible only for rather small description sets) or using an implementation (e.g., ConExp[20] or a self-implemented prototype).

5. Repeat the exploration with this background ontology using a reasoner (like KAON2, Fact++, or Pellet) to cope with the hypotheses brought up by the exploration algorithm in the first place.

6. Compare the two exploration runs in terms of kind and number of the asked questions as well as algorithm runtime.

---

[19]http://www.neon-toolkit.org

[20]http://conexp.sourceforge.net

7. Draw the concept lattice (again either completely manually or using a tool like ToscanaJ[21] or ConExp) based on the formal context resulting from the exploration. What Information can you read from it?

9.8.2   Project 2

Investigate and discuss whether it is possible that a logically consistent and coherent ontology turns inconsistent or incoherent during the exploration process. Please give an explanation for your answer.

9.8.3   Project 3

The purpose of this exercise is to raise awareness for modeling problems that may result from manual or semi-automatic ontology engineering. Choose one or more concepts from your favorite domain and try to find natural language definitions for them. Online glossaries such as Wikipedia[22] or WordNet[23] are good starting points for your search, but you can also ask Google^{TM} (e.g., "define: DNA") or write your own definition if you do not find any on the Web. Once you have the definition, please formalize it by means of a description logic of your choice. The example in Section 9.3 shows how a formal class description of the concept *Reviewer* could look like. However, you do not have to stick closely to the proposed translation approach. Write down the difficulties you encounter and discuss possible solutions.

9.8.4   Project 4

The implementation of RELExO (cf. Section 9.6) relies on two experts who may confirm or deny hypotheses and give counterexamples: a human and an ontology reasoner. An additional expert might be an ontology learning component that tries to obtain evidence for subsumption or class instantiation from external resources such as a collection of text documents. Can you imagine even more kinds of experts? What are their respective advantages and disadvantages?

---

[21]http://toscanaj.sourceforge.net
[22]http://www.wikipedia.org
[23]http://wordnet.princeton.edu

Chapter 10

Faceted Document Navigation

Jon Ducrou

University of Wollongong, Australia

Peter Eklund

University of Wollongong, Australia

10.1   Introduction

The problem of document browsing and navigation attracts the attention of researchers in information retrieval and also in human-computer interface design. Similarly, in conceptual structures research, the use of the concept lattice as an information space for the navigation of documents was one of the pioneering applications of Formal Concept Analysis and maintains the interest of many researchers in the field due to its practicality.

MailSleuth is the extension of previous research work and a number of practical projects preceded it; namely, CEM [Cole and Eklund, 1999, Cole et al., 2000, Cole and Stumme, 2000, Eklund, 2002, Cole et al., 2003] and Rental-FCA [Cole and Eklund, 2001]. Other related and concurrent efforts that are similarly motivated include [Koester, 2006]. **MailSleuth** distinguishes itself from these in that the software has been extensively tested for usability. **MailSleuth** has provided a platform for enhancing the practicability of browsing and navigating information spaces using concept lattices, particularly testing the suitability of the metaphor for non-FCA trained users.

ImageSleuth [Ducrou et al., 2006] was conceived later than **MailSleuth** and similarly underwent an iterative design and usability testing methodology reported in [Ducrou and Eklund, 2007]. **ImageSleuth** is a tool for navigating collections of annotated images and for content-based retrieval using Formal Concept Analysis. It combines methods of FCA for information retrieval with the graphical information conveyed in image thumbnails. In addition to established methods such as keyword search and upper/lower neighbors, a query-by-example function and restrictions to the attribute set are included. Metrics on formal concept similarity are discussed and applied to ranking and automated discovery of relevant concepts: from both concepts and semi-concepts.

In this chapter we trace the evolution of the design of software for document navigation and browsing that exploits Formal Concept Analysis as the metaphor for content navigation. We discuss the idea both in terms of its usability dimension, particularly the suitability of the line diagram in practical systems but also in terms of the integration of Formal Concept Analysis-based systems with traditional information retrieval structures.

10.2   The Email Domain and Formal Concept Analysis: MailSleuth

MailSleuth extends the standard email management paradigm by enhancing an existing email client, Microsoft Outlook^{TM}, with new features based on the theory and practice of Formal Concept Analysis.

Email documents are semi-structured, text-based documents used primarily for interpersonal communication. Structural components within emails are well defined and typically include information on sender and receiver(s), subject, date, and the body, or content, of the email. Traditionally the organizational metaphor for email is not unlike paper-based mail. One mail item has one place in the filing system. There are differing levels of complexity of email filing methods, from simple flat methods to more elaborate hierarchical structures. Simple flat methods have all email collected in a single one-dimensional list. Usually content is maintained chronologically, but often tools for finding and sorting emails to assist retrieval are featured. While this flat method is simple, it does not scale well to large, diverse collections.

Inheritance structures (folders) allow categorization of email into inheritance hierarchies where the root contains folders and emails, and any folder can be a container for further folders and emails. In practice, the folder depth rarely exceeds one or two levels. Folder structure and email list components are traditionally separated within the display as shown in Figure 10.1.

Each folder is a named categorization or sub-categorization that contains email items. Emails physically exist in a single folder, so if an email could be categorized into two folders, it would need to be placed into only one or duplicated in both. For example, if a user has a category for conferences and another for collaboration with a colleague, then where should email from the colleague about a conference go? Placing the mail in one of the categories

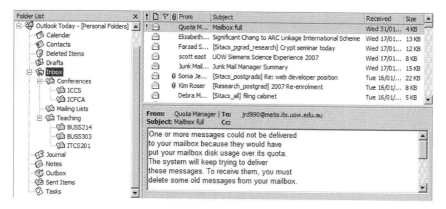

FIGURE 10.1: Example of traditional email categorization. A hierarchical folder list is shown on the left as a tree widget structure. This is complemented with a one-dimensional list of emails shown on the top right. Emails are realized via the preview pane (lower right).

means finding that email may require looking in both folders. Duplication on the other hand creates redundancy.

A more appropriate methodology would allow emails to be found in multiple locations within the system without physical duplication. Namely, an email about joint conference submission with Richard should be in both the "Conference" and "Richard" folders. Some attributes should be able to be derived from others, for instance "AAAI" should automatically imply "Conference." And "Conference" may in turn imply "Research Activities." It should also be possible for the email in each of these folders to be analyzed separately as a sub-collection; allowing analysis to be localized within the overall collection. Formal Concept Analysis naturally supports multiple inheritance and forms a theoretical framework for applying this idea.

10.2.1 Storage and Access for Formal Concept Analysis

To overlay Formal Concept Analysis onto an existing mail system efficiently, appropriate storage and access is required. These components include base-level data structures for the Formal Concept Analysis integration: knowledge base, attribute hierarchy, and visualization. It also includes a discussion of natural interface mechanisms through which these are manipulated in the software.

10.2.1.1 Triple Store Knowledge Base

MailSleuth uses a knowledge base to store all information relating to its operation and state. This includes a full-text index, the attribute hierarchy, user settings, and all other information used by the software. The knowledge base is a triple store, which automatically applies transitive closure operations on new triples with certain relationship types (e.g., asserting "a is-a b" and "b is-a c" will automatically assert "a is-a c"). This design allows partial order operations to be performed automatically and efficiently.

An efficient knowledge base allows **MailSleuth** to create a structured, full text index of the users email collection. This can be thought of as a universal formal context for the **MailSleuth** system. While the context is maintained it is never used in its entirety for the creation of a lattice structure (given the millions of relations created in an email collection such a task is computationally prohibitive). Each email is broken into its component parts (based on SMTP and extended by the email client) and indexed according to these parts. Indexing is based on one of four relation types: "index relation," "before relation," "after relation" and "relative relation." These relation types dictate the behavior of the query engine in returning results from the index. Components that are primarily free-text, such as the email body and its subject, are processed using standard stopword and stemming techniques.

10.2.1.2   Inducing an Attribute Hierarchy

To provide the user with access to multiple inheritance functionality the software allows creation and modification of an attribute hierarchy. The simplest way to represent this without deviating unnecessarily from standard user control types is with a folder tree that enforces a partial order over its structure. This means that some branches of the folder tree will be duplicates of others, however, their content is not duplicated. Each folder's name is a unique identifier indicative of its content and the content of all child folders. If a folder is created with an existing name that does not break the partial order, it will automatically have all child folders associated with that name created under it. This style integrates well with the host software (Outlook^{TM}), which uses a folder view for traditional email management.

The attribute hierarchy is comprised of *labels*. Each label is part of a partial order. For the set G of emails, the set M of labels and the incidence mapping I, the following *compatibility condition* holds:

$$\forall g \in G, m, n \in M : (g, m) \in I, m \leq n \Rightarrow (g, n) \in I$$

The attribute hierarchy is the mechanism that allows the user to create conceptual scales for viewing lattices and is strictly enforced. Users creating, moving, or deleting folders are warned if the action will cause repercussions on other attributes in the partial-order.

Labels can have one or more queries attached to them. A label's queries generate objects associated with that label. When more than one query is used, the result is the intersection of each query result. Each label is a "virtual folder" or "drill-down folder." "Virtual folders" display the email matching the query associated with the label in the traditional one-dimensional list that is

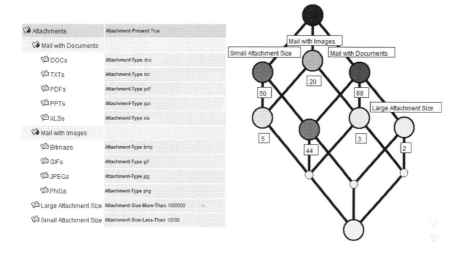

FIGURE 10.2:   The structure and queries used in the "Attachments" drill-down folder and its corresponding lattice. Two of the subfolders are drill-down folders that further specify queries about attachments present with emails.

kept consistent with the collection. Typically, all leaf labels (those without children) are used this way. "Drill-down folders" can be used to view a lattice with its child folders as the scale. When used in a lattice view, all child labels represent the union of their descendants' objects.

10.2.2   Scalable Sub-Lattice Visualization

As mentioned, the complete lattice structure is never created or visualized due to the complexity of an email collection. To control the complexity of the lattice visualization, only points in the virtual folder hierarchy, and therefore the attribute hierarchy, can be used for creating sub-lattices.

Creation of the lattice context for a given label in the attribute hierarchy is performed by taking all direct child labels as attributes. The object set for such a context is the union of all queries from the given label to all its child labels (direct and transitive). Each attribute is assigned objects (emails) that match the union of all queries from that label and its children's labels (both direct and transitive).

For example, in Figure 10.2 the lattice is generated for the label *Attachments*. The attribute set is the direct children, namely *Mail with Documents*, *Mail with Images,Large Attachment Size* and *Small Attachment Size*. The object set will be the combination of the queries from the label *Attachments* and all direct and indirect child labels. Each attribute is then assigned the union of its query and its children's queries. The 'Attachments' folder itself has the query **Attachment-Present True** so that "Attachments" represents all email with attachments, and not just those that match its children's queries.

The visualized lattice offers various manipulations to allow the user to navigate the information presented. Concepts can be arranged, the view can be zoomed, panned, and centered, and font size increased or decreased. The visualized concepts themselves offer the user one of two choices, *extent* or *contingent*, which can be realized as emails for closer inspection.

10.2.3   Interface Metaphors to Email Clients

MailSleuth is then integrated into the user's existing mail system. The user is therefore required to create/adjust the attribute hierarchy to suit their needs and to take advantage of the software. This process is bootstrapped by creating folders with meaningful names and assigning queries or nesting more folders within the created folders. For instance the creation of a "Conferences" Drill-Down Folder, that contains a "AAAI" Virtual Folder with a query "**Email-Body** aaai," and another Virtual Folder "ICFCA" with the query "**Email-Body** icfca." In order to ensure that the "Conferences" label is more correctly represented at higher levels the user attaches the query "**Email-Body** conference." A lattice view of the conference information space can now be seen via the "Conferences" Drill-Down Folder. At a high level in the attribute hierarchy the "Conferences" folder will represent the union of the three queries:

$$\text{\textbf{Email-Body} conference} \bigcup \text{\textbf{Email-Body} aaai} \bigcup \text{\textbf{Email-Body} icfca}$$

These labels can be duplicated within the attribute hierarchy (provided they do not violate the partial order) to combine facets of the information space into new structures. These structures will be kept consistent with the mail collection as new mail enters, old mail is archived, or unwanted mail is deleted.

10.2.4   Usability Evaluation for Iterative Design

MailSleuth was planned as commercial software and in order to forecast its acceptance the software underwent usability testing at the Australian Software Testing Centre (ATC). Testing was split into two components; a comparative functionality review and user-based evaluation sessions. While these outcomes are reported elsewhere [Eklund et al., 2004] they are reiterated in this chapter because they provide important design hints for using line diagrams in practical Formal Concept Analysis-based software systems.

10.2.5   Comparative Functionality and Pre-Defined Scales

The May 2003 version of **MailSleuth** had no preintialized virtual folders when first installed. In FCA terms, **MailSleuth** had no predefined conceptual attribute hierarchy (or conceptual scales). ATC recommended a number of predefined Virtual Folders such as "This Week," and "Attachment" folders for email attachments of different document types and sizes. These form the basis of the Folder List shown to the left of Figure 10.7. This recommendation was followed and in subsequent versions pre-defined Virtual Folders were added including the folders mentioned (various popular document and image attachment types and sizes) and also a "follow-up" folder that tests the Outlook follow-up flag. Predefined virtual folders serve as examples and a useful starting point from which users can extend the Virtual Folder structure (attribute hierarchy) while benefiting immediately from the software. Other comparable products derive Virtual Folders from reading the mailbox but the structure (once built) cannot be modified or extended as with **MailSleuth**. This advantage is highlighted by including an extensible predefined folder structure when the **MailSleuth** program is first installed.

10.2.6 User-Based Evaluation and Iterative Feedback

User-based evaluation of **MailSleuth** involved one-on-one interviews and was intended to evaluate the ease of use and expectations of the user community. Included in the evaluation group were a Librarian, an Insurance Manager, a Financial Analyst, a Recruitment Manager, an Imaging Specialist, and a Personal Assistant. Each session lasted at most 90 minutes and was directed by a usability analyst who observed tasks and recorded relevant data. Each session was analyzed to identify any usability issues and quantitative measures compiled.

Participants were able to learn the basic operations associated with **MailSleuth** and complete the small number of predefined tasks. With a simple orientation script (in the place of a help system, incomplete at the time of testing), participants could quickly learn to use the software. For example, once introduced to the concepts of Virtual Folders and how they are associated with a query (or queries), participants were able to use the application to create their own folders and populate them with appropriate queries. Users indicated they found the interface reasonably intuitive and easy to use.

> *"An encouraging finding was that participants were able to read the lattice diagrams without prompting. Subject six even used the word lattice without it having been mentioned to her. Participants correctly interpreted the major elements – for example, how the "envelope" icons related to the mail folders and how derived vertices represented the intersection of two folders" – ATC Final Report, Usability Analysis, September 2003.*

Based on the usability testing there were a number of design improvements that could be made in order to present the lattice more clearly:

- *The start and end nodes could be removed from the legend, and blue and red arrows added.*

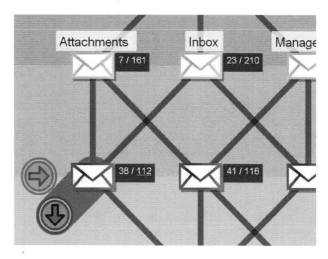

FIGURE 10.3: The left horizontal arrow (colored red) realizes the contingent of the concept) and the left down arrow (colored blue) realizes the extent. These are displayed when the cursor hovers over the envelope icon. The extent and contingent sizes are displayed next to the envelope as a fraction (contingent/extent) and the corresponding numbers are underlined when the user interacts with the arrows.

The introduction of contingent and extent arrows into the lattice diagram of Figure 10.3 highlights the interactive nature of the lattice diagram as a tool for querying emails. This compensates for the fact that only named folders can be accessed via the Folder List. The arrows are clues from the line diagram that the extent and contingent are available and that Derived Folders can be created by manipulating the diagram.

- *For more complicated structures, less emphasis could be placed on regions that are "unmatched." This would reduce visual clutter and further highlight the relationships that do exist.*

This comment refers to the small concept icons that are used to represent unrealized concepts. To solve this, concept representations at unrealized vertices in the line diagram were eliminated. The introduction of a reduced line-diagram was also included as a rendering option.

- *The format for representing total and dispersed emails associated with each folder could be more clearly represented, some users indicated that the present format (using brackets) represented total and "unread" e-mails. A reference to the format could be included in the legend.*

Tying together the textual representation of extent and contingent to arrows and the total (extent) and dispersed (contingent) sizes is achieved by representing them as a fraction (as shown in Figure 10.3).

- *The initial/default node view could be improved – when elements are close their labels can overlap. An interesting finding was that some users found more complicated diagrammatic representations better conveyed the relationships to the left-hand folder list.*

The ability to adjust the highlights and font sizes for diagram labels was included (along with the ability to color the layered shading). The observation that more complex line diagrams more strongly linked the line diagram to the Folder List is because larger line diagrams contain more labels appearing in the Folder List.

Finally, user responses give encouraging indications of an implicit understanding of information visualization using line diagrams. When shown a very large line diagram the librarian found it overwhelming but was certain that there was "value in a lattice of the information space." More specifically, one user said that she preferred a reduced line diagram, namely she saw "no reason that points without corresponding data should be drawn at all."

The term 'Virtual Folder' was also used by another when asked the same question "Drilling down through virtual folders to locate specific emails etc.," this indicates a familiarity with the idea of a 'Virtual Folder', either pre-existing or learned during the 30-40 minutes using the program. Further, the use of the term 'drilling down' in the appropriate context suggests (through correct terminology) comfort among the user group.

10.2.7 Design Improvements to Help Interpret Line Diagrams

The highlighting of adjoining lines is meant to illustrate relationships within the lattice and this could be clearer. There is a hierarchy within the lattice, which could be reinforced through the use of arrows on connecting lines that appear upon rollover — ATC Functional Testing Report, May 2003

A line diagram is a specialized Hasse diagram with several notational extensions. Line diagrams contain vertices and edges with the vertices often labeled dually with the intent (above) and extent (below). Rather than labeling each node in the line diagram with its intent and extent, a reduced labeling scheme is used and each object (and attribute) appears only once. In Toscana-systems [Eklund et al., 2000], a listing of the extent is often replaced with a number representing the cardinality of the extent (and/or the contingent).

In Hasse diagrams, edges (representing the cover relation) are unlabeled. It is well understood in Mathematics that a partially ordered set is transitive, reflexive and anti-symmetric. To simplify the drawing of an ordered set (via its cover relation) the reflexive and transitive edges are removed, and the directional arrows of the relation are dropped. It is therefore understood by

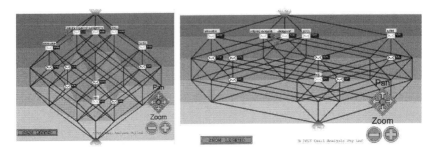

FIGURE 10.4: The interaction of the line diagram layout algorithm and background layer shading can produce an odd effect (left), but with human intervention layer shading can also be used as a guide to adjust the line diagram by hand (right) by moving vertices to align the layers.

convention that the Hasse diagram is hierarchical with the edges as directional relations pointing upward. In other words, if $x < y$ in the partially ordered set then x appears at a clearly lower point than y in the diagram.

The Australian Software Testing Centre suggested arrowheads be used in the line diagram to reinforce its hierarchical character. This is an unacceptable violation of a convention that some other mechanism to reinforce hierarchy in the line diagram had to be found. The idea of a line diagram with shaded layers was conceived: the principle is iterative darkening, dark at the top and light at the bottom; shading progressively lighter from top to bottom. This is shown in Figure 10.5 and Figure 10.4. The top and bottom elements of the lattice have also been replaced with special icons indicating "All Mail" and "No Mail" (when the bottom element is the empty set of objects). In combination, layer shading on iconic shapes highlights a top-to-bottom reading of the line diagram.

Shading does not interfere with the conventions of drawing line diagrams because it is a backdrop to the line diagram. The layers can also be turned off for purists or if the line diagram is to be embedded in a printed document. However, the interaction of the layout algorithm and background layer shading fails (background layers are not aligned) in line diagrams with high dimensionality as shown in Figure 10.4 (left) requiring human intervention to produce something readable (in Figure 10.4 (right)). It is possible to use the alignment of the background layers to guide the manual layout process. Nonetheless, once layer shading was used, it was apparent from test subjects that they were (without prompting) able to read and explain the line diagram from top-to-bottom.

> ... *most nodes in the lattice are depicted using the exact same icon, even though there are a variety of nodes. In particular, the root node, which represents the set of all emails, should be differentiated from all other nodes.* — ATC Report, May 2003

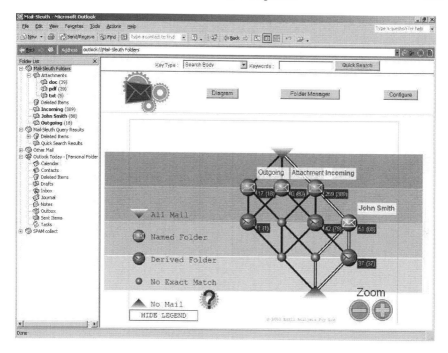

FIGURE 10.5: A line diagram from the August 2003 version of **Mail-Sleuth**. Layer shading is used to suggest a hierarchical reading. Top and bottom elements have been iconified. Unrealized concepts are differentiated. Realized concepts are split into two iconic categories "Named Folders" with an intent label with a white envelope and "Derived Folders," whose intent needs to be *derived* as an orange envelope. Cardinality labels have been replaced with dual labels as "*extent (contingent)*." This version includes a Quick Search bar at the top which provides an entry point for keyword-based search.

In **MailSleuth** the top of the lattice represents all emails in the collection. Some of the vertices shown in the line diagram correspond with actual Virtual Folders that exist in the Folder List to the left, while other vertices represent derivations of the named Virtual Folders. It is useful to indicate, through different icon types, which vertices are named Virtual Folders (appearing in the Folder List) and which are derived. This led to the idea of a "Derived Folder," a type of Virtual Folder that does not appear in the Folder List and whose existence is *derived* from the named Virtual Folders (attribute names) above it in the line diagram.

When drawing "reduced line diagrams," unrealized concepts are excluded but automatic layout can be problematic with this type of diagram. As **Mail-Sleuth** is designed for the non-expert, it is important that the lattice diagram always be as *readable* as possible and therefore reduced-line diagrams are an optional feature. Where elements of a scale are unrealized, the entire label is excluded from the diagram. However, what remains is drawn as a distributive lattice. These issues are covered in detail in [Cole et al., 2006].

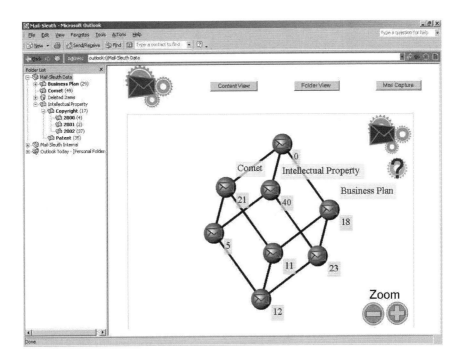

FIGURE 10.6:   A line diagram from the May 2003 version of **MailSleuth**. Most of the usual FCA line diagram labeling conventions are followed with the exception of iconifying vertices with an envelope. There is no obvious search point and limited visual highlighting in the diagram itself. Structural diagrammatic constraints are imposed; for instance, concepts cannot be moved above their superconcepts.

In Figure 10.6, unrealized concepts are the same shape and size as realized concepts, the only difference being the presence or otherwise of an envelope icon on the vertex. To distinguish unrealized from realized concepts, they were reduced in size as shown in Figure 10.5. Top and bottom concepts (when the bottom was an empty set of objects) were also iconified. In addition, realized concepts are identified in two ways. The first is where the intent label matches a "Named Folder" in the Folder List of Outlook (to the left of Figure 10.7). The second is where vertices represent the intent labels of the upper covers; these may have common attribute names ("Named Folder" labels) and are

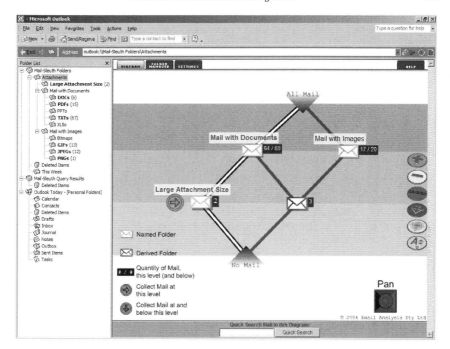

FIGURE 10.7: Screenshot of the final version of Mail-Sleuth in 2003. The line diagram is highly stylized and interactive. Folders "lift" from the view surface and visual clues (red and blue arrows) suggest the queries that can be performed on vertices. Layer colors and other visual features are configurable. Unrealized concepts are not drawn and *derived* Virtual Folders are differentiated from Named Virtual Folders. A high level of integration with the Folder List to the left and the Folder Manager is intended to promote a single-user iterative Formal Concept Analysis task flow using small lattice diagrams. Buttons in Figure 10.6 have been replaced with tabs and the help system is more consistently located to the top right. The Quick Search bar is visually highlighted and placed toward the bottom of the screen.

colored orange. To avoid cluttering the diagram with labels on all vertices, the interface gives scope to query an orange envelope and the result is a new Virtual Folder named after the intent labels of its upper covers appearing in the "**MailSleuth** search results" in the Folder List.

Because we are dealing with concepts that are sets of emails, it is natural to replace stylized vertices (a legacy of the Hasse diagram) with a literal iconic representation relevant to email. In the case where "Derived Folders" are unrealized, no vertex is drawn. Where data is present, an envelope replaces the envelope/ball icon combination as shown in Figure 10.7. Top and (empty) bottom vertices appear at most once in a line diagram and so are removed from the legend (shown in the legend of Figure 10.5 but not in Figure 10.7) and labeled accordingly in the diagram itself (shown in Figure 10.7). The ability to manipulate the line diagram in four directions via the "Pan" widget appears in Figure 10.7, and the envelopes animate by "lifting" on rollover with drop shadowing and zoom. This helps suggest that vertices in the line diagram can be moved and therefore manually adjusted by the user.

Edge highlighting has been used to emphasize relationships in line diagrams in both **ToscanaJ** and in CEM. This idea is mainly used as a method to orient the current vertex in the overall line diagram so that relationships can be identified. **ToscanaJ** allows the edges of the line diagram to be labeled with the ratio of object counts to approximate the idea of "support" in association rules. **ToscanaJ** also uses the size of the extent to determine the color of a vertex. A number of other significant functions for listing, averaging, and visualizing the extent at a vertex are also provided in **ToscanaJ**.

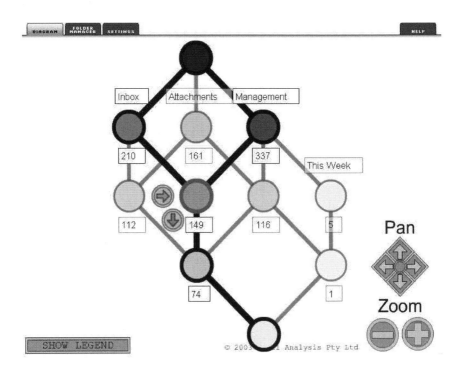

FIGURE 10.8: **MailSleuth** tries to accommodate a new user community to document browsing using Formal Concept Analysis. **HierMail** (shown above) has a stronger conformity to diagrammatic traditions in FCA. It is effectively a version of **MailSleuth** with a **ToscanaJ**-like line diagram.

Creating a new audience for Formal Concept Analysis with software such as **MailSleuth** is an interesting exercise but the original user community also requires support. **HierMail** is a version of **MailSleuth** for the FCA community that conforms to the diagrammatic conventions of **ToscanaJ**. It took only a matter of days to rollback the lessons learned from over four months of usability testing and design refinement with **MailSleuth** to produce **HierMail** as shown in Figure 10.8.

10.2.8   MailSleuth's Evolution

MailSleuth underwent considerable design change mostly in the way the lattice diagrams were presented. Initially, visualization was derived from the pen-and-paper drawing styles from Formal Concept Analysis's mathematical heritage. Deviation from this style removed circles as concepts replacing them with more descriptive icons. Top and bottom concepts were replaced with iconic arrows emphasizing that the relative vertical position of the concepts has meaning. Vertical position is further emphasized with use of layer shading. Highlighting is further used to show the connection between comparable concepts through emboldened labels and edges.

Throughout testing users had no difficulty using and understanding the attribute hierarchy as represented in a folder structure. Using this important part of the system was implicit. **MailSleuth** provided validation of Formal Concept Analysis applicability to wider audiences than researchers. The interpretation of line diagrams was not instantaneous, even with visual clues, but participants found the representation useful once understood.

Following **MailSleuth**, a Web-based version, called *MailStrainer*, was reported and tested in [Domingo and Eklund, 2005]. *MailStrainer* is much the same in its representation of line diagrams as **MailSleuth**, but allowed more freedom with respect to lattice display; lattices of significant complexity could be constructed and drawn. Another substantially larger usability experiment on student subjects concluded that the prospects for novice Formal Concept Analysis users to read and interpret line diagrams using **MailSleuth** and *MailStrainer* showed promise but was not conclusively positive.

This important conclusion leads to a new line of usability evaluation and software development where different mechanisms for navigating the information spaces other than line diagrams are used. This gives rise to the **ImageSleuth** which is presented and discussed in the remainder of this chapter.

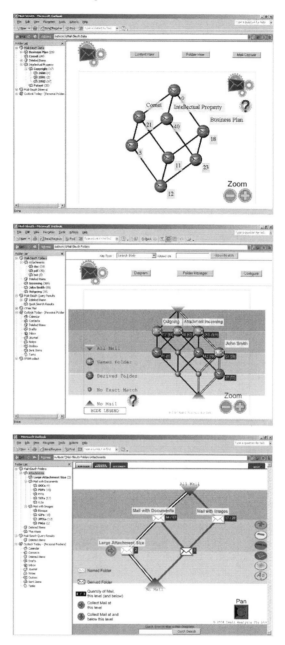

FIGURE 10.9:   Comparison of three versions of **MailSleuth** starting with May 2003 (before initial interface analysis), followed by August 2003 (version used in user testing) and the final, December 2003, design.

10.3   Navigating Images Using FCA: ImageSleuth

10.3.1   The ImageSleuth Image Domain and Test Set

Image collections exist in many forms; from informal collections of digital photographs to formal museum archives and photographic evidence. The established method for browsing collections of images is to display them as thumbnails. Within a collection of thumbnails, each thumbnail usually has the same aspect ratio and is displayed in a two-dimensional layout, sorted by a physical feature of the image's file (e.g., filename, date, filesize, etc.).

The desired outcome of **ImageSleuth** is to design a navigation of thumbnails to best convey the content of an image, and its association with other images, with the advantages of FCA-based document navigation. This requires that the concept lattice representation have a different presentation and navigation paradigm compared to text documents such as emails.

In order to test the usability of **ImageSleuth**, a sample dataset was needed. This was created from the popular computer game *The Sims 2* and features 412 objects of household furniture and fittings, described by 120 attributes that include in-game properties, suggestions for use and automatically extracted color information.

There are 7,516 concepts in the lattice derived from such a context. Each attribute of the context is assigned to one or more "perspectives" (similar to conceptual scales and further described in Section 10.3.2). In this dataset, ten perspectives have been constructed and these are another name for conceptual scales in Toscana-systems [Eklund et al., 2000].

10.3.2   Interface Design and Objectives

Following the approach used for browsing of emails in **MailSleuth**, **ImageSleuth** computes sub-lattices from the full context representing the complete collection of images as objects and their annotated features as attributes. These attributes may be information about the object annotated by human hand or automatically extracted information about shape and color.

In **ImageSleuth** no line diagram of the concept lattice is displayed. Instead, following the usability design of Kim and Compton [Kim and Compton, 2001b,a], the user is always located within the information space at one concept of the concept lattice. This allows thumbnails of the images to be shown as the extent of the present concept and thus to convey most of the graphical information characterizing the concept. The intent is represented as a list of attributes that can be added or removed from the current concept.

As no line diagram of the lattice is shown, lists of upper and lower neighbors are the only representation of the lattice structure surrounding the present concept. Searching and browsing the image collection corresponds to moving from concept-to-concept in the lattice. By including new attributes in the intent, the user moves to a smaller concept where all images in the extent have these features. **ImageSleuth** offers the following possibilities to navigate in the concept lattice:

- Restriction or elaboration of the set of attributes under consideration

- Move to upper/lower neighbors

- Search by attribute(s)

- Search by object(s) (Query-by-example)

- Search for similar concepts (proximity search)

Restricting the set of attributes under consideration allows focus on the features that are relevant for the current navigation needs of the user. Attributes irrelevant to the user's need simply increase the number of concepts; making navigation unnecessarily complex.

ImageSleuth offers predefined sets of attribute combinations (called *perspectives*) covering different aspects of the images. The user may combine these perspectives and include or remove perspectives at any point. Scale attributes are natural candidates for such attribute sets but other sets are allowed (for example, overlapping perspectives and perspectives that are subsets of other perspectives).

10.3.3   Semi-Concepts and Similarity

The option to search for similar concepts requires the use of similarity measures. In order to use similarity together with the normal search or query-by-example, where the user may describe the search by specifying an identical attribute set to a given object (or by identifying an object set from which the attribute set is derived), we want the similarity measure to be defined for semi-concepts as introduced in [Luksch and Wille, 1991] as a generalization of concepts.

A *semiconcept* of a context $\mathbb{K} := (G, M, I)$ is a pair (A, B) consisting of a set of objects $A \subseteq G$ and a set of attributes $B \subseteq M$ such that $A = B'$ or $B = A'$. The set of all semiconcepts of $\mathbb{K}$ is denoted by $\mathfrak{H}(\mathbb{K})$.

Note that every concept is a semi-concept. The underlying structure of **ImageSleuth** is thus:

1. A context $\mathbb{K} := (G, M, I)$ with a collection of images as object set G, possible features as attribute set M and an incidence relation I assigning features to objects.

2. A collection $\mathcal{P}$ of subsets of M called perspectives. Every subset $\mathcal{A} \subseteq \mathcal{P}$ defines a sub-context $\mathbb{K}_{\mathcal{A}} := (G, \bigcup \mathcal{A}, I_{\mathcal{A}})$ with $I_{\mathcal{A}} := I \cap (G \times \bigcup \mathcal{A})$ of $\mathbb{K}$.

3. A similarity measure

$$s : \bigcup_{\mathcal{A} \subseteq \mathcal{P}} \mathfrak{H}(\mathbb{K}_{\mathcal{A}})^2 \to [0,1]$$

assigning to every pair of semi-concepts of a sub-context $\mathbb{K}_{\mathcal{A}}$ a value between 0 and 1 that indicates the degree of similarity.

Since for every $\mathcal{A} \subseteq \mathcal{P}$ the contexts $\mathbb{K}_{\mathcal{A}}$ and $\mathbb{K}$ have the same object set and every attribute of $\mathbb{K}_{\mathcal{A}}$ is an attribute of $\mathbb{K}$, it follows for every $m \in \bigcup \mathcal{A}$ that $m^I = m^{I_{\mathcal{A}}}$. Since for $(A, B) \in \mathfrak{B}(\mathbb{K}_{\mathcal{A}})$ we have

$$A = B^{I_{\mathcal{A}}} = \bigcap \{ m^{I_{\mathcal{A}}} \mid m \in B \} = \bigcap \{ m^I \mid m \in B \}$$

it follows that A is the extent of a concept of $\mathfrak{B}(\mathbb{K})$. Therefore, $\phi(A, B) := (A, A^I)$ defines a map $\phi : \mathfrak{B}(\mathbb{K}_{\mathcal{A}}) \to \mathfrak{B}(\mathbb{K})$ and the image of ϕ is a $\wedge$-subsemi-lattice of $\underline{\mathfrak{B}}(\mathbb{K})$. In the following, the different navigation means based on this structure are described.

10.3.4 Restriction of the Attribute Set

By including different perspectives the user defines a sub-context of $\mathbb{K}$ in which all operations are performed. The user may change this sub-context while browsing, thus obtaining at the present concept further information and search options. If at the concept $(A, A^{I_{\mathcal{A}}})$ the perspective $S \in \mathcal{P}$ is included (i.e., the set of attributes in consideration is increased), then **ImageSleuth** moves to the concept $(A^{I_{\mathcal{A} \cup \{S\}} I_{\mathcal{A} \cup \{S\}}}, A^{I_{\mathcal{A} \cup \{S\}}})$ of $\mathfrak{B}(\mathbb{K}_{\mathcal{A} \cup \{S\}})$. Since for $\mathcal{A} \subseteq \mathcal{P}$ and $S \in \mathcal{P}$ the extent of every concept of $\mathbb{K}_{\mathcal{A}}$ is an extent of $\mathbb{K}_{\mathcal{A} \cup \{S\}}$, we have $A = A^{I_{\mathcal{A} \cup \{S\}} I_{\mathcal{A} \cup \{S\}}}$ and the set of images shown does not need to be updated when a further perspective is included. This allows the addition of perspectives during the search without losing information. A similar strategy is known from Toscana (cp. [Becker et al., 2002]) where the user moves through different scales. At every point the user may also remove a perspective S, which takes them to the concept $(A^{I_{\mathcal{A} \setminus \{S\}}}, A^{I_{\mathcal{A} \setminus \{S\}} I_{\mathcal{A} \setminus \{S\}}})$. If in this way an attribute of $A^{I_{\mathcal{A}}}$ is removed from the current sub-context then the extent may be increased since $A^{I_{\mathcal{A}}} \subseteq A^{I_{\mathcal{A} \setminus \{S\}}}$.

10.3.5 Moving to Upper and Lower Neighbors

ImageSleuth uses most of its interface to show thumbnails of images in the extent of the chosen concept (see Figure 10.10 (top)). The lattice structure around the current concept is represented through the list of upper and lower neighbors that allow the user to move to super- or sub-concepts. For every upper neighbor (C, D) of the current concept (A, B) the user is offered to remove the set $B \setminus D$ of attributes from the current intent. Dually, for every lower neighbor (E, F) the user may include the set $F \setminus B$ of attributes that

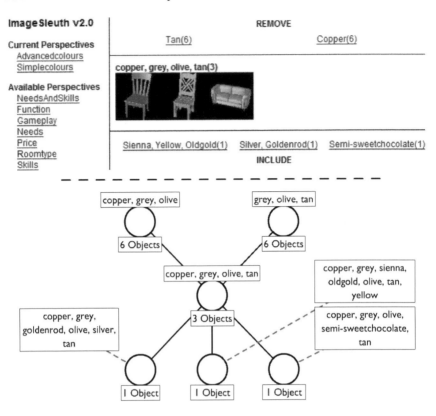

FIGURE 10.10: An example screenshot of **ImageSleuth** and the lattice representation of the corresponding neighborhood. The screenshot shows the four primary navigation functions of **ImageSleuth**. On the left is the listings of current and available perspectives (currently, advanced and simple color perspectives are selected). Top and bottom show the remove and include functions, respectively. The central pane shows the current concept; with intent listed as textual attributes and extent as thumbnailed images. The lattice neighborhood shows the current concept at its center.

takes them to this lower neighbor. By offering the sets $B \setminus D$ and $F \setminus B$ dependencies between these attributes are shown. Moving to the next concept not having a chosen attribute in its intent may imply the removal of a whole set of attributes. In order to ensure that the extent of the given concept is never empty, it is not possible to move to the bottommost concept if its extent is empty.

10.3.6   Search and Query-By-Example

Browsing is achieved by moving to neighboring concepts. In many cases the user will want to go directly to images having a certain set of attributes $B \subseteq \bigcup \mathcal{A}$. This is offered by the search function that computes, for the selected attributes; the concept $(B^{I_{\mathcal{A}}}, B^{I_{\mathcal{A}} I_{\mathcal{A}}})$. Its extent is the set of all images having these attributes, its intent contains all attributes implied by B.

Another type of search is performed by the query-by-example function. Instead of defining a set of attributes, a set of objects A is defined as the sample set. The query-by-example function then computes the common attributes of these images (in the selected sub-context) and returns all other images having these attributes by moving to $(A^{I_{\mathcal{A}} I_{\mathcal{A}}}, A^{I_{\mathcal{A}}})$. In this way, query-by-example is the dual of the search function. While the search for images having certain attributes is not affected by the removal or addition of perspectives to the sub-context, query-by-example depends strongly on the selected sub-context. The more attributes taken into consideration, the smaller the set of images that have exactly the same attributes as the examples.

10.3.7   Concept Similarity

The aim of query-by-example is to find objects that are similar to the objects in a given sample set. This is a narrow understanding of similarity implying equivalence in the considered sub-context; for the query-by-example function two objects g, h are "similar" in a sub-context $\mathbb{K}_{\mathcal{A}}$ if $g^{I_{\mathcal{A}}} = h^{I_{\mathcal{A}}}$. If the objects are uniquely described by the attributes in the chosen sub-context then query-by-example seldom yields new information.

A more general approach is to define a similarity measure for pairs of concepts. In [Lengnink, 2001] several similarity measures on attribute sets are investigated. Similarity of two objects g and h is then described as the similarity of the attribute sets g' and h'. In order to use the grouping of objects provided by the formal concepts, **ImageSleuth** works with a similarity measure on semi-concepts that allows the return of a ranked list of similar concepts. We use semiconcepts since the set of sample images chosen by the user is not necessarily the extent of a concept. The similarity measure is derived from the following metric:

> On the set $\mathfrak{H}(\mathbb{K})$ of semiconcepts of a context $\mathbb{K} := (G, M, I)$ the metric $d : \mathfrak{H}(\mathbb{K}) \times \mathfrak{H}(\mathbb{K}) \to [0, 1]$ is defined as
>
> $$d((A, B), (C, D)) := \frac{1}{2} \left(\frac{|A \setminus C| + |C \setminus A|}{|G|} + \frac{|B \setminus D| + |D \setminus B|}{|M|} \right).$$

This definition formalizes the idea that two semi-concepts are close if there are few objects and attributes belonging to only one of them. In order to compare the number of objects and the number of attributes where the two objects differ, these numbers are set in relation to the total number of objects or attributes. Semi-concepts with small distance are considered similar. **ImageSleuth** uses $1 - d((A, B), (C, D))$ for scoring purposes – so that smaller distances equate to larger scores.

For a similar purpose, a related similarity measure was introduced in [Saquer and Deogun, 2001] as

$$s((A, B), (C, D)) := \frac{1}{2} \left(\frac{|A \cap C|}{|A \cup C|} + \frac{|B \cap D|}{|B \cup D|} \right).$$

This definition of extends to semi-concepts (A, B), (C, D) if $A \cup C \neq \emptyset$ and $B \cup D \neq \emptyset$. In particular, the similarity $s((A, A'), (C, D)))$ is defined for every nonempty set A of objects and every concept $(C, D) \neq (G, \emptyset)$. For a sample set A of images, **ImageSleuth** uses a combination of both measures to return a ranked list of concepts similar to the semi-concept (A, A^{I_A}). An example of the system output is shown in Figure 10.13.

The metric on semi-concepts has two advantages. First, it allows the return of a list of similar concepts rather than just a list of images. This provides a reasonable grouping of the similar images and, since the attributes of the concepts are displayed, it shows the way the images relate to the sample set.

Second, in contrast to other approaches such as distance in the lattice, the number of different objects of two concepts is taken into account. Instead of counting only the attributes in which two concept intents differ, we assume that the significance of this difference is reflected in the difference of their corresponding attribute sets. If (A, B) is a concept and (C, D), (E, F) are upper neighbors of (A, B) with $|C| \leq |E|$, then the attributes in $B \setminus F$ are considered as more characteristic for the concept (A, B) than the attributes in $B \setminus D$. Thus, if $|D| = |F|$ then (C, D) is closer to (A, B) than (E, F) even though they differ from (A, B) in the same number of attributes. In this way, even an incomparable concept may be closest. This contradicts the intuition that, for a concept, its sub- and super-concepts should be closest but upper and lower neighbors are directly accessible by other navigation means. The advantage of the search for similar concepts for a given concept is that it offers a selection of (in the lattice order) incomparable but close concepts that are otherwise invisible.

As the original query-by-example function described above is the dual of a search, this approach can be used for the search function, too. If a search is carried out for a set of attributes B, and if B' is empty, then the concept (B', B'') contains only the information that these attributes do not occur together. No images are returned as a result of this search, since there are no images having the required attributes. In this case, the user may be shown a list of concepts similar to the semi-concept (B', B).

The most common solution to concept searches in FCA, that result in an empty extent, is to offer attributes that can be removed from the search to supply a more general answer that meets a majority of search attributes. Most other forms of search (for example, text search) do not work this way. Instead they supply the user with a list of results that are ranked by a relevance to the query. **ImageSleuth** addresses this using the semi-concept search result and a combination of distance and similarity measures (see Section 10.3.7). When a search is performed that would return the concept with an empty extent, the user can opt to allow the system to find and rank conceptually relevant concepts. This process is achieved by finding possible neighbors of the semi-concept and performing a bounded traversal that ranks the traversed concepts. These possible neighbors become the first concepts traversed. Each concept visited has its relevance calculated and stored. A test is applied to each concept visited to calculate whether it is to be used for further traversal. The test condition is based on the distance metric compared to a weighted average of the query concepts intent and extent size. The condition is represented as:

$$Dist((A, B), (C, D)) \times SearchWidth < \tfrac{1}{2}(|A|/|G| + |B|/|M|)$$

where (A, B) is the query concept and (C, D) is the current concept of the traversal. $SearchWidth$ is a modifier to allow the search to be made wider or narrower. If the traversal is to continue, the concept's neighborhood is added to the traversal list, the concept is marked as visited and the process continues. This algorithm is shown as psuedocode in Figure 10.11.

Relevance is calculated as the average of the similarity scores that is presented to the user as a percentage.

The following shows a simple example of **ImageSleuth's** semi-concept searching within the example collection. This example uses two perspectives, *Function* and *RoomType*, which have 20 attributes in total. The *Function* perspective is a simple nominal scale with each object having one *function* attribute. The *RoomType* perspective, on the other hand, is more complex with each object having zero or more *room type* attributes. With this context the complete lattice has 194 concepts.

The query for this example will be *Appliances, Electronics, Study*, the first two attributes from the *Function* perspective and the remaining one from *RoomType*. *Function* being nominally scaled, the inclusion of two attributes from this perspective means that if the concept was completed it would result in the empty extent concept or $(\emptyset, M)$. Although this result is technically correct, it does not suit the query's intention.

To identify a concept that is more representative, a concept traversal is started using the semiconcept, $(\emptyset, (Appliances, Electronics, Study))$. In this example, the traversal visits 12 concepts, four of which are conceptually close enough to extend the traversal. Consequently, only 6.19% of the total lattice is computed. The first three of five rankings are shown in Figure 10.13. Relevance is shown as a large percentage, while individual distance and similarity scores are displayed below. Each result is displayed as a list of attributes representing the intent and a collection of thumbnails representing the extent. The highest ranking concept, with relevance 64.92%, has the intent (*Electronics, Study*), which is two of the three original query attributes. Following

Search Traversal

Input: Concept (X, Y), Number $Width$

Output: Ranked List of Concept Set (not shown)

```
1.      RankedConceptSets := ∅
2.      Candidates := UN((X,Y)) ∪ LN((X,Y))
3.      Visited := ((X,Y))
4.      while |Candidates| > 0
5.        Concept := Candidates[0]
6.        Visited := Visited ∪ Concept
..
..        compute and store rank information for Concept.
..
7.        if Dist((X,Y),Concept) × Width < ½(|X|/|G| + |Y|/|M|)
8.          Candidates := Candidates ∪ upperNeigh(Concept)
9.          Candidates := Candidates ∪ lowerNeigh(Concept)
10.         Candidates := Candidates/Visited
11.       end if
12.     end while
```

FIGURE 10.11: Pseudocode representation of search traversal. Parameters are the starting concept or semiconcept and a numeric value used to modify the width of the search. Output is a ranked list of concept sets – not shown in this example for brevity.

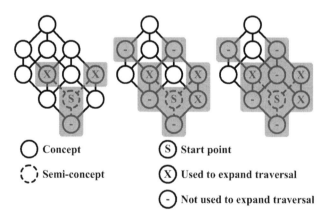

FIGURE 10.12: An example of lattice traversal starting from a semi-concept. The traversal in this example is complete in 3 steps. The shaded area shows the computed concepts at each step.

64.92%

Distance: 0.965189 Similarity: 0.333333

Electronics, Study(7)

55.74%

Distance: 0.914985 Similarity: 0.2

Bedroom, Electronics, LivingRoom, Study(5)

54.42%

Distance: 0.921883 Similarity: 0.166667

Appliances(21)

Distance: 0.921883 Similarity: 0.166667

Electronics(21)

FIGURE 10.13: Results of a concept traversal from the query *Appliances, Electronics, Study* using the perspectives *Function, RoomType*.

that, at 55.74%, is the concept with the intent (*Bedroom, Electronics, LivingRoom, Study*). The third ranking, at 54.42% relevance, has two concepts, with the intents (*Appliances*) and (*Electronics*), which represent the mutually exclusive elements of the original query.

The version of **ImageSleuth** presented here is the second version. The original prototype used concept neighborhoods and include/remove attributes, but was limited to traversal between three mutually exclusive subcontexts via single objects. It underwent user-evaluation to test functionality and usability of the navigation paradigm. Results of that study are presented in [Ducrou and Eklund, 2007] and indicate that concept neighborhoods offered a useful navigation method.

10.4   Conclusion and Summary

The research presented in this chapter covers almost a decade of research and development. The idea of a cluster of documents as concepts based on the analysis of keywords as attributes was foundational to the field and presented an early and promising thread for the integration of Formal Concept Analysis with information retrieval [Godin et al., 1989, Carpineto and Romano, 1996]. However, the approach was not without its challenges: the complexity of computing the complete lattice when dealing with 10,000 objects and thousands attributes mean that the resulting line diagrams are formidable and of limited value as a visualization of the information space. The solution proposed in tools like CEM [Cole et al., 2003] is to reduce the complexity of the rendering of the line diagram via attribute scaling and object zooming. The resulting lattice diagrams are more manageable but the question of the readability of line diagrams for the uninitiated has never been systematically tested until **MailSleuth**. Experiments with different styles of line diagrams, different integration points, and software manipulators, proves the value of Formal Concept Analysis applied to document browsing in low-dimensional faceted data.

ImageSleuth takes a different interface design approach that reduces the view of concept lattice as the interaction paradigm to rely more heavily on a traditional keyword search approach to the discovery of formal concepts. It then emphasizes navigation from a concept to other formal concepts by the manipulation of attributes (as facets) in the conceptual neighborhood. When considered in this way, tools like **ImageSleuth** raise a different set of problems: the problem of moving to similar but incomparable formal concepts (moving across anti-chains of formal concepts in the lattice), the problem of query-by-example, and the problem associated with partial matches or formal concept similarity. Query-by-example is easily done by taking an object's attribute set, computing the formal concept and then looking at the formal concepts extent. Fortunately also, a practical approach to similarity measurement can be implemented from the theoretical literature on Formal Concept Analysis to provide a principled way of developing relevance orders (rankings) of formal concepts. **ImageSleuth** therefore demonstrates the complete integration of FCA ideas in a content-based retrieval application and contains all the analog elements of vector-space models for information retrieval.

10.5   Student Projects

10.5.1   Project 1

Section 10.2.1.1 describes the idea of *transitive closure* on an attribute hierarchy namely asserting "a is-a b" and "b is-a c" will automatically infer that "a is-a c." Characterize transitive closure mathematically and discuss the role of transitivity in a concept lattice. What other basic mathematical properties hold for concept lattices and discuss the difference between these ideas to those that hold for inheritance hierarchies?

10.5.2   Project 2

What is SMTP? Find and describe the schema for SMTP. What extensions to SMTP are implemented in Microsoft Outlook that are not part of the *de facto* SMTP standard?

10.5.3   Project 3

What is the difference between the term *extent* and *contingent* used in concept lattices? Give an example and develop a mathematical characterization of the terms.

10.5.4   Project 4

Regarding the design of a usability experiment, what types or task would you include in the usability test for an email client?

10.5.5   Project 5

Figure 10.9 shows the evolution of the line diagram used to display the concept lattice in the mail-sleuth project. Discuss step-by-step the changes that lead from the traditional line diagram of a concept lattice to the final stylized version of the line diagram shown in Figure 10.7. Can you suggest other graphical elements and features that would make reading line diagrams easier for novice users?

10.5.6   Project 6

Section 10.3.6 describes the idea of query-by-example in the context of image search. Research other popular techniques that implement query-by-example and discuss how these might be deployed to help solve the problem of image retrieval.

10.5.7   Project 7

In Section 10.3.7 the idea of concept similarity and distance is defined. Research other ways that vector space models of information retrieval implement distance and similarity. Start with text retrieval and consider how these ideas from text retrieval might be applied to distance and similarity when dealing with MPEG-7 descriptions of images?

Chapter 11

Optimizing Social Software System Design

Aldo de Moor

CommunitySense, Tilburg, The Netherlands

11.1   Introduction

Collaborative communities are important catalysts of research, economic, and social processes. In these communities, many stakeholders collaborate on joint goals, although often having partially conflicting interests. Examples are research communities, knowledge management teams, innovation platforms, and environmental campaign networks. These communities make use of an ever-widening range of social software tools, including many types of discussion fora, blogs, content management systems, and advanced knowledge analysis and processing tools. These tools increasingly come with Web 2.0 functionalities like tagging and reputation management systems.

Collaborative communities are complex and rapidly evolving socio-technical systems. The design of these systems includes the communal specification of communication and information requirements, as well as the selection, configuration, and linking of the software tools that best satisfy these requirements. Supporting the effective and efficient community-driven design of such complex and dynamic systems is not trivial.

To represent and reason about the system design specifications we use conceptual graph theory. We do so because the knowledge representation language of choice must be rich enough to allow the efficient expression of complex definitions. Also, since design specifications derive from complex real-world domains and community members themselves are actively involved in specification processes, a close mapping of knowledge definitions to natural language expressions and vice versa is useful. Finally, the representation language must be sufficiently formal and constrained for powerful knowledge operations to be constructed. Conceptual graph theory has all of these properties.

In this chapter, we explore how conceptual graphs can be used to:

1. model the core elements of such socio-technical systems and their design processes.

2. specify communication and information requirements and match these with social software functionalities.

We illustrate these design processes with examples from a realistic scenario and end with a number of suggestions for student projects to extend the ideas proposed in this chapter.

Section 11.2 summarizes our view of social software, which we describe as an evolving socio-technical system consisting of a *tool system* embedded in a *usage context*. Using these notions, in Section 11.3 we present a conceptual model of functionality matching in collaborative communities. Section 11.4 shows how the functionality matching process can be supported using these conceptual definitions. Section 11.5 contains a discussion and conclusion. Section 11.6 lists a set of student projects expanding the material introduced.

11.2   Social Software: From Tools to Systems

Collaborative communities, in which people work together to accomplish common goals, make use of a wide variety of social software tools to support their information processing, communication, and coordination needs. A typical configuration would include a content management system, some mailing lists, a bulletin board, and increasingly more sophisticated Web 2.0 tools like LinkedIn[1] for maintaining professional contact networks, Digg[2] for filtering relevant Web documents, and so on. In [de Moor, 2007], we presented a conceptual model of a tool system and its usage context as a way to describe such socio-technical systems. Here we briefly summarize the main elements.

We define a *tool system* as the set of integrated and customized information and communication tools tailored to the specific information, communication, and coordination requirements of a collaborative community. There are numerous, partially overlapping implementations of such functionalities.

---

[1]http://www.linkedin.com

[2]http://www.digg.org

In addition, many of these tools are built on top of an emerging cyberinfrastructure of organizational practices, technical infrastructure, and social norms [Edwards et al., 2007]. Furthermore, each community has its own, unique way of using these functionalities. Finally, the requirements and technologies used are in constant flux. In all, this makes it extremely complicated to come up with standardized prescriptions for the best tool system for a particular community at a particular moment in time. New forms of analysis,, roles in software development, and the meaning of use and maintenance need to evolve [Sawyer, 2001], which applies even more to the case of community information systems development.

To design useful information systems by selecting, linking, and configuring the right components, available tool system functionalities in the form of modules and services need to be evaluated in their context of use by the communities of use themselves. Effective use is key here, which can be defined as the capacity and opportunity to successfully integrate ICTs into the accomplishment of self or collaboratively identified goals [Gurstein, 2003]. Evaluating the usefulness of a functionality can be defined as the evaluation of the extent to which users can translate their intentions into effective actions to access the functionality [Gaines et al., 1997]. Such evaluation should ultimately contribute the purpose of the community of use [Preece, 2000]. Such an approach requires the continuous comparison of tool functionalities with usage context requirements, for which this chapter provides some building blocks.

We first present a hypothetical scenario of a knowledge-driven topic community on climate change, which will be used to illustrate the ideas introduced in this chapter. We then present a conceptual model of both the tool system and the usage context, together making up the socio-technical system of a collaborative community.

11.2.1 Scenario: Building a Knowledge-Driven Topic Community on Climate Change

Scientific consensus is growing rapidly that climate change is indeed happening and will have a serious societal impact. However, the causes and effects of this phenomenon are still ill-understood and need urgent and concerted analysis by many stakeholders from all over the world. Let us assume that the United Nations Environment Programme (UNEP) has been commissioned to—as quickly as possible—create a range of task groups to examine scientific consensus on the causes and effects of climate change and to make appropriate policy recommendations. The task groups are to be broad in scope, including scientific experts as well as opinion leaders from business, governments, NGOs, and the general public. Over time, these *ad hoc* task groups should grow into well-established, collaborative topic communities, able to rapidly evaluate highly complex and contradictory results and to efficiently inform policy makers and other stakeholders all over the world.

Jane, a senior UNEP official, is in charge of growing these topic communities. Having selected an initial topic community charged with examining causes and effects of ice cap melting,[3], she now faces the problem of defining the appropriate tool system to support their collaboration.

The first task of the community is to take stock of the relevant climate change entries in Wikipedia,[4] as this is a well-established repository of basic knowledge on the main issues, with contributions made by numerous (self-appointed) experts and stakeholders worldwide. As these authors represent a wide range of, often contradictory, opinions, and since Wikipedia has such a high visibility, the knowledge represented in the Wikipedia entries forms a good starting point for more in-depth analysis and dissemination by the topic community.

The Wikipedia network of hyperlinked entries forms an informal *concept network*. However, many of the nodes contain irrelevant or wrong information from the point of view of the topic community. Furthermore, other interesting links between items within Wikipedia and with information resources on the Internet at large could be conceived. Finally, this resource only provides the raw information for the topic community's goal: advising on policy and decision making. What is needed is a sensemaking, filtering, and knowledge recombination and extension *discourse* between various topic community members. We call these discussion foci *discourse topics*, which in this case include causes and effects of and policy recommendations for addressing ice cap melting. The discourse topics thus are orthogonal to the knowledge entries, one topic potentially linking to many entries and vice versa. Several domain and discussion roles are defined as well. Representatives of relevant societal stakeholders should be included in the topic community. Topics are to be discussed by discussants, facilitators should keep the complex discourse process on track, and summarizers should formulate consensus positions for dissemination to the general public. To support this discourse process, Jane opts for the Compendium tool.[5] This is a visual discourse-supporting hypermedia tool for sensemaking that provides a variety of mechanisms for analysts to tag and connect media assets in relation to issues, ideas, and arguments. Figure 11.1 outlines the socio-technical system of the topic community.

Having decided on these principal tools, Jane now faces the design problem of determining how the workflows should look that these tools are to support, which functionality components of the tools are to support what workflow steps, and how these functionality components should be integrated or at least aligned.

---

[3]In [de Moor and Anjewierden, 2007] we presented a socio-technical approach for the initial stage of this process: the selection of candidate members for these topic communities by using the tOKo tool for text analysis to mine the blogosphere.

[4]e.g., http://en.wikipedia.org/wiki/Climate_change

[5]http://www.CompendiumInstitute.org

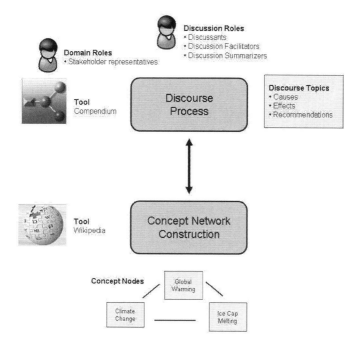

FIGURE 11.1: Outline of the topic community socio-technical system.

11.2.2   The Tool System

A functionality is a set of functions and their specified properties that satisfy stated or implied needs.[6] Functionalities come in different forms of aggregation, from simple word processing functions to complete modules and applications supporting particular workflows or business processes. When evaluating functionality, the right level of granularity should be chosen.

In [Christiaens and de Moor, 2006], we introduced a basic model of a *tool system*, consisting of components of different levels of granularity. At the lowest level of granularity, we distinguish *systems* of tools or services. The next level consists of the *tools* or services themselves. Then come the *modules* comprising the tools or services. Finally, we distinguish the particular *functions* grouped in a module.

Functionality components of different levels of granularity can be linked in many ways. The *interface* between two functionality components A and B is described by the externally visible representation of the knowledge structures and processes used in the interoperating components A and B. It is important only to model such high-level conceptual structures when designing the tool

---

[6]Definition by the Software Engineering Insitute Open Systems Glossary:
http://www.sei.cmu.edu/opensystems/glossary.html

system, to prevent unnecessary modeling and interpretation efforts in this initial stage. Only when the total system architecture has been agreed upon, should further modeling efforts be made.[7]

In [Christiaens and de Moor, 2006], we focused on the communication aspects of interoperating functionality components. We abstracted from the internal *information processes*. These are operations on information objects within a functionality component, such as the creation, modification, or deletion of an object, such as editing an HTML page. For tool selection, such processes may be important, however, so we include them in our conceptual model. Again, note that we are only interested in those information processes visible to a user, so we abstract from all internal computational processes.

Example:

Wikipedia's main unit of analysis is the *page*. Each page contains a textual description of the topic, and can contain *links* to other pages as well as *categories* indexing the content. Two functionality modules of particular interest are the *Search Pages* and *Edit Page*. Functions within the Edit Page module include *Edit Text*, *Edit Link*, and *Edit Category*.

Compendium has many different functionality modules. The one of particular interest to our purpose concerns *Create Node*, with functions like *Create Question*, *Create Answer*, *Create Argument*, *Create Decision*, *Create Reference*, etc.

11.2.3   The Usage Context

In [de Moor, 2007], we distinguished three important categories of ontological elements making up the usage context: goals, actors, and various domain factors. We summarize these elements here.

11.2.3.1   Goals

Goals or objectives are crucial in the pragmatic view. Everything starts with goals. Goals give a sense of purpose, drive people and processes, and can be used as evaluation criteria.

We distinguish two types of goals. First, there are *activities*, such as "writing an advisory report" or "making a climate change argumentation analysis."

---

[7]Note that in this chapter, we first discuss the tool system, then the usage context. Strictly speaking, one should first analyze the requirements determined by the context, and only then determine which functionalities match best. However, we express requirements in terms of mappings to functionalities. Furthermore, our approach is an evolutionary one, which means that legacy functionalities often predetermine the degrees of freedom in functionality selection. Socio-technical systems analysis is a continuous and delicate balancing act between requirements and functionalities, which in practice is much more a parallel than a serial process.

Such activities in fact are operationalized goals: processes with a concrete deliverable as an outcome, often in the form of an information object such as a report or a message. However, many goals are abstract and cut right across processes and structures. Examples of such goals are non-functional requirements and quality criteria, like "efficiency," "participation rate," and so on. We call such abstract goals *aspects*.

Although activities can be viewed as workflows, we abstract from their design and implementation details, such as concurrency and sequence. Although definitely important in the final construction of the information system, for the purpose of functionality component selection they add unnecessary complexity. This initial stage, the focus of this chapter, concerns itself with the selection of *potential* functionalities, not their actual configurations for workflow support.

Example:

In the scenario, the following *activities* are distinguished:

1. Select relevant concepts and their entries in the knowledge system

2. For each discourse topic, conduct a discussion among relevant stakeholders, resulting in a consensus position.

3. Disseminate the consensus position to the general public.

One quality *aspect* which should be ensured by the tool system is *legitimacy*. This is defined as that all actors should be involved in those and only those activities that are relevant to their discussion or domain roles.

11.2.3.2 Actors

Many stakeholders are involved in tool system design and use. For example, in the domain of courseware evaluation, students, teachers, the computer center, etc. all have their specific, often partially incompatible interests, needs, and goals. Students prefer easy-to-use functionalities, whereas main concerns of the computer center include security and reusability [de Moor, 2007]. In order to ensure that all requirements are captured, contrasted, and balanced, it is not sufficient to examine "The User." Instead, an inventory carefully needs to be made of the *actor roles* that the various stakeholders play. Making roles explicit is gaining prominence in, for instance, the Role-Based Access Control paradigm[8] as a way to systematically determine actor responsibilities in workflows and access to functionalities and information resources. Most role classifications are quite abstract and technology-focused (administrator, facilitator, member, etc.). However, many other typologies are possible, often

---

[8]http://csrc.nist.gov/rbac/

quite specific to a particular domain. Roles could, for instance, be based on
workflow process (author, editor, reviewer, ...), on organizational structure
(secretary, manager, ...), or on the main stakeholders in a particular domain
(UNEP, Corporation, NGO, ...). Precisely defining the responsibilities, per-
missions, and prohibitions attached to these roles helps in creating better
design patterns of information systems. Besides having their normal operational status, such actor roles should also be explicitly involved in the design
and optimization of the socio-technical system-in-use.

Example:

As the goal of the topic community is to provide policy advice, it is imper-
ative that voices across society have their say. The *domain role* of representative is therefore introduced. Further analysis would need to specify which
stakeholders need representation: science, business, government, NGOs, the
general public, and so on. To ensure the efficiency and effectiveness of the
required discussion processes, the following *discussion roles* are necessary:
discussant, discussion facilitator, and discussion summarizer.

11.2.3.3 Domain

A third layer of the model is the domain in which the collaborative community using the tool system (inter)operates. The domain is an important, but
still ill-understood element in tool system design and evaluation. Aspects that
influence design processes and functionalities of the tool system include issues
of structure and size; for example, is it a distributed network or centralized
organization, a large or a small organization? What setting is the tool being
used in: an academic, a corporate, a governmental, or a non-governmental
setting? What is the financial situation: are there resources for acquisition
or customization of software, or is off-the-shelf, open source software the only
option? Are there political alliances and commitments that force or preclude
the use of certain software? In other words, the domain determines how many
degrees of freedom there are in the design process.

Such domain issues are not likely to be translated into goals like activities
and aspects directly, but they do help to define the scope of the project and
the general affordances and constraints in which the tool system needs to (inter)operate.

Example:

The topic community concerns a distributed, medium-sized network, in
which the mode of working in general will be asynchronous and dispersed due
to the global nature of the community and the almost impossible matching
of agendas. It concerns an intergovernmental setting, with much attention
to procedure, legitimacy, participation, and accountability, features that the

tool system should reflect. As with many (inter)governmental bodies, there is some money for customization and process facilitation, but no budget to buy expensive, proprietary tools. Also, the UN has a strong interest in promoting free software and prefers to lead by example. Well-documented, reliable, and interoperable open source tools are therefore strongly preferred.

11.3 A Conceptual Model of Functionality Matching in Collaborative Communities

In this section, we formalize and extend the notions introduced so far. First, we present a concept type hierarchy of the socio-technical system. We then use this hierarchy to define some key functionality mappings using conceptual graphs. Then, we present a worked example of the functionality matching process based on these mappings.

11.3.1 A Concept Type Hierarchy of the Socio-Technical System

Based on the scenario, the following concept type hierarchy can be created to model the socio-technical system of the topic community.

Actors either play discussion or domain roles. *Discussion roles* include *Discussants*, *Discussion Facilitators*, and *Discussion Summarizers*. *Domain roles* include *Representatives* of various stakeholders, which (as of yet) do not need to be formalized in more detail.

Functionality Components of different levels of granularity are distinguished. The *Topic Community Support System* is the focus of attention. The two main *Tools* used are *Compendium* and *Wikipedia*. Each tool contains one or more *Modules*, such as *Create Node*, *Edit Wiki*, or Search Pages. Each of these modules contains a number of *Functions*, which themselves are further grouped in categories like *Create a Text Element*, *Create Links* to and from text elements, or *Create Indices* of text elements.

Goals of the system are the *Activities* of *Selecting Relevant Concepts* and their entries in the knowledge system, *Conducting a Discussion* among relevant stakeholders, and *Disseminating the (consensus) Position* to the general public. One important Aspect is the *Legitimacy* of participation in the activities.

Mappings within and between elements of the tool system and usage context include *Enabled Functionalities*, *Functionality Requirements*, *Required Implementations*, and *Support-definitions*.

```
T >                             Module >
  Actor >                         Create_Node
    Disc_Role >                   Edit_Wiki
```

```
    Discussant                    Search_Pages
    Disc_Facilitator            System >
    Disc_Summarizer               Topic_Community_Support_Sys
  Domain_Role >                 Tool >
    Representative                Compendium
Funct_Comp >                      Wikipedia
  Function >                    Goal >
    Create_Index >                Activity >
      Create_Category               Conduct_Disc
      Create_Tag                    Disseminate_Pos
    Create_Link >                   Sel_Rel_Concepts
      Create_Reference            Aspect >
    Create_Text >                   Legitimacy
      Create_Argument           Mapping >
        Create_Arg_Con            Enable
        Create_Arg_Pro           Funct_Req
      Create_Position >          Req_Impl
        Create_Answer            Support
        Create_Decision
        Create_Idea
      Create_Issue >
        Create_Question
      Create_Note
      Edit_Disc_Page
      Edit_Page
```

11.3.2   Functionality Mappings

Functionality mappings define relations within and between various elements of the usage context and the tool system, and are used in the functionality matching process. In [de Moor and van den Heuvel, 2001], we presented a functionality matching meta-model and process for virtual communities, using conceptual graphs to assess the match between required and enabled functionalities. Here, we present an adapted version of that approach.[9]

We first propose a set of effective tool functionality axioms, which help establish the link between the tool system and usage context.

Effective Tool Functionality Axioms

- A functionality component can enable one or more functions.
  Example: the Compendium *Create Node*-module allows a user to create an issue, a position, an argument, etc.

- Different functionality components may have partially *overlapping* functionality, i.e., each enabling the same functions, while also each enabling different functions at the same time.
  Example: Both Compendium and Wikipedia allow for links between knowledge

---

[9]The most important changes are: including roles instead of specific users in the definitions, simplifying the workflow mappings by leaving out the interaction layer, using functions instead of information/communication processes as the atomic unit of functionality, and not including access-relations. The reader is referred to the original paper for more detail.

nodes to be created. However, only Compendium also allows for visually mapping the structure of a debate, while only Wikipedia provides users with the capabilities to jointly efficiently edit and review complex wiki knowledge nodes.

- All community members involved in a functionality requirement must have at least one *enabling functionality component* at their disposal.

  Example: a community member may need to be able to create links between knowledge nodes. Both Compendium and Wikipedia provide their users with this functionality.

Enabled Functionality

Any function enabled by some functionality component for a particular actor role is called an *enabled functionality*. Such a mapping is represented as a definition, which conforms to a specialization of the following definition of the enabled functionality-mapping[10]:

[Enable : *x] → (Def) → [Mapping :?x]−
   (Inst) → [Tool] → (Part) → [Module] → (Obj) → [Function]
  (Agnt) → [Actor].

Example:

The following definition says that creating an argument is enabled by the Compendium *Create Node*-module for any user of the component:

[Enable : #114]−
   (Inst) → [Compendium] → (Part) → [Create_Node]−
    (Obj) → [Create_Argument]
  (Agnt) → [Actor].

Required Functionality

Functionality requirements are defined by functions in their usage context, which we define as the activity for which a function is used and the actor role involved in that activity:

[Funct_Req : *x] → (Def) → [Mapping :?x]−
   (Obj) → [Activity] → (Agnt) → [Actor]
  (Inst) → [Function].

---

[10]First a type definition of each concept is given. The (Def) relation connects the defined type on the left with the genus (supertype) on the right, while the rest of the definition comprises the differentia, the subgraph that specialises the genus to the defined type. The knowledge definitions in the examples are specializations of defined type plus differentia.

Example:

The following definition says that for conducting a discussion a discussion summarizer needs to be able to create a position.

[Funct_Req : #156]−
    (Obj) → [Conduct_Disc] → (Agnt) → [Disc_Summarizer]
    (Inst) → [Create_Position].

Assigned Functionality

For each functionality requirement, at least one enabling functionality component needs to be assigned. The actual selection of the functionality components that are to enable functionality requirements is not automated in our approach. The reason is that there may be many domain and nonfunctional requirements constraining the choice of implementation, which require human interpretation. Two types of assigned functionality include support-mappings and required implementation-mappings.

The Support-Mapping

A *support*-mapping represents a function supporting a functionality requirement.

[Support : ∗x] → (Def) → [Mapping :?x]−
    (Obj) → [Funct_Req]
    (Inst) → [Tool] → (Part) → [Module] → (Obj) → [Function].

Example:

The following support-relation says that for a discussion summarizer to be able to create a position (Functionality Requirement #156) she can use the *Create Decision*-function of the *Create Node*-module of Compendium.

[Support : #212]−
    (Obj) → [Funct_Req : #156]
    (Inst) → [Compendium] → (Part) → [Create_Node]−
        (Obj) → [Create_Decision].

The Required Implementation-Mapping

Legacy systems and the need to standardize implementations for maintenance efficiency often lead to the demand for certain functionality requirements to be supported by a specific tool. Such a *required implementation* puts an additional constraint on possible support-relations.

[Req_Impl : *x] → (Def) → [Mapping :?x]−
   (Obj) → [Funct_Req] → (Obj) → [Activity]
   (Inst) → [Tool].

Note that for readability we repeat the activity-concept that is already part of the functionality requirement.

Example:

The following required implementation-mapping says that for any functionality requirement concerning the conducting of discussions, the Compendium tool needs to be used.

[Req_Impl : #186]−
   (Obj) → [Funct_Req] → (Obj) → [Cond_Disc]
   (Inst) → [Compendium].

11.4 The Functionality Matching Process

Using the functionality mappings, we now examine the steps of the *process* in which required and enabled functionalities can be matched in practice.

Already with these very basic functionality mappings, essential functionality matching operations can be performed. The real power of conceptual graphs, however, is that on top of these definitions, a wide range of (meta)-constraints can be defined, at various (meta)-levels of analysis [Mineau et al., 2000]. Such constraints are especially important when operationalizing abstract quality aspects, such as efficiency, extensibility, security, and legitimacy. [11] Bootstrapping the complex definition of such requirements with conceptual graphs could significantly help optimize the design process.

The matching process consists of three stages: (1) creating a knowledge base of socio-technical system specifications, (2) proposing some change to the specifications, and (3) performing the match (which includes formulating a set of functionality matching criteria, calculating the match, and interpreting the results).

---

[11] We gave legitimacy as an example of a goal-aspect. We do not work out its operationalization here, but have defined it as a student project at the end of this chapter.

11.4.1 Define System Specifications

Collaborative communities are continuously evolving socio-technical systems. At t = 0, before a change is proposed, we assume that the current system specifications are properly matched, i.e., that all required functionalities are enabled, and that all required implementation-constraints are satisfied.

Example:

Assume the concept type hierarchy and definitions given above. Furthermore, the following graphs are given.

Enabled functionality:

This definition captures (part of) the functionality provided by the Compendium *Create Node*-module:

[Enable : #115]—
   (Inst) → [Compendium] → (Part) → [Create_Node]—
      (Obj) → [Create_Answer]
      (Obj) → [Create_Idea]
      (Obj) → [Create_Argument]
      (Obj) → [Create_Arg_Con]
      (Obj) → [Create_Arg_Pro]
      (Obj) → [Create_Decision]
      (Obj) → [Create_Note]
      (Obj) → [Create_Question]
      (Obj) → [Create_Reference]
   (Agnt) → [Actor].

The next definition captures (part of) the functionality provided by the Wikipedia *Edit Text* and *Search Pages*-modules. [12]

[Enable : #116]—
   (Inst) → [Wikipedia]—
      (Part) → [Edit_Wiki]—
         (Obj) → [Create_Category]
         (Obj) → [Create_Link]
         (Obj) → [Edit_Disc_Page]
         (Obj) → [Edit_Page]
      (Part) → [Search_Pages].
   (Agnt) → [Actor].

---

[12]Note that the functions of the *Search Pages*-module has not been defined. However, including this module-concept in the definition shows that there is at least *some* searching-functionality, which could be investigated and explicated further if required.

Assigned functionality:

Support-definition #212 stated that Compendium's Create Position-function is to be used to support the Discussion Summarizer in her participating in the discussion. Required Implementation-definition #186 said that every functionality requirement having to do with conducting discussions has to be supported by Compendium. In addition, the following definition expresses that disseminating the final position to the general public debate should be supported by Wikipedia:

[Req_Impl : #187]−
   (Obj) → [Funct_Req] → (Obj) → [Disseminate_Pos]
   (Inst) → [Wikipedia].

11.4.2   Propose Specification Changes

At t = 1, one or more specification changes are proposed by one of the users. Such a change concerns the creation, modification, or deletion of one or more specification knowledge definitions like the ones presented so far. In collaborative communities, it is essential *who* has the legitimate authority to make such changes: in [de Moor and Weigand, 2007], we show how such legitimacy can be ensured by calculating these authorizations using the set of *applicable composition norms*. The calculation says which community members may, must, or may not be involved in the specification of particular parts of their socio-technical system. In this chapter, we unfortunately do not have the space to say more about this highly relevant, but complex topic.

We illustrate the processing of a newly specified (legitimate) functionality requirement to show how conceptual graphs can help optimize the design process.

Example

A discussion summarizer, besides having to be able to create a position during the conduct of the discussion also must be able to disseminate the position in the right form to the general public. As it is not clear yet in what form this dissemination best take place, all that can be said for now is that it should be some form of text creation, to be specified in more detail later, represented by this new functionality requirement #158:

[Funct_Req : #158]−
   (Obj) → [Disseminate_Pos] → (Agnt) → [Disc_Summarizer]
   (Inst) → [Create_Text].

The question is how to assign the appropriate enabled functionality? To answer this question, we need to match required and enabled functionalities.

11.4.3 Perform the Functionality Matching

Functionality matching can help to correctly process specification changes, such as new functionality requirements, ensuring the legitimacy, consistency and completeness of the definitions of the evolving socio-technical system.

Many different forms of functionality matching are conceivable. *Matching criteria graphs* (or constraints) need to be specified on which the matching steps are to be performed. The matching criteria should at least partially be expressed in terms of the functionality mappings. These graphs are the CG queries necessary for retrieving the knowledge definitions that satisfy the matching criteria. Often, a functionality matching process is complex, consisting of a sequence of matching graphs to be projected, joined, etc. The outcome of this process will be an answer to the question whether a specification change does have adverse effects on the socio-technical system, or provide suggestions for new specification changes. For example, if a new functionality requirement is introduced, there may not yet be the right enabling functionality. In that case, either the functionality requirement needs to be modified, a new functionality component needs to be installed, or at least a procedure needs to be in place with instructions for how to deal with the functionality problem.

In our example, the initial matching criteria graphs are the actor-concept (the concept in the graph that is the specialization of [Actor]) and the function-concept (the concept in the graph that is the specialization of [Function]) of each enabled functionality ef in the total set of enabled functionalities EF. To enable a particular functionality requirement fr, the function-concept of at least one $ef_i \in EF$ should be a subtype of its counterpart in the functionality requirement fr, as it should be some implementation of the required function-concept. However, the actor-concept of that ef_i should also be a *super*type of the actor-concept of fr, as it must at least be enabled for the required role.

11.4.3.1 Functionality Matching Steps

For a particular functionality requirement fr, the functionality matching steps are as follows:

- Determine the set of potentially enabling functionalities EF_{pot} of fr. To do so, for each enabled functionality $ef_i \in EF$:
    - If the function-concept of fr projects into the function-concept of ef_i and the actor-concept of ef_i projects into the actor-concept of fr, then ef_i is in EF_{pot}.

- Determine the set of relevant required implementation-mappings RI, which are those required implementation-mappings of which the activity-concept projects into the activity-concept of fr.

- Determine the set of acceptable enabling functionalities EF_{acc}, which equals EF_{pot} minus those enabled functionalities ef_i EF_{pot} where none of the $ri \in RI$ has a tool-concept that projects into ef_i. In other words, if a particular activity (as defined in fr) is to be supported by one or more specific tools, but the potentially enabling functionality does not contain a specialization of any of those, that functionality is not acceptable.

- Create the support-mapping that defines which function enables fr by selecting one or more acceptable enabling functionalities from EF_{acc}.

Example

We now conduct the functionality matches, using the CG projection operation. One tool that supports these operations is Prolog+CG, part of the Amine platform. [13]

We need to define a support-mapping for functionality requirement fr #158. The function-concept of fr is [Create_Text], its actor-concept is [Disc_Summarizer].

Projecting [Create_Text] into all $ef \in EF$, Prolog+CG returns the following enabled functionality-(sub)graphs[14]:

```
?- cg(_,g1),subsume([Enable]-inst->[Tool]-part->[Module]
     -obj->[Create_Text],g1,g2).

{_=FREE, g1=[Enable : nr114] -
  -inst->[Compendium]-part->[Create_Node]-obj->[Create_Argument],
  -agnt->[Actor],
  g2=[Enable : nr114]-inst->[Compendium]-part->[Create_Node]
    -obj->[Create_Argument]}

{_=FREE, g1=[Enable : nr115a]
  -inst->[Compendium]-part->[Create_Node] -obj->[Create_Answer],
  -agnt->[Actor];,
```

---

[13] Prolog+CG currently does not allow for nonfunctional relations to be processed. This means that graphs with more than one instance of the same relation emerging from a concept node, are not processed correctly. In a future version of Amine, the tool environment embedding the newest versions of Prolog+CG, such nonfunctional relations will be interpreted correctly [Kabbaj, 2007]. For now, we need to flatten graphs like enabled functionalities #115 and #116 into their separate subgraphs with only a single (Part)-relation per subgraph.

[14] cg(_,g1) retrieves a graph from the knowledge base. The subsume(g0, g1, g2) operation then checks that g0 subsumes g1 and returns in g2 the image of g0 in g1 (the subgraph of g1 that is isomorph to g0.

```
g2=[Enable : nr115a]-inst->[Compendium]-part->[Create_Node]
  -obj->[Create_Answer]}
```

[...]

```
{_=FREE, g1=[Enable : nr116d]
  -inst->[Wikipedia]-part->[Edit_Wiki] -obj->[Edit_Page],
  -agnt->[Actor];,
  g2=[Enable : nr116d]-inst->[Wikipedia]-part->[Edit_Wiki]
    -obj->[Edit_Page]}
```

Enabled functionality-subgraphs #115i, #116a, and #116b are not returned, as their enabled functions (resp. `Create_Reference`, `Create_Link`
and `Create_Category`) are not specializations of the `Create_Text`-concept
of the functionality requirement fr to be supported. Since the actor-concept
of all ef is [`Actor`] (the most generic role), all projections obtained just now
are in EF_{pot}.

The activity-concept of fr is [`Disseminate_Pos`]. Required implementation #187 is the only one where the activity-concept (also [`Disseminate_Pos`])
projects into the activity-concept of fr, and is therefore the only element in
the set of relevant required implementation-mappings RI.

To calculate the set of acceptable enabling functionalities, EF_{acc}, we need
to find those $ef_i \in EF_{pot}$ where the tool-concept is a specialization of the
tool-concept [`Wikipedia`] of required implementation #187:

```
?-cg(_,g1),subsume([Enable]-inst->[Tool]-part->[Module]-obj->
    [Create_Text],g1,g2),subsume([Wikipedia]-part->
      [Module]-obj->[Function],g2,g3).
```

```
{_=FREE, g1=[Enable : nr116c]-inst->[Wikipedia]-part->[Edit_Wiki] -
  -obj->[Edit_Disc_Page],
  -agnt->[Actor];, g2=[Enable : nr116c]-inst->[Wikipedia]-part->
    [Edit_Wiki]-obj->[Edit_Disc_Page], g3=[Wikipedia]-part->
      [Edit_Wiki]-obj->[Edit_Disc_Page]}
```

```
{_=FREE, g1=[Enable : nr116d]-inst->[Wikipedia]-part->[Edit_Wiki] -
  -obj->[Edit_Page],
  -agnt->[Actor];, g2=[Enable : nr116d]-inst->[Wikipedia]-part->
    [Edit_Wiki]-obj->[Edit_Page], g3=[Wikipedia]-part->
      [Edit_Wiki]-obj->[Edit_Page]}
```

The resulting acceptable enabling functionalities for fr #158 give the specifier two possible functions to implement the requirement: *Edit Page* and *Edit
Discussion Page*. It is now up to the specifier to select the appropriate one,
based on her experience with the domain. Jane chooses to have the position
described on the climate change wiki page itself, not on the discussion page
of that page, as the wiki page is what the general public, not interested in

the meta-discussion, will most likely read. She therefore defines the following support definition:

```
[Support : #213]−
    (Obj) → [Funct_Req : #158]
    (Inst) → [Wikipedia] → (Part) → [Edit_Wiki] → (Obj) → [Edit_Page]
```

11.5   Discussion and Conclusion

In this chapter, we examined how to optimize the design of tool systems for collaborative communities using conceptual graphs. Ever more, such communities are supported by rapidly evolving systems of (social) software tools. However, there are many dependencies within and between the tool system and the usage context. Such dependencies can be captured with what we called functionality mappings. Using these mappings, tailored functionality matching processes can be set up that help optimize the design process. Using the power of conceptual graph theory, in particular the properties of graph generalization hierarchies and basic projection operations, many of these matching processes can be semiautomated.

The role and definition of functionality matching processes is not at all trivial. In this chapter, we only gave an illustration of how such processes could be implemented. More sophisticated, complete, and sound approaches are conceivable, but our goal here was modest: introducing a way of conceptual thinking about social software design for collaborative communities, showing the need for functionality matching, and firing an opening shot in the direction how this could be operationalized in practice. Future research should lead to more systematic approaches of conceptualizing, implementing, and applying functionality matching processes in the design of social software systems.

One direction in which this work could be extended is by expanding the conceptualization of the design process that embeds the kind of functionality matching processes introduced here. Work on testbed development for collaboratories, such as pragmatic methodologies for knowledge representation tools research and development, could inform such conceptualizations [Keeler and Pfeiffer, 2006]. Successful collaboratory development requires (1) a system architecture and integration to explore ways that people and machines can use component technologies most effectively, (2) a research program to study the conditions required for collaboration, and (3) user-oriented rapid-prototyping testbeds, to understand the impact of the technologies used [Computer Science and Telecommunications Board, 1993]. Pattern languages can help to capture communication knowledge of large distributed communities [Schuler,

2002], particularly useful in the design of collaboratories. In [de Moor, 2004], we used conceptual graphs to model collaboratory improvement as a set of layered testbed processes: information and communication processes enabled by tools, workflows through which the community accomplishes its goals, design processes in which they change their socio-technical system, and improvement processes in which the design processes themselves are made more effective and efficient. We showed how conceptual graphs can be used to optimize these processes. Combining that approach with the detailed functionality matching processes proposed in this chapter, could be one useful way to go about optimizing social software design using conceptual graphs.

11.6   Student Projects

11.6.1   Project 1

This chapter introduced a basic concept type hierarchy. Find a number of state-of-the-art research articles on online communities, Web services design, etc. Using these articles, refine and expand the hierarchy presented in this chapter. Which newly found concept types are universal, which ones are domain or case-dependent?

11.6.2   Project 2

The functionality mappings presented in this chapter are only very basic. Improve the definitions given, e.g., currently functionality requirements concern individual functions, but probably aggregates of functions could be useful as well. Required implementation-mappings currently concern full tools, but might also be defined at the module or even function-level. What impact do your revised definitions have on the functionality matching processes?

11.6.3   Project 3

Select two cases of (online) collaborative communities, for example e-learning communities, gaming communities, or corporate communities. Using the (revised) concept type hierarchy and functionality mappings introduced in this chapter, try to characterize the socio-technical systems in the respective cases. If necessary, further modify the concept type hierarchy and functionality mappings.

11.6.4 Project 4

The example of a functionality matching process given in this chapter was that of implementing a functionality requirement. However, there are many more applications of such processes. One example not worked out is that of checking software quality aspects, like the effectiveness, efficiency, security, and legitimacy of software (components). Select one such quality aspect and operationalize it in terms of functionality mappings and matching processes. Apply this process to your case description of the previous assignment.

11.6.5 Project 5

Much advanced Web services-related research is focusing on standards like the Universal Description, Discovery, and Integration (UDDI) protocol,[15] Business Process Modeling Notation (BPMN),[16] and Business Process Execution Language (BPEL).[17] Find a project description using these standards and model selected representations and processes using the notions introduced in this chapter as a starting point. How could our functionality matching approach help address problems identified in the selected project description?

11.6.6 Project 6

We have used the projection operation for the matching process. What roles could other core CG operations (e.g., maxJoin) play in developing new ways of functionality matching and design? How could you use these operations to make the example process presented in this chapter more efficient?

11.6.7 Project 7

Amine is a powerful open source platform for knowledge system implementation,[18] with a strong grounding in conceptual graphs. In particular its Prolog+CG module is very useful for CG representation and reasoning, as it combines a range of CG representations with the reasoning power of Prolog. Make an Amine implementation of the functionality matching process presented in this chapter.

---

[15]http://www.uddi.org/

[16]http://www.bpmn.org/

[17]http://en.wikipedia.org/wiki/BPEL

[18]http://amine-platform.sourceforge.net/ See http://www.huminf.aau.dk/cg/ for an online course.

Part VI

Intelligent Systems

Chapter 12

Semantic Annotations and Localization of Resources with Labelled Graphs

Michel Chein

C.N.R.S. and University Montpellier 2, France

12.1 Introduction

Resources and metadata

Two basic sorts of objects are considered in this chapter: resources and metadata.

A resource can be an electronic document (e.g., a jpeg image, a software component, a relational database, a text file, a Web resource, etc.), or a non-electronic document (e.g., a book, an argentic picture, etc.), or even any "object" (e.g., a historical building, a tourist resort, a drawing, etc.). Thus, a resource is a very general notion, and we do not restrict it to documents. On the other hand, we greatly restrict metadata.

The metadata considered in this chapter are computer data structures or, more precisely, mathematical objects that can be implemented as computational data. A metadata is always associated with a resource. Thus, we consider two sets. A set of resources, sometimes called primary data when electronic data are considered, and a set of metadata associated with these resources, sometimes called secondary data.

Each resource is identified by an identifier, which is used in the metadata base for referencing the resource. A resource can be decomposed, e.g., a book can be decomposed into chapters. But hereafter a part of a resource is considered as a resource, i.e., we do not necessarily take into account, in the annotation base, the structure of a decomposed resource.

Objective and subjective metadata

The resource address (e.g., a URL, the class number of a book in a library, the ground address of an historical building), the resource type and subtype (e.g., html page, jpeg image, DivX video, text file, book, tourist spot), the document size, the authors, the rights attached to the document, etc., are examples of metadata.

Metadata that can be attached to a resource depend on the resource type. Nevertheless, for any kind of resource, metadata can be roughly clustered into two classes: objective metadata and subjective metadata. The resource address, authors name, and the document size are some examples of objective, or external, metadata. The second class of metadata is called subjective because it aims at representing information generally depending on the author of the metadata. The most usual example of such subjective metadata is the representation of the content of a resource, i.e., an indexation of the resource. Besides indexation, we also consider metadata that aims at representing a comment, note, or remark about a resource, e.g., the historical importance of a given painting or a viewpoint about a movie. Such metadata are usually called (semantic) *annotations*. Thus, an annotation is simply a piece of knowledge associated with a resource (cf. [Reeve and Han, 2005] for a recent survey of annotation platforms).

Queries and answers

A query is a search task in a metadata base. We assume hereafter that a query is expressed by metadata properties. Otherwise, for instance, if the query is a natural language expression, it must be transformed into an expression built with metadata properties before being processed by the search engine. Thus, basically, the structure of the metadata base determines queries which can be made on this metadata base.

The answer to a query is built from the set of metadata which satisfy the properties represented in the query. Metadata that satisfy a query are obtained by matching algorithms. A fundamental point to mention is that, in the model presented here, we assume that *the resources cannot be directly accessed* by the search engine. The resources can only be selected by searching metadata bases. Even if the resources are electronic resources, we assume that the queried resource base cannot be directly accessed but can only be accessed through a metadata base. In other words, the resources are only known by their metadata. For instance, if a resource is structured, the structure can be used to search resources only if the structure is represented as a metadata. Otherwise said, a sub-resource R' of a resource R is simply a resource, and if one wants to use this information in the query/answering mechanism the fact that R' is a sub-resource of R has to be represented in a metadata associated with R' or/and with R.

A privileged application

A subtask of many applications involves the use of metadata for selecting and localizing resources. Information retrieval is the most classical and important one. Due to the large amount of electronic documents and to the easiness of decomposing an electronic document, as well as enriching it with metadata, *authoring document* is a rapidly growing application. More precisely, the problem is to edit a new document composed of parts of existing documents. The metadata are used to select parts of existing documents that are relevant to the objectives of the new document. This authoring process by resource recontextualization (cf. [Gaillard et al., 2007]) is a key thread of the present chapter.

Why conceptual structures enter the picture?

Until now there is nothing new! Some data are described in a data language (the metadata language), queries concerning these data are expressed in a query language, and a query is answered by matching algorithms between the query and the metadata base.

So it is natural to ask why conceptual structures are worth using, and, more generally, why Artificial Intelligence techniques are used? The main reason is that, in this chapter, we are only interested in subjective metadata that are usually called (semantic) annotations. We do not consider objective metadata that fall within the scope of the database domain.

Among annotations, one can distinguish: indexation (a general description of the content of a document used for retrieving documents having a given content), analysis (e.g., explanation of a law, rhetorical analysis of a political speech), comments or judgments (e.g., this historical building is interesting to visit for such and such reason), etc.

Annotations can be used for different purposes, for example, as already said, to help in the construction of publications using annotated resources (cf. [Nanard et al., 2006] and [Gaillard et al., 2007]), or to improve a resource (e.g., modifying the resource taking all negative comments concerning the resource into account, such as a law project), or as a collaborative tool for characterizing or discussing a set of resources, etc. In any case, annotating a resource enriches this resource (e.g., cf. [Petridis et al., 2006], [Schreiber et al., 2001], [Aubert et al., 2006b], [Hollink et al., 2003], [Handschuh and Staab, 2003]). Let us once again note that, in this chapter, "subjective metadata" and "annotation" are considered as synonymous expressions.

The subjective nature of the searchable annotation base means that they cannot be considered as universally exact as the objective metadata are (or have to be ...). This has an important consequence for search algorithms. Contrary to the database domain, exact search (return all the correct answers and only them) is not sufficient. Considering that annotation bases imply *flexibility for the search algorithms*, one has to build approximate as well as exact search methods. Depending on the user knowledge about the exact content of the annotation base, and depending on the quality of the annotation base, the search process can be more or less exact or approximate.

The previous considerations lead to the conclusion that an annotation base attached to a resource set is (a kind of) a knowledge base, and this explains why we put forward the hypothesis that Artificial Intelligence techniques should be at the core of an annotation system. The main components of the annotation system presented in this chapter are briefly described as follows.

- In order to search by (semantic) queries annotated resources, we first need *a data language for annotations*. The proposed annotation language uses "conceptual structures," i.e., structures richer than a weighted set of terms. Roughly said (precise definitions are given later), an annotation is a set of terms related by relations, i.e., it is a labelled graph.

  An annotation is built from an ontology that is basically composed of three hierarchies. A hierarchy of concepts (or terms), a hierarchy of relations between concepts, and a hierarchy of nestings for representing the different kinds of annotations.

  An ontology can also contain representations of other knowledge. Relation signatures indicate the types of relation arguments. Rules can represent knowledge of the form "if a piece of information is present (within an annotation) then further information can be added (to this annotation)." Thus, rules can be used to automatically complete an annotation by implicit and general knowledge. Constraints can represent knowledge of the form "if a piece of information is present (within an annotation) then other information cannot be present (in the annotation)." Signatures and constraints are used to avoid the construction of absurd annotations. Prototypes of annotations can also be used to help annotators. Rules, constraints, signatures, and prototypes are all represented by labelled graphs.

- Secondly, a *query language* is used for searching annotation bases. Elementary queries are graphs having the same form as the annotations. Elementary queries can be combined with the usual boolean connectors (and, or, not).

- Thirdly, graph operations are used for *computing answers*. Graph homomorphisms are used for computing exact answers, i.e., annotations which are specializations of the query. The (graph) annotations have logical semantics in first-order logic (FOL), and the homomorphism mechanism is equivalent to deduction in FOL. Thus, the exact answers to a query are annotations that entail the query. Instead of using probabilistic techniques (cf., for instance, [van Rijsbergen, 1979] or [Fuhr, 2000]), versatile matching algorithms for computing approximate answers can be based on graph transformations. A general framework for building approximate answers is proposed.

- Last but not least, as the intended users are not necessarily computer scientists or engineers, *friendly tools* for building and interrogating annotation bases have to be developed, as well as easily explainable answering mechanisms.

The model presented in this chapter is based on the conceptual graph model as developed at LIRMM since 1992 (cf. [Chein and Mugnier, 1992] and [Chein and Mugnier, 2008]). The main difference from the initial general model of Sowa [Sowa, 1984] is that it is a *graph-based* knowledge representation and reasoning language. Graphs are used for representing knowledge, and graph operations are used for reasonings. Logic is used as denotational semantics. In addition to efficient graph algorithms, graphs are useful for explaining to end-users how the system works because knowledge and reasonings can be visualized in a way that is easily interpretable in natural language. Indeed, a knowledge representation system should allow users to have maximal control over each step of the process:

- entering and understanding the data (labelled graphs),

- understanding the results given by the system (labelled graphs),

- understanding how the system computes the results (simple graph operations).

Chapter organization

Section 2 briefly presents the main notions of an annotation language. Section 3 deals with some notions (e.g., modules and prototypes) introduced to help and guide the annotation construction process (cf. [Moreau et al., 2007]). Section 4 concerns searching algorithms. It first briefly presents exact search, which is based on graph homomorphisms and is logically founded. Then, it is shown how approximate searching algorithms based on van Risjbergen's uncertainty principle (cf. [J.Y. Nie, 1998]) can be built thanks to graph formalism (cf. [Genest and Chein, 2005]).
The main advantages and weaknesses of the proposed language are discussed in the conclusion.
The present chapter has benefited from many applications that have been developed for information retrieval tasks using conceptual graphs (cf. [Martin, 1997], [Martin and Eklund, 1999], [Corby et al., 2000] and [Chein and Genest, 2000] for older references), and especially from the development of two ongoing applications Saphir [SAPHIR, 2008] and Logos [LOGOS, 2008].

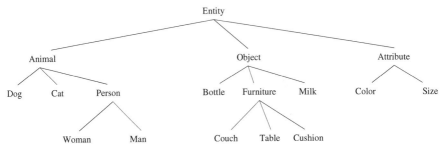

FIGURE 12.1: A primitive concept type set.

12.2 An Annotation Language

Mathematical definitions of the main notions of the model are briefly reviewed here.

12.2.1 Annotation

First we present simple (conceptual) graphs built over a vocabulary. Then typed nested graphs, i.e., the notion used for representing annotations, are introduced.

12.2.1.1 Simple Graphs

A *simple conceptual graph (SG)* is a bipartite multigraph. "Bipartite" means that a BG has two disjoint sets of nodes — concept nodes and relation nodes — and that any edge links two nodes from different sets, i.e., a concept node and a relation node. "Multigraph" means that a pair of nodes may be linked by several edges (see also Chapter 2).

Concept nodes are used to represent entities, and relation nodes are used to represent relations between entities. A relation can be of any arity ≥ 1. If a concept c is the i-th argument of a relation r, then there is an edge between r and c, and this edge is labelled i.

The node labels are built from a simple ontology called a *vocabulary*. A vocabulary is composed of a hierarchy of primitive concept types, a set of basic banned conjunctive types, a hierarchy of relation symbols, and a hierarchy of nesting symbols.

A hierarchy of acceptable concept types is built as follows from a hierarchy of primitive concept types, and from a set of banned conjunctive types. A *primitive concept type set* is an ordered set T with a greatest element. An example of a primitive concept type set is given Figure 12.1.

Conjunctive types, which are subsets of T, are provided with a natural partial order that extends the order defined between primitive types. More precisely:

DEFINITION 12.1 Conjunctive concept type set   *Let T be a set of primitive concept types. A* conjunctive concept type *is given by a (non-empty) set of pairwise incomparable types $\{t_1, ..., t_n\}$. $T^\sqcap$ denotes the set of all conjunctive types over T. $T^\sqcap$ is provided with the following partial order, which extends the partial order on T: given two types $t = \{t_1, \ldots, t_n\}$ and $s = \{s_1, \ldots, s_p\}$, $s \geq t$ if for every $s_j \in s$, $1 \leq j \leq p$, there is a $t_i \in t$, $1 \leq i \leq n$, such that $s_j \geq t_i$.*

To any (non-empty) subset A of T, one can associate the conjunctive type $min(A)$, which is the set of minimal elements of A. A primitive type t is identified with the conjunctive type $\{t\}$.

Generally, not all conjunctions of types have a meaning. Thus we need a way of expressing that a set of types cannot have a common subtype. A classical way of doing this is to add a special `Absurd` type below types with an empty intersection. But this technique is not always precise enough. For instance, t_1, t_2, t_3 being direct supertypes of `Absurd` means that the intersection of any two of them is empty. We cannot express that t_1 and t_2 are disjoint as well as t_1 and t_3, but that there can be an entity of type t_2 and t_3 (e.g., t_1 = `Animal`, t_2 = `Ship`, t_3 = `Robot`, all being subtypes of `MobileEntity`). The introduction of *banned* type sets allows to overcome this difficulty. Giving in extension all acceptable types is not conceivable in practice; therefore we define the set of acceptable conjunctive types by the primitive type set and the (maximal) banned conjunctive types, and we fulfill the following principle: "everything that is not banned is permitted." Theoretically, the number of banned conjunctive types can be exponential in the number of primitive types, but, practically, considering banned types of cardinality two seem to be sufficient and their number is at most quadratic in the number of primitive types.

Banned types are defined as follows.

DEFINITION 12.2 Banned type set   *Let B be a set of conjunctive types. An element of $T^\sqcap$ is said to be* banned *w.r.t. B if it is less than or equal to (at least) an element of B. B^* denotes the set of all banned types: $B^* = \{t \in T^\sqcap \mid \exists t' \in B, t \leq t'\}$.*

For example, in Figure 12.2 the conjunctive types $\{Animal, Object\}$ and $\{Color, Size\}$ are indicated as banned.

We can now define a concept type hierarchy.

DEFINITION 12.3 Concept type hierarchy   *A concept type hierarchy*

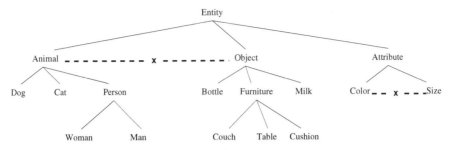

FIGURE 12.2:   Example of banned conjunctive concept types.

T_C is defined from a pair (T, B), where:

- T, the set of primitive concept types, is a partially ordered set, with a greatest element, denoted by $\top$

- B, the set of basic banned conjunctive types, is composed of conjunctive types over T

- B complies with T, i.e., for all $b \in B$, there is no type $t \in T$ with $t \le b$ i.e., $B^* \cap T = \emptyset$

T_C is defined as the set $T^{\sqcap} \setminus B^*$. $T^{\sqcap}$ is thus partitioned into acceptable types T_C and banned types B^*.

A vocabulary is composed of three hierarchies.

DEFINITION 12.4 Vocabulary   *A* vocabulary *is a triple of pairwise disjoint partially ordered sets* (T_C, T_R, T_N), *where:*

- T_C, *is a concept type hierarchy*

- T_R, *the set of* relation symbols, *is a partially ordered set partitioned into subsets* $T_R^1, \ldots, T_R^k$ *of relation symbols of arity* $1, \ldots, k$, *respectively. The arity of a relation* r *is denoted* $arity(r)$. *Any two relations with different arities are not comparable.*

- T_N *is a hierarchy of nesting symbols with a greatest element*

An example of a relation type hierarchy is given in Figure 12.3.

An example of a nesting type hierarchy, which represents different sorts of annotations, is given in Figure 12.4.

A simple (conceptual) graph (with conjunctive concept types and coreference links) built over a vocabulary is precisely defined as follows (see also a slighlty different formalization in Chapter 2).

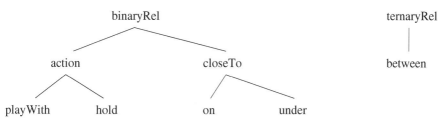

FIGURE 12.3:   A relation type hierarchy.

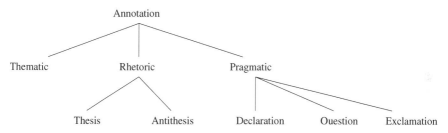

FIGURE 12.4:   A nesting type hierarchy.

DEFINITION 12.5 Simple conceptual graph   *A simple conceptual graph (in short SG) over a vocabulary $\mathcal{V}$ is a 5-tuple $(C, R, E, l, coref)$, such that:*

- (C, R, E) *is a finite, undirected and bipartite multigraph. C is the concept node set, R is the relation node set, and E is the family of edges.*

- *l is a labelling function of nodes and edges that satisfies:*

    1. *A concept node c is labelled by a pair $(\text{type}(c), \text{marker}(c))$, where $\text{type}(c)$ is a type of T_C, i.e., an acceptable conjunctive type, and $\text{marker}(c) \in \mathcal{I} \cup \{*\}$, where $\mathcal{I}$ is a set called the individual marker set, and $\{*\}$ is called the generic marker.*

    2. *A relation node r is labelled by $l(r) \in T_R$. $l(r)$ is also called the type of r and is denoted by $\text{type}(r)$. The degree of a relation node r is equal to the arity of $\text{type}(r)$.*

    3. *Edges incident to a relation node r are totally ordered, and they are labelled from 1 to $arity(\text{type}(r))$.*

    4. *For any individual marker m, the set of all concepts with marker m is compatible, i.e., the union of their types is an acceptable conjunctive type.*

- *$coref$ is an equivalence relation over C, such that:*

    any coref class is compatible,

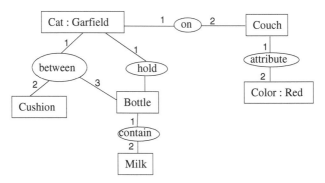

FIGURE 12.5: A simple graph.

> *if two individual concepts have the same marker, then they are in the*
> *same coref class.*

The coreference relation represents the equality relation, but note that we consider the open-world assumption, which means that if two concept nodes are not coreferent, i.e., are not in the same coreference class, they nevertheless may represent the same entity. The SG given in Figure 12.5 is an annotation of a picture representing Garfield holding a bottle of milk and sitting on a red couch between a cushion and the bottle of milk.

12.2.1.2   Positive Nested Graph

The nested conceptual graph model presented is a simple extension of SGs that is used to represent hierarchically structured knowledge (cf. [Chein and Mugnier, 1997]). Nestings are represented by boxes, and *positive* is used to differentiate these boxes from those used in general conceptual graphs, where a box represents a negation (cf. [Dau, 2003a]). In this chapter, only positive nested graphs are considered, thus we omit "positive." Nested graphs are used here to represent complex annotations built with different viewpoints. Simple Nested Graphs are first defined, then we introduce a typing of the nestings.

DEFINITION 12.6 Simple Nested CG

1. *An elementary SNG is obtained from a simple graph (SG) by adding to the label of each concept node c, a third field, denoted by $Desc(c)$, equal to ∗∗ (∗∗ can be considered as the empty description). There is a trivial bijection between elementary SNGs and SGs.*

2. *Let G be an elementary SNG, let $c_1, c_2, \ldots, c_k$ be concept nodes of G, and $G_1, G_2, \ldots, G_k$ be SNGs. The graph obtained by substituting G_i to the description ∗∗ of c_i for $i = 1, 2, \ldots, k$ is an SNG.*

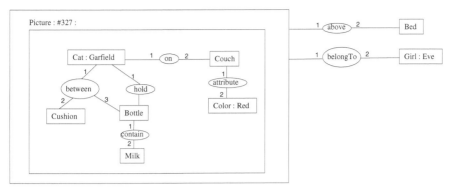

FIGURE 12.6: A simple nested graph.

3. *coref is an equivalence relation defined on the set X_G of all the concept nodes in G that satisfies:*

   (a) *each class of the coreference relation of any SGs of G is included in coref*

   (b) *the concepts of a coref class are compatible*

   (c) *individual concepts having the same marker are in the same coref class*

An example of an SNG is given in Figure 12.6. It represents the fact that the picture number 327, whose description is given by the graph in Figure 12.5, belongs to the girl Eve and is located above a bed. Other examples are given in Chapter 2.

DEFINITION 12.7 Typed Nested Graph

1. *An elementary TNG is an elementary SNG.*

2. *Let G be an elementary TNG, let $c_1, c_2, \ldots, c_k$ be concept nodes of G, $n_1, n_2, \ldots, n_k$ be a set of nesting types, and $G_1, G_2, \ldots, G_k$ be TNGs. The graph obtained by substituting the set $\{(n_1, G_1), \ldots, (n_k, G_k)\}$ to ** in the label of c_i is a TNG.*

An example of a TNG is given Figure 12.7.

Finally, *annotations are represented by TNGs*, i.e., the annotation language involves TNGs built upon a vocabulary.

12.2.2 Rules and Constraints

A rule R is first composed of a couple (A, B) of SGs, A is called the hypothesis of R and B is called the conclusion of R and, secondly, of l generic

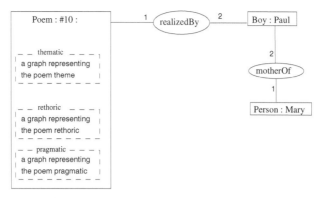

FIGURE 12.7: An annotation = a typed nested graph.

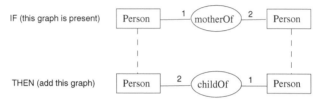

FIGURE 12.8: A rule graph.

concepts $c_1, \ldots, c_l$ of A, which are associated with l generic concepts $c'_1, \ldots, c'_l$ of B, and for all i c_i and c'_i have the same type. The distinguished generic concepts c_i and c'_i are called connection nodes.

Let R be a rule having the form "if A then B," then it is interpreted as follows: if a graph G contains A, then B can be added to G.

The rule representing the knowledge that "if A is the mother of B then B is a child of A" is represented as a graph in Figure 12.8. Thus, rules are useful for representing implicit (common sense) knowledge. An annotator does not need to put knowledge represented by a rule into an annotation, the system will automatically do that (if someone has entered the rule into the system).

Positive and negative constraints can be defined using the same formalism as a rule. Roughly said, a positive constraint "if A then B" means that if a graph G contains A, then B must be present in G, and a negative constraint "if A then B" means that if a graph G contains A, then B cannot be present in G.

A hierarchy of family of conceptual graphs using rules and constraints, and called the $\mathcal{SG}$ family is defined and studied in [Baget and Mugnier, 2002a].

A constraint is represented in the same way as a rule, i.e., a pair of simple graphs with a pair of connection node sets. The positive constraint representing the fact that "if there is a picture then the person who took this picture must be given" is represented as a graph in Figure 12.9.

IF (this graph is present)

THEN (this one must be present)

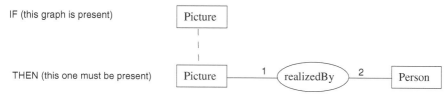

FIGURE 12.9: A positive constraint graph.

12.3 Construction of Annotations

On top of a vocabulary and a set of rules and constraints, different notions can be used to facilitate, and also to direct, the construction of relevant annotations.

12.3.1 Mandatory Constraints

Signatures of relations and typing of individuals, are two classical notions useful for avoiding absurd annotations. The signature function is a mapping σ that assigns a tuple of concept types to each relation. The length of the tuple is the arity of the relation and the i-th element of $\sigma(r)$ is the greatest type of any i-th argument of r. For example, the signature of the relation *realizedBy* can be (*Object, Person*), and this leads to forbid *realizedBy(Car, Flower)*. The typing of individuals is a function from the set of individual markers to the set of concept types. For example, in some application, saying that the type of the individual *Mercure* is *God* induces the use of another individual for denotating the *Mercure hotel*. The examples previously given show how positive constraints can be used to force the annotations to contain necessary information, and how the negative constraint can be used to force the annotations to not contain absurd statements.

12.3.2 Modules

The module notion restricts the vocabulary to a subset dedicated to represent a specific kind of annotation. An annotation can be constructed from scratch, i.e., from an empty graph. If there are several annotators, there may be great variety among annotations of similar resources. The module notion, and the notions of pattern graphs, prototype graphs, and individual graphs aim to direct the construction of an annotation by proposing annotation templates. Precise definitions are now given. A *modular vocabulary* is a vocabulary composed of *modules*. A module is a subset of the global vocabulary that is associated with a nesting type.

DEFINITION 12.8 Module   A module *of a vocabulary*

$$V = (T_C(T, B), T_R, T_N),$$

where $T_C(T, B)$ *is the set of acceptable conjunctive types generated by the primitive type set* T *and the basic banned type set* B, *is a triple* $m = (T'_C(T', B'), T'_R, T'_N)$ *where:*

- T' *is a subset of* T, *and* B' *is the set of maximal elements of* $B^\sqcap \cap T'$

- T'_R *is a subset of* T_R

- T'_N *is a subset of* T_N

T'_C, T'_R, T'_N *are partially ordered by the partial orders on* T_C, T_R *and* T_N, *respectively.*

Note that if a vocabulary is equipped with a signature function, then a module is not necessarily a (sub)vocabulary because the concept types of the signature of a relation in T'_R can be outside T'_C. The module notion is used to define modular vocabularies and annotation dimensions.

DEFINITION 12.9 Modular vocabulary   A modular vocabulary *is a triple* $W = (V, M, D)$, *where* $V = (T_C, T_R, T_N)$ *is a vocabulary,* M *is a set of modules of* V, *and* D *is a mapping from* T_N *to* M. *A dimension is a pair* $d = (n, m)$, *where* $n \in T_N$ *is a nesting type, and* $m \in M$ *is the module* $D(n)$ *representing the point of view that* n *designates.*

The partial order on nesting types induces a specialization relation between dimensions. A dimension $d' = (n', m')$ is said to be more specific than a dimension $d = (n, m)$ if $n' \leq n$. The hierarchization of dimensions leads to constraints on the associated modules. A type used in a specific dimension must be interpretable in a more general dimension. Let us suppose that $m = (T_C, T_R, T_N)$ and $m' = (T'_C, T'_R, T'_N)$. Then, two kinds of constraints can be considered:

- m' is *included* in m if $T'_C \subseteq T_C$, $T'_R \subseteq T_R$, and $T'_N \subseteq T_N$;

- m' is a *specialization* of m if for every $t' \in T'_C$, there is $t \in T_C$ such that $t' \leq t$, and the same conditions hold for relation types and nesting types.

The specialization constraint is weaker than the inclusion constraint, i.e., it allows, in a specific module, a concept type that does not appear in the more general module.

A TNG G built over V respects a modular vocabulary $W = (V, M, D)$ if it fulfills:

1. the elementary NG from which G is constructed, i.e., composed of all nodes not contained in a nesting box, is an elementary NG built over V

2. for each typed nested graph (n_i, G_i) in G, the graph G_i is built over m_i, i.e., the module that is associated with the nested type n_i.

Thus, a modular vocabulary allows restriction of the terms used when constructing the part of an annotation corresponding to a given viewpoint. It is easy to check if an annotation respects a modular vocabulary, or the annotator tool can even simply allow the construction of annotations respecting a modular vocabulary.

12.3.3 Pattern Graphs, Prototype Graphs, Individual Graphs

A *pattern graph* is a TNG associated with a dimension d, which represents a starting point when annotating a resource with respect to dimension d. Such a pattern graph consists of general and frequent notions appearing when annotating a resource with respect to d. A pattern graph for d has to respect the module of d. Several pattern graphs can be associated with a given dimension d, e.g., when a dimension is rather general and contains several objectives. Figure 12.10 is an example of a pattern graph, used in the Saphir project, associated with the *rhetoric* dimension. It is used to describe the use of a language.

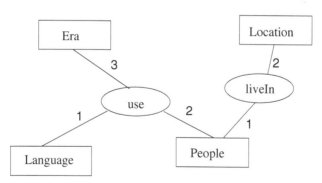

FIGURE 12.10: A pattern graph associated with the *Rhetoric* dimension.

A *prototypical graph* can be associated with a concept type or with a relation symbol. A prototypical graph defines a usual context of a concept (or a relation) of some type. In a prototypical graph, there is a special node, called the *head* of the graph, with the same type (or relation) as that from which it is the prototype. Nodes of an annotation whose types have prototypical graphs are called extension nodes. Such a node n of type t can be used to extend the annotation by merging the head of a prototypical graph of t with n. A (concept or relation) type can have several prototypes, with each prototype representing a way to describe an entity or a relation. Prototypical graphs are not directly related to dimensions, but during the annotation process one can only access prototypes whose types are all in the module of the current dimension. Figure 12.11 shows an example of a prototypical graph for the concept type *linguistic sign system.*

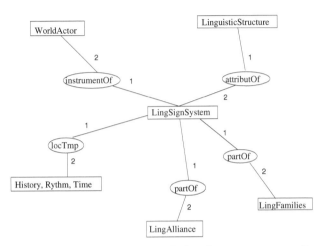

FIGURE 12.11: A prototypical graph for the concept type *linguistic sign system.*

An *individual graph* looks like a prototypical graph, but its head is necessarily an individual concept and it is not attached to a type. An individual graph for the individual marker m represents general knowledge concerning m. The individual concepts, in an annotation, having associated individual graphs are used as extension nodes. Indeed, an annotation A can be extended by merging the head of an individual graph for m with an individual concept of marker m in A. An individual marker m can have several individual graphs that represent different ways to describe the entity denoted by m. Individual graphs are not directly related to dimensions, but during the annotation process one can only access individual graphs whose types are all in the module of the current dimension.

FIGURE 12.12: A query graph.

12.4 Search Methods

We first briefly consider exact answers to a query, then we explain a method to compute approximate answers. Approximate answers are important because vagueness, incompleteness, and impreciseness are intrinsic to the problem discussed here. Indeed, annotations are subjective metadata about resources, so annotations cannot be assumed to be exact and complete data as is the case in the database domain. Furthermore, it is the same for queries that cannot always be assumed to be exact and complete. It is essential to have a model allowing flexible searches in order to cope with the intrinsic difficulties of our problem.

12.4.1 Exact Search

Let $\mathcal{A}$ be an annotation base (of a resource base) composed of TNGs built over a vocabulary $\mathcal{V}$. Let Q be a query graph (also built over $\mathcal{V}$). The *exact answer* of Q over $\mathcal{A}$ consists of all graphs in $\mathcal{A}$, which are specializations of Q. For example, let us consider the query graph Q in Figure 12.12. The graph in Figure 12.5 is an answer to Q since there is a mapping from Q to G, which maps: the concept [Animal] to the concept [Cat : Garfield], the relation (closeTo) to the relation (on) and the concept [Furniture] to the concept [Couch]. One can say that H is a specialization of G if there is a mapping from G to H, which preserves the structure and can restrict the labels. This is the definition of the homomorphism notion between simple graphs that is reviewed below (homomorphisms between CGs are traditionally called "projections" in the CG community), and this notion can be simply extended to nested graphs (cf., for instance, [Chein et al., 1998]).

DEFINITION 12.10 Homomorphism   *If $G=(C_G, R_G, E_G, l_G, coref_G)$ and $H = (C_H, R_H, E_H, l_H, coref_H)$ are two SGs defined on the same vocabulary, then a homomorphism π from G to H is a mapping from C_G to C_H and from R_G to R_H, which preserves edges and may decrease concept and relation labels, that is:*

1. *$\forall (r, i, c) \in G, (\pi(r), i, \pi(c)) \in H$*

2. *$\forall e \in C_G \cup R_G,\ l_H(\pi(e)) \leq l_G(e)$*

3. *$\forall x, y \in C_G, coref_G(x, y) \Rightarrow coref_H(\pi(x), \pi(y))$*

Actually, the sought answers are not annotations but are resources whose annotations are specializations of the query. As each annotation contains a reference to the resource it annotates, this set of resources can be simply obtained from annotations answering a query.

TNGs can be equipped with first-order logical semantics, which is an extension of the FOL semantics Φ introduced for simple graphs in [Sowa, 1984], and the homomorphism notion between TNGs is sound and complete with respect to Φ. Simply stated, there is a homomorphism from G to H if and only if $\Phi(G)$ is a logical consequence of $\Phi(H)$ and $\Phi(\mathcal{V})$, where $\Phi(\mathcal{V})$ is a set of FOL formulas representing the order relations over the type sets (cf., for instance, [Chein et al., 1998]).

This soundness and completeness result means, firstly, that exact answers can be computed with graph homomorphism algorithms and, secondly, that graph transformations can be used to define and compute approximate answers.

Let us insist on the importance to have a model that enables approximate search in order to cope with the intrinsic difficulties of our problem, i.e., impreciseness and incompleteness of annotations and queries.

Let Q be a query graph, $\mathcal{A}$ be a graph annotation base, and $ans(Q, \mathcal{A})$ be the answer of Q on $\mathcal{A}$, i.e., the set of graphs in $\mathcal{A}$, which are specializations of Q. It is easy to extend the elementary exact search by the classical boolean operators AND, OR, and NOT as follows.

- $ans(Q \text{ AND } Q', \mathcal{A}) = ans(Q, \mathcal{A}) \cap ans(Q', \mathcal{A})$

- $ans(Q \text{ OR } Q', \mathcal{A}) = ans(Q, \mathcal{A}) \cup ans(Q', \mathcal{A})$

- $ans(Q \text{ AND NOT } Q', \mathcal{A}) = ans(Q, \mathcal{A}) \setminus ans(Q', \mathcal{A})$

12.4.2   Approximate Search

Probabilistic or numerical approaches are commonly used to develop approximate search methods. In this section, we present a combinatorial approach based on graph homomorphisms and simple graph transformations, which is easily visualizable and understandable by nonspecialist users (see also [Genest and Chein, 2005]).

This approach is an instantiation of the following form of van Rijsbergen's Uncertainty Principle: *Let A and B be two formulas, a measure of the uncertainty of $A \to Q$ relative to a knowledge base is determined by the minimal transformation from A to A' such that $A' \to Q$ holds.* (cf. [J.Y. Nie, 1998]).

In our case, A is an annotation graph and Q is a query graph. Therefore we consider transformations from an annotation graph A to an annotation graph A', although usually in Information Retrieval transformations of queries (query reformulation) are considered. The two approaches can be considered as equivalent because dual operations can be defined to transform the queries and it can be shown that $A' \to Q$ holds if and only if $A \to Q'$ holds, where Q' is obtained from Q by a dual transformation of a transformation from A to A'.

The advantage of modifying annotations rather than queries is that the annotation base can be enriched in different ways by using the system. A modified annotation obtained when the system finds approximate answers to a query can be proposed to be included in the annotation base (if the user is satisfied by the answer). Modifying annotations makes the annotation base a "living" base.

Graph transformations have to be simple in order to be understood by end-users. There are three kinds of elementary transformations: label substitutions, concept node identifications, and node adjunctions. Let us consider an annotation graph A and a query graph Q. A label substitution consists of replacing a concept or a relation node label by a non-smaller label (replacing a label by a smaller label is taken into account by the homomorphism definition). A homomorphism is a mapping from the node set of Q to the node set of A, by identifying two concept nodes in A or by adding concept or relation nodes in A, the number of homomorphisms from Q to A can be increased.

In case of approximate search, it is mandatory to rank the answers. Ranking procedures can be proposed by the way of a total preorder relation defined over the transformation set. The admissible transformation sequences and the total preorder among them are application-dependent. They are especially dependent on the vocabulary upon which the annotations are built. In [Genest and Chein, 2005] three predicates are used to define admissible transformation sequences.

- *substitute-term*(x, y) asserts that the concept type x can be a substitute for the concept type y.

- *substitute-relation*(x, y) asserts that the relation type x can be a substitute for the relation type y.

- *acceptable-sequence*(s) asserts that a sequence s of transformations is acceptable. Replacing y by x when *substitute-term*(x, y) is a simple example of an acceptable transformation.

These predicates must be defined in such a way that if A is an annotation of a resource R, then a graph A' obtained from A by an acceptable sequence of transformations is an acceptable annotation of R.

A fourth predicate defining a total preorder $s \leq s'$ between sequences of acceptable transformations must be defined in such a way that it can be used as a ranking function for answers to a query. Usually these predicates are defined using parameters. For instance, *substitute-term*(x, y) holds if and only if the distance from x to y in the ordered concept type set is less than some integer number. A sequence s of substitutions and identifications is admissible for a query Q if s does not contain too many identifications relative to the number of nodes in Q (i.e., the structure of the annotation is only slightly modified). A ratio a can be chosen by the user. The previous total preorder $\leq$ must be extended to the set of sequences of substitutions and identifications. As an identification can substantially change the graph structure, this transformation is usually considered more distancing than the substitutions.

A sequence s of substitutions and identifications and node adds is acceptable for a query Q if it is acceptable for the substitutions, and if there are not too many node adjunctions. Parameters can be considered for controlling these transformations, i.e., parameter b concerning the ratio number of new concept nodes, and c concerning the ratio of new relation nodes. The previous total preorder $\leq$ must be extended to the set of sequences of label substitutions and identifications and node adjunctions.

12.5   Conclusion

The task tackled in this chapter can be defined as an enrichment of a set of resources by semantic annotations that can be automatically processed. This task is fundamental and interesting itself since it is the basic task of any content information retrieval system. Furthermore, as said in the introduction, such a task is a critical component of editing applications.

12.5.1   Advantages

The annotation model presented in this chapter has good properties that make it relevant to semantically annotated resources:

- it is a kind of semantic network, thus it allows a graphical representation, and thus an intuitive interpretation, of the knowledge;

- it allows structuring and contextualization of knowledge, thanks to nested graphs;

- it allows the representation of different kinds of knowledge: annotation, patterns, rules, constraints (*cf. SG* family [Baget and Mugnier, 2002a]);

- it is logically founded, allowing to define notions of consistency (for an annotation and an annotation base), and deduction from annotations;

- it has exact as well as approximate reasoning mechanisms based on graph theory. Using graph models allows to use existing efficient graph algorithms, and to graphically visualize inconsistencies as well as answers;

- it is close to RDF, the annotation standard language of the Semantic Web (*cf.* equivalence properties in [Baget, 2005a]), and to Topic Maps, the ISO specification of document descriptions (*cf.* transformations in [Carloni et al., 2006]), thus allowing the semantic interoperability of built annotations.

12.5.2 Limits

A first fundamental limit of the proposed model is that building a shared ontology is a difficult task.

A second fundamental limit is that annotations cannot be automatically built from the resources, even if the resources are electronic. Such annotations have to be manually built. Manually building has two main drawbacks, first it is a time-consuming process, and secondly it is an unstable process, i.e., two persons (or even the same person at different time) can build different annotations of the same resource.

These limits are not intrinsic to the graph model proposed; indeed they concern any model equivalent to a nontrivial fragment of the first-order logic. Nevertheless, any approach based on complex conceptual structures can be worthwhile in specific situations, such as:

- the resources are not in electronic form,

- the resources are electronic images or videos for which high-level semantic annotations are needed and there is no automatic mechanism for directly extracting them from the resources (e.g., cf. [Schreiber et al., 2001], [Hollink et al., 2003], [Aubert et al., 2006b]),

- the resources are (multimedia) documents and one wants to use them in other contexts than their first publication context for editing new documents composed of annotated parts of a document base.

12.5.3 Further Developments

Many improvements can be considered within the same graph-based approach and some of them are described as follows.

- theoretical and algorithmic results have been obtained concerning extension of the model with simple forms of negation (cf. [Mugnier and Leclère, 2007] and [Leclère and Mugnier, 2007]), and they have to be included in the annotation model and in the tools

- classification techniques can be used for structuring a large annotation base in order to improve the search algorithms (cf. [Ellis, 1995])

- hybrid models combining a relational data base for the objective metadata with the presented model for the subjective metadata can be developed

- hybrid models combining the approach presented here with a description logic for representing the vocabulary can be developed

- automatic learning techniques for building prototypical graphs can be used

- approximate search algorithms using prototypical graphs can be developed

- a set of resources is generally dynamic, i.e., resources can appear or be removed. Thus, the vocabulary should be able to evolve. If new terms appear, the existing annotations are still correct but they can use terms that are too general relative to the new vocabulary. If terms are removed, all existing annotations containing removed terms are no longer valid.

12.6   Student Projects

Two sets of exercises are proposed. The first one concerns an application and the second one deals with the conceptual graph model used for constructing semantic annotations.

12.6.1   An Example

In this section we will consider the problem of describing photos representing children who are playing. The objective is to build annotations in order to help the management of a set of such photos.

1. Construct a set of primitive concept types (about 50) relevant to the "children playing" application.

2. Construct a set of banned types relevant to the primitive concept type set built previously.

3. Construct a set of relation-types (about 20) with signatures.

4. Propose some rules and constraints. First, describe them by sentences in English, then represent these sentences by graph rules and constraints.

5. Choose 2 or 3 photos concerning children playing and represent them by simple conceptual graphs.

6. What are the main difficulties encountered while studying the previous questions?

12.6.2 Questions about the Formal Model

1. Is it possible to consider conjunctive relation-types in the same way as conjunctive concept types?

2. Let G and H be two SGs and π a homomorphism from G to H. What are the properties of two relation nodes r and r' in G such that $\pi(r) = \pi(r')$? At what conditions can two relations nodes of an SG have the same image by a homomorphism?

3. Propose a mechanism for concept type definitions. Give a FOL semantics concerning your proposal. What are the questions to be solved concerning this concept type definition mechanism?

4. Define a *partial* homomorphism from an SG to an SG. How could this notion be used for approximate reasoning?

Chapter 13

An Overview of Amine: An Integrated Development Environment for Symbolic Programming, Intelligent System Programming and Intelligent Agents Programming

Adil Kabbaj

INSEA, Rabat, Morocco

13.1 Introduction

Amine can be viewed as a modular integrated development environment (IDE) for symbolic programming, intelligent system programming, and intelligent agents programming.

Amine is a Java open source multi-layer platform, based on Conceptual Graph theory. Amine is a synthesis of 20 years of work done by the author [Kabbaj, 1987, 1993, 1995, 1996, 1999a,b, 2006, Moulin and Kabbaj, 1990,

Kabbaj et al., 1994, Kabbaj and Janta-Polczynski, 2000, Kabbaj et al., 2001]
to investigate several aspects of Conceptual Graph theory (see the Chapter
by John Sowa in this volume for a historical background about CG theory).

This chapter presents an overview of Amine[1] and it is organized as follows:
Section 13.2 introduces briefly the architecture of Amine platform. Sections
13.3-13.7 introduce the Ontology/KB layers, the Algebraic layer, the Memory-
based inference and learning strategies layer, the Programming layer and the
Multi-Agents Systems layer respectively. Section 13.8 specifies briefly how
Amine can be used as an IDE for symbolic programming. Section 13.9 illustrates the use of Amine for intelligent system programming. The illustration is
done with Natural Language Processing applications. Section 13.10 specifies
briefly how Amine can be used for intelligent agents programming and multi-
agents systems. Section 13.11 provides a comparison of Amine with other CG
tools. Section 13.12 outlines some current and future works. Section 13.13
presents a conclusion. Section 13.14 provides a list of some suggested projects.

13.2   Amine Architecture

Amine is composed of seven layers (Figure 13.1):

1. *Ontology layer:* It concerns the creation, edition, and manipulation of
   multi-lingua ontologies , which provides the "conceptual vocabulary" and
   the semantic of a domain.

2. *Knowledge Base layer:* it concerns the creation, edition and manipulation of Knowledge Bases (KB).

3. *Algebraic layer:* this layer provides several types of structures and opera-
   tions. Amine offers *elementary data types* (AmineInteger, AmineDouble,
   String, Boolean, etc.), *structured descriptions* (AmineSet, AmineList,
   Term, Concept, Relation and Conceptual Graph) and *structured patterns* (structured descriptions that are generic; they contain variables).
   In addition to operations that are specific to each kind of structure,
   Amine provides a set of basic common operations (clear, clone, toString,
   etc.) and various *matching-based operations* (match, equal, unify, subsume, maximalJoin, and generalize) common to all structures. Operations on structured patterns take into account *variable binding* (the
   association of a value to a variable) and *binding context* that specifies
   the programming context that determines how variable binding should

---

[1]http://sourceforge.net/projects/amine-platform

be interpreted and resolved, i.e., how to associate a value to a variable and how to get the value of a variable.

Amine structures and operations (including CG structure and CG operations) are APIs that can be used by any Java application. They are also "inherited" by the higher layers of Amine.

Amine uses "Java interfaces" to guaranty genericity and openness of Amine to Java: new structures, that implement these Java interfaces, can be integrated as new structures to Amine.

4. *Memory-Based Inference and Learning Strategies Layer*: Ontology and KB constitute the memory of the system. Amine provides some inference and learning strategies that are defined as memory-based processes.

5. *Programming layer:* Three complementary programming paradigms are provided by Amine:

    (a) *pattern-matching and rule-based programming paradigm*, embedded in PROLOG+CG language, which is an object-based and CG-based extension of PROLOG language,

    (b) *activation and propagation-based programming paradigm*, embedded in SYNERGY language and based on *executable CG*, and

    (c) *ontology or memory-based programming paradigm*, which is based on memory-based inference and learning strategies.

6. *Agents and Multi-Agents System layer:* Amine can be used in conjunction with a Java Agent Development Environment to develop multi-agents systems. Amine does not provide the basic level for the development of multi-agents systems (i.e., implementation of agents and their communication capabilities using network protocols) since this level is already offered by other open-source platforms (like JADE[2]). For instance, JADE is currently used in Amine for this purpose.

7. *Application layer:* various kinds of applications can be developed, using Amine (or a subset of Amine): Knowledge-Based Systems, Ontology-based applications, natural language processing applications, problem-solving applications, planning applications, reasoning based applications, learning based applications, multi-agents applications, etc.

Amine also provides several graphical user interfaces (GUIs): Ontology/KB editor GUI, CG editor GUI, CG Operations GUI, Ontology processes GUI, Dynamic Ontology GUI, Inference and learning strategies GUI, Prolog+CG GUI, and Synergy GUI. *Amine Suite Panel* provides an access to all these GUIs, as well as an access to some ontology examples and to several tests

---

[2]http://jade.tilab.com

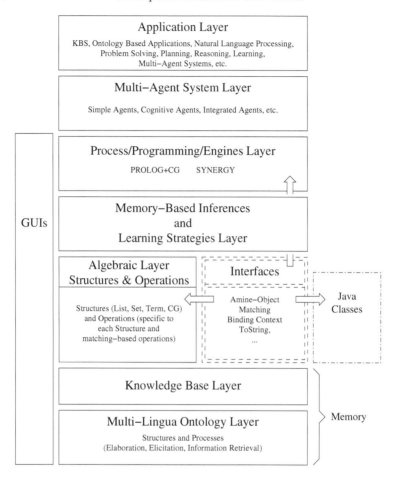

FIGURE 13.1: Amine architecture.

that illustrate the use of Amine structures and their APIs. Amine also has a Website, with samples and a growing documentation.

Amine seven layers form a *hierarchy*: each layer is built on top of and uses the lower layers (i.e., the algebraic layer inherits the ontology layer; the KB layer is supported by the ontology layer, the programming layer inherits the ontology and the algebraic layers, etc.). However, a lower layer can be used without the higher layers: the ontology layer (i.e., with the associated APIs) can be used directly in any Java application without the other layers. Algebraic layer (i.e., with the associated APIs) can be used directly, too, etc.

Among the goals (and constraints) that have influenced the design and implementation of Amine was the goal to achieve a higher level of modularity and independence between Amine components/layers.

Beside the documentation provided by the Amine Website, the group of Peter Øhrstrøm and Henrik Schärfe has developed on-line courses that cover some parts of Prolog+CG and Amine Platform[3] [Petersen et al., 2001-2008].

Amine is used by the author to teach Artificial Intelligence (AI) courses. Amine is suited for the development of projects in various domains of AI (i.e., natural language processing, problem solving, planning, reasoning, case-based systems, learning, multi-agents systems, etc.). Different groups are actually using Amine in their projects.

Please consult the Amine Website[4] for more detail. Source code, samples, documentation and Amine Website can be downloaded from SourceForge.net site[5].

13.3   Ontology and KB Layers

In Metaphysics and Philosophy, Ontology is not concerned by our "Commonsense Knowledge about the World," but by "the quest of what is necessarily true about the Physical World" [Smith, 1995, 2003, 2004, Guarino, 1995, Farrar and Bateman, 2004]. Knowledge should be discarded in favor of a truthful and formal description of the "essence" of the World.

In the last decade, Artificial Intelligence (AI) and Information Sciences scientists started with "an analogical use" of the term Ontology; viewed as a Terminological/Definitional Knowledge Base that guarantees an efficient sharing of (Large) Knowledge Bases (KB). Ontology is therefore viewed as a "terminological" support to (large) KB [Guarino, 1995, Brewter and O'Hara, 2004].

This "analogical use" of the term/concept Ontology has been criticized by philosophers [Smith, 1995, 2003, 2004, Guarino, 1995, Farrar and Bateman, 2004] who propose instead the use of "Formal Ontology" and a methodology suited for ontology design [Guarino, 1995, Smith, 1995].

"Applied Ontology," based on Formal Ontology, is now emerging as a new multi-disciplinary field, with Semantic Web as the application domain "par excellence" [Brewter and O'Hara, 2004].

In general, Formal Ontology is proposed as a new approach to Computer-Human Interaction/Communication (CHI/C). However, in [Kabbaj et al., 2006] we argue that "Full/Natural" (i.e., uncontrolled, unrestricted, unlimited) CHI/C is also a central topic in Cognitive Science in general and in AI in particular. It is well established, from these sciences, that Commonsense

---

[3]http://www.huminf.aau.dk/cg/

[4]http://amine-platform.sourceforge.net

[5]http://sourceforge.net/projects/amine-platform

Knowledge Base is a (if not the) vital aspect in natural CHI/C. Commonsense KB is a definitional and schematic/situational KB. Also, "cognitive ontology"; ontology of a human or of a soft agent is not a formal ontology but an ontology that is more like a Commonsense KB [Dölling, 1993]. Cognitive ontology of an agent is fully integrated to the other kinds of agent's knowledge in his dynamic memory.

For these reasons, Amine provides two layers (Ontology and KB layers) and adopts a generic Ontology/KB meta-model that covers all the above cases: Ontology in Amine is defined on a large "Knowledge Base spectrum" with "formal ontology" and "Commonsense KB" as the lower and upper limits respectively. Formal ontology is considered in this scheme as a rigorous, systematic, definitional, and truthful commonsense KB.

With its two layers and its generic Ontology/KB meta-model, Amine allows for the construction of different kinds of intelligent systems: intelligent system with only a formal ontology, intelligent system with an ontology augmented by commonsense knowledge, and intelligent system with an ontology and various KB (i.e., commonsense or specific knowledge are kept separate from ontological knowledge).

13.3.1 Amine's Ontology/KB Meta-Model

Amine does not provide a specific ontology, but provides "epistemological constructs" that determine the organization of "multi-lingua conceptual ontology" and "multi-lingua KB." It is the responsibility of the user/designer to use "adequately" these constructs to model her/his ontology and KB according to her/his domain semantics.

A *multi-lingua conceptual ontology* is an ontology with one or many lexicons (for different languages). An ontology is a hierarchical knowledge base that provides the "meaning" of the conceptual vocabulary (concept types, relations and individuals) of domain semantics.

In Amine, the "meaning" of a concept type is not restricted (closed) to the (classical) type definition, it is "open" to the whole commonsense knowledge (CK) associated to the type. The CK for each type is organized in terms of different *Conceptual Structures* (CS).

More specifically, Amine's ontology meta-model proposes three *kinds* of CS for three *kinds* of CK (CK that is related to each type):

1. *Definition of a type:* this CS enables the expression of definitional knowledge about a specific type.

2. *Canon for a type:* this CS enables the expression of constraint and axiomatic knowledge about a specific type. It is currently limited to the expression of canonical constraints [Sowa, 1984].

3. *Schematic cluster (schemata) for a type:* schematic cluster is an open-end extendible and dynamic set of schemata related to a specific type.

In some cases, we cannot reduce the meaning of a concept to a closed and situation-free definition (assuming that such a definition is possible). A concept is like a "coin with two sides": a possible closed-fixed-static side (the classical definition) and an open-extendible-dynamic side (the growing schematic cluster). Schematic cluster constitutes a fundamental part in the specification of the meaning of a concept. Schema is strongly related to other fundamental cognitive notions like "Patterns," "Conceptual Network," "Mental Models," "Situations," "Contexts" and "Experiences" (chunk of Knowledge). There is also various implementations of schemata in AI (frames, scripts, situation, schema, MOP, TOP, TAU, etc.).

At the structural level, ontology in Amine is a generalization graph where nodes represent Conceptual Structures (CSs). Currently, there are five types of nodes: Type node, RelationType node, Individual node, Schema (or Situation) node, and Rule node:

1. *Type and RelationType nodes:* nodes that represent concept type and relation type, respectively. These nodes contain the *definition* of the type (if provided) and/or the *canon* of the type (if provided).

2. *Individual node:* a node that represents an individual (an instance) of a concept type. This node contains the *description of the individual* (if provided).

3. *Schema/Situation node:* a node that contains the description of a schema-/situation.

4. *Rule node:* a node that contains the description of a rule.

There are three types of links used in the composition of an ontology (in Amine):

1. *Specialization link (s):* A type (node) can be (s)pecialized by other types (related to them by the (s)pecialization/subtype link), a schema/situation can be specialized by other schemata/situations, and a rule can be specialized by other rules.

2. *Instantiation link (i):* the (i)ndividual link relates an individual (an instance) to its type.

3. *Utilization link (u):* In general, a schema/situation is not indexed under all types contained in it, but only to some of them (determined by the user or by a process that interacts with the ontology). The schema is related to these types with (u)se links.

None of the above Conceptual Structures are mandatory: a "minimal" definition of a type is the specification of its super-types (the equivalent of primitive concepts in KL-ONE family). Then, if a definition is available, it can be specified. Also, schemata/situations can be added to the ontology as more commonsense knowledge (CK) is allowed, acquired and required. Also, the content of a canon can be augmented by more constraints and axiomatic rules as needed.

By constraining what kind of CK to consider, Amine's user can define and use different *kinds* of ontology. For instance, the simple kind of ontology is "taxonomic ontology"; ontology composed of primitive-types hierarchy. Another kind of Ontology is "terminological ontology" where the hierarchy is enriched by type definitions (and optionally axioms). Lastly, user can define and use "Commonsense Knowledge-based Ontology," where "full" CK is considered; with the use of schematic clusters and rules.

Amine Ontology Layer can be used to construct and develop all these kinds of ontologies.

Amine-lingua KB has an ontology as a support and it corresponds to a generalization graph composed mainly of situations and rules nodes.

13.3.2　Ontology/KB modelling language

Amine ontology and KB Layers are not committed to a specific knowledge modelling language; any language (CG, KIF, XML, RDF, Frame-Like, etc.) that is interfaced with Java may be used as a modelling language for the description of CS. Indeed, at the implementation level, a description of a CS is declared as "Object" (the root class in Java). However, since Conceptual Graph (CG) is the most developed structured description in Amine and since CG is used as the basic modelling language in the other layers of Amine (algebraic, programming and multi-agents layers), we use also CG as a modelling language in our ontology and KB examples.

To be able to create/edit/update ontology/KB in Amine, the developer may either (i) use ontology/KB layer's APIs, from a Java program, or (ii) use directly Amine's ontology/KB GUI. The Amine Website provides examples of ontology and KB creation and update using related APIs from Java programs. The next section presents briefly Amine's ontology/KB GUI.

13.3.3　Amine's Ontology/KB GUI

Amine's ontology/KB GUI is a multi-view editor (a tree view editor, a drawing view editor and a browser view editor) that allows the creation, consultation, browsing, edition and update of ontology and KB. This section presents briefly these three views. Figure 13.2 is a snapshot of an ontology edited using the tree view editor. The ontology is visualized in the main frame as a hierarchy. Tree nodes represent CSs nodes of the ontology (each type of node is represented with a different color). As noted before, a type node can be specialized by other types (and related to them by the (**s**)pecialization link). For instance, "Pelican" is a specialization of "Bird." Also, a type can be related to several individuals (by (**i**)ndividual link) and it can be related to several schemata/situations (by (**u**)se link). For instance, situation SIT#1 is associated to the type "Open." Also, a situation can be specialized by other situations.

User can select a specific language (from the language list associated to the ontology/KB) to get the ontology/KB according to the selected language. Selection of a new language leads to the reformulation of the ontology/KB in the selected language.

If required by the user, the content of a node is visualized in an auxiliary frame. Some auxiliary frames are shown in Figure 13.2: one frame shows the content of the type "Pelican" and two other frames show contents of two situations. A Type node contains type's definition (if provided) and canon (if provided). For instance, "Pelican" has only a definition and type "Drink" only a canon. The description of a CS is a CG that is displayed in multi-lingua multi-notations CG editor (see the four auxiliary frames in Figure 13.2): CG can be visualized according to the selected language (one tab for every language) and according to the selected CG notation (one tab for every notation: Linear form, CGIF, and Graphical form).

Amine's Ontology/KB GUI also provides the possibility to display the super types of a given type. For example, the frame "Super Types" shows the super types of the type "Pelican." It is also possible to edit synonyms in all available languages for a given type.

The second possibility is to edit/consult ontology/KB in "Draw View" Editor. Using graphical components (graphical nodes for CSs and arrows for links), this view enables the use of the same functions as those available in tree view mode. Of course, the classical graphical operations are also available (edition, selection, move, cut/copy/past, etc.) and any change (addition, suppression, modification of nodes and/or links) made on either modes directly affect the ontology/KB and thus is visible in both modes. Also, user can zoom in/out the ontology/KB, she/he can change the vertical spacing between ontology/KB nodes and she/he can locate the position of a specific type or an individual node in the ontology/KB.

To produce an automatic drawing view of the ontology/KB from its tree view, we use currently the basic Sugiyama algorithm that enables automatic drawing of a hierarchical graph. An improvement of ontology/KB automatic drawing is underway.

For a large ontology/KB, user may have trouble reading the ontology/KB from the drawing view. To accommodate this problem, Amine provides a "Browser View Editor" that allows the user to focus on a particular area of the ontology/KB: a) the Browser Editor allows getting the neighborhood of a particular node in the ontology/KB, b) then, functions "expand" and "collapse" allow user to explore neighborhood of any node in the Browser

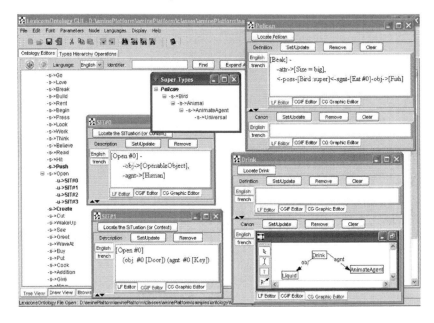

FIGURE 13.2: Ontology/KB Editors (Tree View Editor).

view. With expand/collapse functions. user can perform selective browsing of her/his ontology/KB.

13.4 Algebraic Layer

As noted in Section 13.2, algebraic layer provides several types of structured descriptions/patterns and related operations, as well as various matching-based operations. This section focuses on the most important structure in Amine; namely, Conceptual Graph (CG) and related editors and operations.

13.4.1 Conceptual Graph (CG)

Actually, CGs in Amine are restricted to CGs with dyadic relations only, and CG operations are defined on "simple and composed functional CGs." Any CG is considered according to the current ontology that constitutes its semantic support.

13.4.2   Multi-lingua and Multi-notations CG Editor

Amine provides *a multi-lingua and multi-notations CG editor*: a CG can be expressed in different languages associated to the current ontology and in different CG notations (see Figures 13.2 to 13.3 for embedded multi-lingua and multi-notations CG editors). Indeed, since Amine supports multi-lingua ontology, Amine is able to translate a CG from one language to another (from French to English or to Spanish, for instance). Also, Amine supports three CG notations: Linear, CGIF, and Graphic CG notations (with a procedure for automatic CG drawing). User can specify her/his CGs in any one of the three notations. Amine parses the specified CG notation and produces an internal representation for the CG (i.e., a Java object). Also, user can ask Amine to display a CG in LF, CGIF, or in Graphic notation. As a consequence of this flexibility, user can provide, for instance, a CG in LF and asks Amine to return the same CG in CGIF or in Graphic notation.

13.4.3   CG operations

Apart from operations specified in AmineObject and Matching interfaces (*match, equal, unify, subsume, maximalJoin, and generalize*) that are implemented by CG and by the other structures, CG (in Amine) has a large set of specific operations: *getters and finders* to get and find concepts and/or relations, *checkers* to check various constraints, *adders* and *removers* to add/remove concepts and/or relations, *setters* to set values for specific parameters, *file operations* (Load/Save) to load/save a CG from/in a file, *I/O operations* to read/write a CG according to the different CG notations, etc. There is also a set of operations that implement the *canonical formation rules*: restrict-Type(), restrictDesignator(), restrictDescriptor(), generalizeType(), generalizeDesignator(), generalizeDescriptor(), analogizeType() and join().

CG provides also a set of specific CG operations that are either variants or derived from the Matching operations:

- *specialize():* specialize a CG g1 by a maximal join with another CG g2. Unlike maximalJoin(), specialize does not create a new CG but modifies g1 itself.

- *generalize():* generalize a CG g1 by a generalization with another CG g2. Unlike generalize(), generalize does not create a new CG but modifies g1 itself.

- *isCanonic():* check if the specified CG g is canonic, i.e., if canons of concept types and relation types used in g, subsume g.

- *expand():* two versions of expand are provided, one for concept type expansion and another for relation type expansion.

- *contract()*: contract a CG from another CG taking into account the connectivity and the information specificity of the two graphs. Different versions are provided for the contraction operation.

- *compare()*: compare two CGs and return the detail of the comparison: a) the result of the comparison: the first CG is either equal, or more general, more specific, have common information, or have nothing in common with the second CG, b) the common CG (if the two CGs have common information), c) the matched part of the first CG, d) the matched part of the second CG, e) the specific parts of the first CG, and d) the specific parts of the second CG.

- *analogy()*: this operation has similarities with the analogy engine developed by Falkenhainer, Forbus, and Gentner [Falkenhainer et al., 1989, Forbus et al., 1994]. The operation analogy (sourceCG, targetCG) determines the "best" structural matching of the two CGs and then transfers, to the targetCG, all elements that are specific to the source CG and that respect the canonicity of the targetCG. An example is provided below (Figure 13.3).

13.4.4   CG Operations GUI

To familiarize users with matching-based CG operations and to enable a quick test bed for these operations, Amine provides a *CG Operations GUI*. This GUI provides also a possibility to see an animation of the operation, i.e., a possibility to see the application of the operation step, by step using the drawing of the two CG (and to see the construction, step by step of the third CG that could result from the operation). Amine visualizes, in this way, how the progressive matching of concepts and relations of the two inputs CG produces the new CG that results from the matching.

13.5   Memory-Based Inference and Learning Strategies Layer

Amine provides memory-based deductive, abductive and analogical inference strategies. It also provides two knowledge integration strategies: classification-based integration and generalization-based integration. Classification-based integration uses subsumption operation to classify a description in an ontology or in a KB. Generalization-based integration involves similarity identification and generalization construction [Kabbaj and Frasson, 1993, Kabbaj, 1995, 1996]. Generalization-based integration is related to concept learning

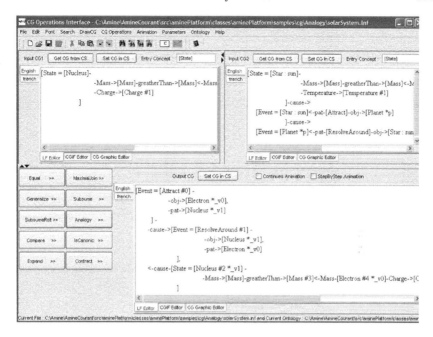

FIGURE 13.3: CG Operations GUI.

and machine learning in general and to dynamic memory models in particular [Kolodner and Riesbeck, 1986, Schank, 1982, 1991].

Beside these memory-based inference and learning strategies, Amine provides three basic memory (Ontology/KB) processes: Information Retrieval, Elicitation, and Elaboration. Due to limitations in space, we present only a brief description of these processes:

- *Information retrieval process (IR)* uses the classification process and searches to know if the specified description is contained or not in the ontology/KB. The aim of IR is not to answer "yes" or "no," but rather to situate the specified description in the ontology/KB; to determine its neighborhood: which nodes are "fathers" (minimal generalizations of the specified description), which nodes are "children" (minimal specialization of the specified description) and which node is equal to the specified description. Here is an example that illustrates the use of IR from Prolog+CG program/console:

```
?- ask([Robot]<-agnt-[Wash]-thme->[Inanimate], [Wash]).

The description is : EQUAL to/than the known
[Wash #0] -
        -Agnt->[Robot],
```

```
        -Thme->[Inanimate]

The description is : MORE_GENERAL to/than the known
[Wash #0] -
        -Thme->[Truck],
        -Agnt->[Robot]

The description is : MORE_GENERAL to/than the known
[Wash #0] -
        -Thme->[Car],
        -Agnt->[Robot]
 yes
?-
```

- *Elaboration process* uses the inheritance mechanism to provide more information about a description D. We differentiate between "deductive elaboration," which uses inheritance of type definitions and "inductive/plausible elaboration," which uses inheritance of schemata/situations ("plausible elaboration" because a situation provides only typical/plausible information about a type, contrary to a definition). A mixture of the two kinds of elaboration is possible.

- *Elicitation process* is an interactive process that helps a user to make his description D more precise and more explicit.

13.6   Programming Layer

This section presents a brief account of Amine's programming layer: PRO-LOG+CG language and SYNERGY language.

13.6.1   PROLOG+CG

Prolog+CG was developed by the author as a standalone programming language [Kabbaj et al., 1994, Kabbaj, 1996, Kabbaj and Janta-Polczynski, 2000, Kabbaj et al., 2001]. The group of Peter Øhrstrøm developed a very good on-line course on some aspects of Prolog+CG. Let us recall three key features of previous versions of Prolog+CG:

- CG (simple and compound CGs) is a basic and primitive structure in Prolog+CG, like list and term. And like a term, a CG can be used as a structure and/or as a representation of a goal. Unification operation of Prolog has been extended to include CG unification. CG matching-based operations are provided as primitive operations.

- By a supplementary indexation mechanism of rules, Prolog+CG offers an object-based extension of Prolog.

- Prolog+CG provides an interface with Java: Java objects can be created and methods can be called from a Prolog+CG program. Also, Prolog+CG can be activated from Java classes.

The interpreter of Prolog+CG, that takes into account these features (and others) has been developed and implemented in Java by the author.

The above three key features are still present in the new version of Prolog+CG but the re-engineering of Prolog+CG, which was necessary for its integration in Amine platform [Kabbaj, 2006], involved many changes in the language (and in its interpreter). Five main changes are of interest:

- Type hierarchy and Conceptual Structures (CSs) are no more described in a Prolog+CG program. Prolog+CG programs are now interpreted according to a specified ontology that includes type hierarchy and CSs.

- Prolog+CG program has the current ontology as " a semantic support": for each symbol in a Prolog+CG program, the interpreter checks first if the symbol is an identifier of an element (type, individual, or relation) in the ontology. If no such identifier is found, then the symbol is considered as an atomic identifier (without any underlying semantic).

- Project of Prolog+CG programs is introduced: user can consult several programs (not only one) that share the same ontology.

- Prolog+CG inherits from the lower layers of Amine: all Amine structures and operations are also Prolog+CG structures and operations. And of course, Prolog+CG user can manipulate the current ontology and the associated lexicons according to their APIs.

- The interface between the new version of Prolog+CG and Java is simpler and more "natural" in comparison with previous interfaces.

- Interoperability between Amine components: Prolog+CG can be used in conjunction with other components of Amine.

Section 13.9 illustrates the use of Prolog+CG in the implementation of Natural Language Processing (NLP) applications.

13.6.2 SYNERGY

SYNERGY is a visual multi-paradigm programming language based on *CG activation mechanism*, which is a computation model for *executable CG* [Kabbaj, 1993, 1996, 1999a,b]. Synergy integrates functional, procedural, process, reactive, object-oriented, and concurrent object-oriented paradigms. The integration of these paradigms is done using CG as the basic knowledge structure, without actors or other external notation. Activation-based computation is an approach used in visual programming, simulation, and system analysis where graphs are used to describe and simulate sequential and/or parallel tasks of different kinds: functional, procedural, process, event-driven, logical, and object oriented tasks. Activation-based interpretation of CG is based on *concept lifecycle, relation propagation rules*, and *referent/designator instantiation*. Each concept is considered as a process that evolves according to its lifecycle. Concept lifecycle is defined as a transition state network where nodes represent the possible states of the concept. Thus, in addition to concept type, concept designator, and concept descriptor, a concept (in Amine) has a state that changes according to its lifecycle.

Concept lifecycle is similar to process lifecycle (in process programming) and to active-object lifecycle (in concurrent object-oriented programming), while relation propagation rules are similar to propagation or firing rules of procedural graphs, in dataflow graphs and Petri Nets.

Previous versions of Synergy have been presented in [Kabbaj, 1999a,b]. Synergy was developed as a standalone programming environment. The integration of Synergy in Amine required re-engineering work, involving some changes and extensions to the language and to its interpreter [Kabbaj, 2006]. New features of Synergy include:

- Long-term memory introduced in previous definitions of Synergy corresponds now to ontology that plays the role of a support to Synergy program and request/expression.

- Previous versions of Synergy did not have an interface with Java. The new version of Synergy includes such an interface; Java objects can be created and methods activated from Synergy. This is an important feature since user is not restricted to (re)write and to define everything in CGs. Also, primitive operations are no more restricted to a fixed set of operations.

- The new version of Synergy has an access to the lower layers of Amine. Also and since Prolog+CG, Synergy and dynamic ontology formation process are integrated in the same platform and share the same underlying implementation; it is now possible to develop applications that require all these components. We sketch an example of this synergy in the next section.

- Another new feature of Synergy is the integration of *dynamic programming*, i.e., dynamic formation-and-execution of program. We focus on this feature in the rest of this section.

13.6.3 Dynamic Programming with SYNERGY: An Example

To illustrate what we mean by "dynamic programming," let us start with the idea of database inference proposed by Sowa [Sowa, 1984, p. 312] that combines the user's query with background information about the database to compute the answer. Background information is represented as type definitions and schemata. Sowa stressed the need for an inference engine to determine what virtual relations to access. By joining schemata and doing type expansions, the inference engine expands the query graph to a working graph (WG) that incorporates additional background information. Actors bound to the schemata determine which database relations to access and which functions and procedures to execute. According to Sowa, his inference engine can support a dynamic way of deriving dataflow graphs [Sowa, 1984, p. 312]. In other words, his inference engine can be considered as a basis for *a dynamic programming approach* (recall that dataflow graphs, Petri Nets, executable CG and other similar notations have been used to develop visual programming languages). Indeed, his inference engine is not restricted to database, it can be extended to other domains and can be considered as an approach to dynamic programming.

The inference engine proposed by Sowa has been integrated to Synergy. Figure 13.5 illustrates the implementation in Synergy, of Sowa's example. Background information (procedural knowledge in terms of strategies, methods, procedures, functions, tasks, etc.) is stored in ontology as situations associated to concept types (Figure 13.5.a). Let us consider the interpretation/execution of the working graph (WG) that is initiated to the request (Figure 13.5): if a concept needs a value that cannot be computed from the actual content of the WG (Figure 13.5.b), then Synergy looks, in the ontology, for the best situation that can compute the value (i.e., the descriptor) of the concept. The situation is then joined to the WG (Figure 13.5.c) and Synergy resumes its execution. In this way, the program (i.e., the WG) is dynamically composed and adapted during its execution (Figure 13.5).

This simple example illustrates the advantage of Amine as an Integrated Development Environment (IDE); it illustrates how various components of Amine (ontology, CG operations, Prolog+CG, Synergy) can be easily used in one application: semantic analysis of the request can be done by a Prolog+CG program. The result (an executable CG), will be provided to Synergy that illustrates the visual execution of the "dynamic program." After the termination of the execution, the final CG can be an input for a text generation program (that can be implemented in Prolog+CG) to provide a text that paraphrases the composed "program" responsible for the result.

13.7   Multi-Agents Layer

Instead of developing a specific multi-agents layer for Amine from "scratch," we decided to use available Java open-source Agent Development Environments in conjunction with Amine. We started with JADE. Jade allows the creation of Agents (i.e., it offers a Java class *Agent* and manages the underlying network protocols), the use of different kinds of behaviors and the communication with ACL according to FIPA specification. Currently, we use Jade to handle the lower level of the Agent (i.e., creation and communication between agents) and Amine for the higher level (i.e., cognitive and reactive capabilities of the agents are implemented using Currently).

Amine, the Multi-Agent System (MAS) layer of Amine contains one plug-in (AmineJade package). Other plugs-in (i.e., other packages) could be added as other combinations of Amine with other "Java Agent Development Environments" are considered. Using Amine and Jade, we have defined a hierarchy of different kinds of AmineJade agents. The most important for the moment is "PPCGAmineJadeAgent"; an agent of this type incorporates a PROLOG+CG interpreter that plays the role of inference engine, problem-solving engine, planning engine and NLP engine of the agent. Section 13.10 describes briefly the MAS Renaldo that was developed using Amine's MAS layer.

13.8   Amine as an IDE for Symbolic Programming

"Symbolic programming" is the main programming paradigm in Artificial Intelligence (AI). Functional programming (incorporated in programming languages like LISP and SCHEME) and declarative/logic programming (incorporated in programming languages like PROLOG) are two families of symbolic programming. These two families are present in Amine; SYNERGY can be used as a visual/graphical functional language and PROLOG+CG as a declarative/logic programming language.

Basic constructs in symbolic programming include: a) the possibility to construct complex structured descriptions and patterns, b) the use of symbols and variables as elements of structured descriptions/patterns, c) the possibility to define and/or use pattern-matching operations, like unification in PROLOG (which is used as a primitive in PROLOG). These constructs are provided in Amine at the algebraic layer.

But symbols, in usual symbolic programming languages like LISP and PRO-LOG, have no associated meaning; no underlying semantic. This semantic/ontological level is missing in these usual symbolic programming languages. Ontology layer in Amine provides this missing level.

Also, the algebraic layer of Amine provides the possibility to construct much more complex structured descriptions and patterns, with the use of List, Term, and especially CG. Also, Amine provides various kinds of pattern-matching operations (unification, maximalJoin, generalize, subsume, analogy, etc.).

These features, and others, make Amine a very suitable IDE for symbolic programming.

13.9 Amine as an IDE for Intelligent System Programming: Natural Language Processing with Prolog+CG as an Example

As stressed in [Kabbaj and Janta-Polczynski, 2000, Kabbaj et al., 2001], several features of Prolog+CG makes it a suitable language for the development of natural language processing (NLP) applications: a) Prolog+CG is an extension of Prolog, which is suited for NLP, b) CG, both simple and compound, is provided as a basic data structure. The usefulness of CG for NLP is well known (see chapter of John Sowa), c) Prolog+CG allows CG with variables (variable as concept type, concept designator, concept descriptor, or as relation type). This feature of Prolog+CG is very useful for NLP and other intelligent systems programming, d) several CG matching-based operations are provided (maximalJoin, generalize, subsume, contract, expand, analogy, etc.), e) CG basic operations are available (find a concept or a relation in a CG that verify some constraints, etc.), f) the possibility to construct and update a CG (by adding more concepts and relations). All these features are very useful for NLP. To illustrate this point, let us consider briefly two tasks in NLP: semantic analysis and question/answering.

13.9.1 Semantic Analysis with Prolog+CG

The following rules provide an example of how the above features of Prolog+CG can be exploited to develop a semantic analysis process.

```
declarative_sentence(P,
   [T_TypeOfSentence=G]-modalityOf->[Modality=declarative]) :-

   NP(P, P1, G_NP, E_NP),
   stativeOrActiveVP(P1, ["."], G_NP, E_NP,
T_TypeOfSentence, G).
```

```
. . .
stativePart([A|P1], P1, G_NP, E_NP, G) :-
   Adj(A, R1, T1, V1), !,
   E_NP = [N : A1],
   G = [N : A1]-R1->[T1 = V1],
   branchOfCG(B, [N : A1]-R1->[T1 = V1], G),
   E_N is B:getSourceConcept(),

   G:specialize(E_N, G_NP, E_NP).
```

The above two rules illustrate : a) the use of variables as concept type, concept designator, concept descriptor, and relation type, b) the construction of a concept (E_NP = [N : A1]), c) the construction of a CG (G = [N : A1]-R1->[T1 = V1]), d) the use of the primitive branchOfCG that locates a branch B in the CG G so that B unifies with the pattern given as the second argument of branchOfCG, e) the use of the first two layers of Amine: branch (i.e., a relation with its source and target concepts) and CG are two structures of Amine, these structures with their methods can be used directly in a Prolog+CG program. In our example, we have a call to the method getSourceConcept() that returns the source of the branch/relation and a call to the method specialize() that specializes a CG by the maximal join of another CG.

Let us consider now, as a second example, the formulation of the lexicon: in previous versions of Prolog+CG, the semantic of the words should be specified in the Prolog+CG program itself. For instance, consider the word "open" with some of its different meanings:

```
lexicon("open", verb,
[Human]<-agnt-[Open]-obj>[OpenableObject]).
lexicon("open", verb, [Key]<-agnt-[Open]-obj->[Door]).
lexicon("open", verb, [Open_Box]<-agnt-[Open]-obj->[Box]).
lexicon("open", verb, [Shop]<-pat-[Open]-
                                    -obj->[Door],
                                    -ptime->[Time]).
```

With the new version of Prolog+CG, another formulation is now possible: the above different meanings can be considered as background information, stored in the used ontology as situations associated to the type "Open." User can access the ontology to get background information (definition, canon, situation, etc.) for a specific type or individual. This change in the formulation of the lexicon in Prolog+CG leads to the following reformulation of the above example:

```
lexicon("open", verb, Open).
lexicon(_verb, verb, _type, _sem) :-
   lexicon(_verb, verb, _type),
   getSemantic(_type, _sem).
```

Definition of the goal getSemantic/2 is provided below. It searches, from the ontology, the background information for a specific type or individual. The first rule looks for the canon. The second rule looks for the situations that are associated with the specified type. With backtracking, it will search for a situation that can be joined with the working CG. Note the call to the method getCanon() that returns the canon of the type (or returns null if the type has no canon) and the call to the method getSituationsDescription() that returns, in a list, all situations descriptions that are associated to the specified type.

```
getSemantic(_Type, _Sem) :-
    _Sem is  _Type:getCanon(),
    dif(_Sem, null).

getSemantic(_Type, _Sem) :-
    _EnumSitDescr is _Type:getSituationsDescription(),
    dif(_EnumSitDescr, null),
    _ListSitDescr is
                    "aminePlatform.util.AmineObjects":
enumeration2AmineList(_EnumSitDescr),
    member(_Sem, _ListSitDescr).
```

Word disambiguation is performed in the current version of our semantic analysis process by using backtracking of Prolog+CG: if the maximal join of the working graph with the semantic of the current word fails, Prolog+CG backtracks and resatisfies the goal getSemantic/2, which returns another meaning (i.e., another conceptual structure) for the current word.

Here is an example of semantic analysis of a simple sentence:

```
?- nlp.
|: john drive to the store.
[Action =
    [Drive #0] -
        -dest->[Shop],
        -agnt->[Boy:John]]-modalityOf->[Modality = declarative]
 yes
```

13.9.2 Question/Answering

Semantic analysis of a (short) story would produce a compound CG (see the fragment below). Let us call it CGStory. In our example, CGStory is a fusion of three networks: a) temporal network composed by "after" relations that specify the temporal succession of actions, events, and states, b) causal network composed by "cause" relations, and c) intentional network composed by "motivationOf" and "reason" relations.

```
story(
[Action #act1 =
[Time : Early]<-time-[WakeUp]-pat->[Man: John]]-after->
[State #stt1 = [Hungry]-pat->[Man: John]]-after->
...
[State #stt1]<-reason-[Goal=
[Action=[Food]<-obj-[Eat]-agnt->[Man:John]]]-
                                    <-reason-[Action #act2],
                                    <-reason-[Action #act5],
                                    <-reason-[Action #act7],
                                    <-reason-[Action #act8]
[Action #act3]<-motivationOf-[Goal =
[Action = [Man:John]<-dest-[Greet]-agnt->[Woman: Mary]]
]<-reason-[Action #act4]
[Event #evt1]-
           -cause->[State #stt2 = [ParkingLot]<-pat-[Slick]],
           <-cause-[Event #evt2]
                         ).
```

Semantic analysis process is applied also to questions and for each type of question there is a specific strategy responsible for the search and the composition of the answer [Graesser et al., 1992]. Here is the formulation in Prolog+CG of the strategy for answering "why" question. It concerns the intentional network: the strategy locates in CGStory the branch/relation with relation type "reason" or "motivationOf," then the source concept of the branch should unify with the content of the request. The recursive definition of the goal reason/2 provides the possibility to follow an "intentional path" to get the reason of the reason, etc.

```
answerWhy(A, Y) :-
   story(_story),
   member(R, [reason, motivationOf]),
   branchOfCG(B, [T = G]<-R-[T2 = A], _story),
   reason([T = G], Y).

reason(X, X).
reason([T = G], Y) :-
   story(_story),
   member(R, [reason, motivationOf]),
   branchOfCG(B, [T1 = G1]<-R-[T = G], _story),
   reason([T1 = G1], Y).
```

For instance, to the question "why did john drive to the store?", the question/answering program returns:

```
?- questionAnswering("why did john drive to the store ?",
```

```
_answer).
{_answer = [Goal = [Action = [Eat #0] -

                                    -agnt->[Man :John],
                                    -obj->[Food]
                ]
            ]};
{_answer = [State = [Hungry]-pat->[Man :John]]};
 no
?-
```

Of course, the above definition of "why-strategy" is very simplistic; the aim of the example is to show how Prolog+CG constitutes a suitable programming environment for CG manipulation and for the development of NLP applications.

13.10   Amine as an IDE for Intelligent Agents Programming and Multi-Agents Systems (MAS)

Amine platform can be used as a modular IDE for symbolic programming and intelligent systems programming. It can be used also as the kernel; the basic architecture of an *intelligent agent*: a) the ontology/KB layers can be used to implement the dynamic memory of the agent, b) the algebraic layer, with its various structures and operations, can be used as the "knowledge representation capability" of the agent, c) the memory-based inference and learning strategies layer can be used for the formulation and development of many inference and learning strategies (induction, deduction, abduction, analogy), d) c) the programming layer (i.e., Prolog+CG and Synergy) can be used for the formulation of agent's cognitive processes (reasoning, problem solving, planning, natural language processing, dynamic memory, learning, etc.), e) Synergy language can be used to implement the reactive and event-driven behavior of the agent, f) agent layer can be used to implement the whole agent architecture.

One long-term goal/project of the author is to use Amine, in conjunction with Java Agent Development Environments (like Jade), to build various kinds of intelligent agents, with multi-strategy learning, inferences and other cognitive capabilities.

To investigate in this direction and to illustrate the use of Amine's MAS layer, we have developed a specific MAS called "Renaldo," that concerns the simulation of a child story. The setting of the story is a forest; it corresponds to the environment of Renaldo. The characters of the story (the bear John, the bird Arthur, the bee Betty, etc.) are the agents of Renaldo. Each type of agents (Bear, Bird, Bee, etc.) has a set of attributes, knowledge, goals, plans, and actions that are specified as Prolog+CG programs. A specific agent can have, in addition, specific attributes, knowledge, goals, plans and actions, specified also as Prolog+CG programs.

Please consult Amine Website for more detail on MAS layer of Amine and on Renaldo.

13.11   Related Work

Philip Martin provides [Martin et al., 2007] a detailed comparison of several available CG tools[6] (Amine, CharGer, CGWorld, CoGITaNT, Corese, CPE, Notio, WebKB). CGWorld is a Web-based workbench for joint distributed development of a large KB of CG, which resides on a central server. CGWorld is no more developed. Corese is a Semantic Web search engine based on CG. WebKB is a KB annotation tool and a large-scale KB server. CoGITaNT is an IDE for CG applications. Notio is not a tool but a Java API specification for CG and CG operations. It is no more developed. It is re-used, however, by CharGer and Corese. CharGer is a drawing CG editor with the possibility to execute primitive actors and to perform matching operation. CPE has been developed as a standalone application. Currently, CPE is being upgraded to a set of component modules (to render CPE an IDE for CG applications). Its author announces that CGIF and basic CG operations (projection and maximal join) are coming soon. The new upgraded version of CPE is underway and it is not yet available. CGWorld, Notio and CPE will not be considered in our comparison of available (and active) CG tools.

In his comparison, Philip focuses mainly on the "ontology-server dimension," which constitutes the specificity of his tool (WebKB); he did not consider other dimensions, i.e., other classes of CG tools. Indeed, CG tools can be classified under at least 8 categories of tools: CG editors, executable CG tools, algebraic tools (tools that provides CG operations), ontology/KB tools, ontology server tools, CG-based programming languages, IDE tools for CG applications and, agents/MAS tools.

The category "IDE for CG applications" means a set of APIs and hopefully of GUIs that allow user to construct and manipulate CGs and to develop various CG applications. Only Amine and CoGITaNT belong to this category. The category "CG-based programming language" concerns any CG tools that provide a programming language with CG and related operations as basic constructs. Only Amine belongs to this category, with its two programming languages: PROLOG+CG and SYNERGY. The category "Agents/MAS Architecture" concerns CG tools that allow the construction and execution of

---

[6]These tools are listed also at http://www.conceptualgraphs.org, [Delugach, 2007]

	Amine	CharGer	CoGITaNT	Corese	WebKB
CG Editor(s)	++	++	++	-	-
Exec. CG	++	+	-	-	-
Algebraic	++	+	+	/	/
KB/Ontology	++	-	?	+	+
Ont. Server	-	-	-	-	++
IDE	++	/	+	-	-
Programming	++	-	-	-	-
Multi-Agent	+	-	-	-	-

FIGURE 13.4: Comparison of available CG tools.

intelligent agents (with cognitive and reactive capabilities) and multi-agents systems (MAS). As illustrated in this chapter, Amine, in conjunction with a Java Agent Development Environment, can be classified under this category.

13.11.1 Symbols Used in the Table:

- "++": the tool offers different features concerning the associated category. For instance, Amine provides multi-lingua and multi-notations CG editors. The same for executable CG: it offers not only the equivalent of actors, as CharGer does, but a programming language based on executable CG. This chapter illustrates in addition a new feature of Synergy: dynamic programming. The same for "Ontology/KB" category: Amine provides a rich ontology API, ontology editors and various basic ontology processes. And the same for "Programming" category: Amine provides two CG based programming languages (i.e., PROLOG+CG and SYNERGY).

- "+": the tool can be classified under the associated category.

- "-": the tool can not be classified under the associated category.

- "/": the tool is not intended to belong to the specified category but it uses some aspects of the category. For instance, "Web ontology tools" like Corese and WebKB are not intended to be used as "algebraic tools" even if they use some CG operations (like projection and generalization).

13.12 Future Work

All layers of Amine are under continuous update and development. For instance:

- enhance CG and Ontology editors and graph (automatic) drawing,

- enhance Prolog+CG: consider the development of a better editor and provide a set of primitives similar to the set of standard Prolog,

- enhance Synergy: Consider a full implementation of Synergy. Currently, only a subset of Synergy has been implemented and integrated to Amine; composed coreference and object-oriented aspects of Synergy were not considered. Also, Synergy should provide a rich set of primitives,

- take into account interoperability with other CG tools and consider for that purpose the CGIF version used by CharGer,

- Current and future works concern also the development of applications in various domains: reasoning, expert systems, ontology-based applications, natural language processing, problem solving and planning, case-based systems, multi-strategy learning systems, multi-agents systems, intelligent tutoring systems, etc.

13.13   Conclusion

This chapter presents an overview of Amine; a multi-layer Java open source platform, based on CG theory and suited for symbolic programming, intelligent system programming, and intelligent agents programming. The multi-layer architecture of Amine and its openness to Java enable great modularity and genericity.

With the Ontology layer, Amine provides not only powerful tools for Ontology-based applications, but also the semantic support for symbolic programming, intelligent system programming and agents programming. The great variety of structures and operations of the Algebraic layer, especially CG and CG operations, is valuable for various applications in different domains. Programming layer, with Prolog+CG, Synergy and Ontology-based processes, makes Amine a suitable IDE for symbolic programming and intelligent system programming. These layers with the agent layer provide the possibility to develop different kinds of agents and multi-agents systems.

We hope that Amine will federate works and efforts in CG community (and elsewhere) to develop a robust and mature modular IDE, suited for symbolic programming, intelligent system programming, and intelligent agents programming.

13.14   Student Projects

13.14.1   Project 1

Ontology: To enable interoperability with other Ontology tools, like Protégé, consider the development of a mapping between Amine's Ontology/CG and OWL/RDF.

13.14.2   Project 2

CG and CG operations: Consider CG with n-adic relations and update CG editor to include this extension. Enhance CG operations to consider the case of CG with n-adic relations and to consider the case of general CG (not only functional CG).

13.14.3   Project 3

Develop a Natural Language Processing project (semantic analysis, question/answering, generation) around a subset of natural language (i.e., a Controlled Language like ACE).

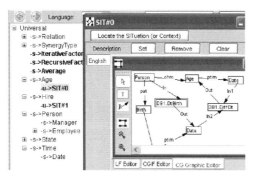

(a) Snapshot of the ontology

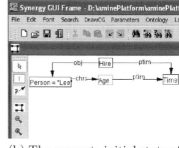

(b) The request: initial state of the WG

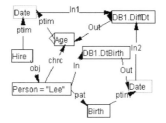

(c) WG after first maximalJoin

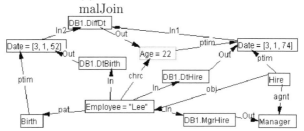

(d) WG after second maximalJoin and termination of execution

FIGURE 13.5:   Example of dynamic programming with Synergy.

Chapter 14

Active Knowledge Systems

Harry S. Delugach

University of Alabama in Huntsville, U.S.A.

Conventional knowledge bases comprise essentially static specifications of some model. Traditional logic methods are quite adequate for reasoning within such systems, since the model's assertions are monotonic and therefore unchanging. There are two main limitations to this approach: (i) knowledge about the (ever-changing) world must reflect its ever-changing nature and (ii) the symbols and relationships within a model must correspond to individuals, objects, and relationships in the world.

We assume that there are at least three broad purposes in developing and maintaining a knowledge-based computer system.

1. To capture and represent knowledge about the world.

2. To reason about the world and draw appropriate conclusions.

3. To adapt to changing knowledge in the world.

Conventional systems are best suited for the second purpose. The first purpose, however, is generally performed in an ad-hoc and informal manner, although it is certainly possible to produce a useful model this way. And by their nature conventional systems are able to address the third purpose only with much difficulty. Active knowledge systems serve all three purposes, as will be explained in this chapter. It is the third purpose where active knowledge systems are clearly different from conventional knowledge bases.

This chapter introduces the reader to the notion that knowledge bases are not static collections of information, but instead are constantly changing in response to their changing environment. We call them *active knowledge systems*. In effect, we propose providing knowledge bases with "eyes" and "ears" to inter-*act* (not just inter-*face*) with the world around them. This gives us two capabilities:

- Knowledge bases can accurately represent the world around them, so answers to queries can reflect what is known at the time they are asked.

- Knowledge bases are thereby *grounded* – meaning that the symbols in a knowledge base are explicitly associated with the things they represent; namely, entities, relations and concepts in the so-called "real world" (the quotes will be explained later).

There are at least two major issues involved in such systems:

- How does knowledge of the changing world become captured into the knowledge base's representation?

- How does the knowledge base ensure that the acquired knowledge is consistent (e.g., does not render previous conclusions invalid).

This chapter is primarily concerned with the first issue; later, some ideas will be offered about directions toward the second.

This work is related to other work in dealing with non-monotonic knowledge, such as the situation calculus [Levesque et al., 1998]. That approach is concerned with describing actions that change the state of a system. The current work is more focused on how to establish formal connections between a model and the environment being modeled; i.e., the effect of an action is not necessarily formally modeled – instead it is interpreted by observing (or interacting with) the environment. Future work in active knowledge systems may allow incorporating more formal action semantics. Another topic of interest is that of model checking [Edmund M. Clarke et al., 1999], whereby conditions may be reasoned about over changing states of a system; again, this approach is more concerned with the closed-world assumption that states and transitions are formally specified. Active knowledge systems can certainly accommodate these reasoning techniques, but that is not their main contribution.

14.1   What Is an Active Knowledge System?

This chapter describes the notion of *active knowledge systems* in conceptual systems, and illustrate them with a few examples. This section introduces the basic terms and ideas around which the work is based. Section 14.2 explains how active knowledge systems provide a grounding (interpretation) for a given model. We then illustrate the approach with some example applications in section 14.3. We end with some concluding thoughts.

Aspects of this work have been described previously in [Delugach, 2003], [Delugach, 2006]. We use the formalism of conceptual graphs [Sowa, 1984] to

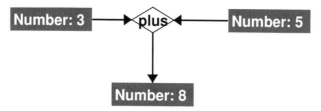

FIGURE 14.1: Sample actor.

illustrate some of the characteristics of active knowledge systems; however, the principles are applicable in a wide range of formalisms.

Any formal system (including a computer system) is composed of symbols and a set of rules for transforming them. Of course, the form or effect of the rules (and their constraints) depends on what kind of system is being developed. In a logic system, the rules are logical relationships and transformations governing conclusions that can be derived from assumptions about the original symbols. So if we have the premises $(p \implies q)$ and p, we can conclude q by using a rule called *modus ponens*. For a more extensive description of basic logic principles, see Chapter 2.

Even in this simple example, there is a big difficulty: how do we come to know that the premises are true? The answer is: we do not! The premises are initial assumptions, taken on faith as it were, sometimes even called axioms. If we assume that $p =$ *"Harry is a millionaire"* and $q =$ *"Harry is happy"*, and we assume that Harry is a millionaire, we thereby infer that Harry is happy, but then there is a more fundamental problem: Harry is not a millionaire (at least the one writing this chapter!). That bigger problem is not a concern of the logic itself (which only considers symbols and rules), but it is a major concern for the users of a knowledge system. So how do we attach actual meaning to the premises, and hence to their conclusions? The answer lies in a process called *grounding*, which logicians call *interpretation*. Grounding can be conceptualized as a procedure that maps individual symbols in a model to actual individuals in some domain of discourse, as well as a set of procedures that map the results of functions with arguments in the model to actual functions whose arguments represent such actual individuals.

The formalities of such interpretations are part of model theory and beyond the scope of this chapter (see [Hodges, 1997] for the formal details). A key requirement is that a system have the capability to access its environment. Using conceptual graphs, this capability can be realized through the use of *actors*. An actor in a conceptual graph can be treated as an active relation; i.e., a function that provides a mapping from zero or more input concepts to zero or more output concepts. (Actors with zero outputs may not seem useful, but some examples are provided later in the chapter.) A simple actor is shown in Figure 14.1. It shows a **plus** actor that establishes a mathematical addition relation between the numbers 3, 5, and 8.

Actors are well described in [Sowa, 1984], with further elaborations in [Delugach, 1992], [Raban and Delugach, 1997], and [Delugach, 2003]. In brief, an actor operates much as a node in a Petri Net [Peterson, 1981]: when its input concepts are ready, the actor "fires," possibly changing its output concepts' referents (i.e., the individuals to which they refer). If the output concept is in fact a context (see Figure 14.15), then the actor is able to create or modify an entire sub-graph within that context; in that situation, the context itself is treated as s single concept whose referent is its enclosed graph. This is similar to the semantics of a *demon* that will be explained below.

Note how the actor `plus` is different from a relation `plus`. We could assert any *plus* relation we want; e.g., between the numbers 10, 100, and 1000, but it would not be true with respect to the "real world" of integers. We therefore require some procedure (outside of the reasoning formalism) to provide us with the "real world's" interpretation of what *plus* means. The actor supplies this procedure.

A conceptual graph whose concepts denote values or identities obtained by actors in this way is called an *actively grounded conceptual graph*. Such graphs are a necessary ingredient in building true active knowledge systems. Figure 14.2 is an example of an actively grounded graph – in this case, a graph that represents the sugar molecule whose formula is $C_6H_{12}O_6$. There is a particular relationship between a molecule and its molecular weight, a relationship that is determined by specific experimental values for the atomic weights of its constituent atoms. The significance of the example is that a knowledge system cannot determine these values logically; the atomic weight values just happen to be what they are.

Actors in a graph provide "hooks" to things *outside* the sheet of assertion, that allow the model to reflect those outside things. Figure 14.2 shows several kinds of actors that may be used to provide meaning to the elements of a model. The `plus` actor represents a functional relationship between one or more input concepts' referent and an output referent that will always represent the mathematical sum obtained by adding the input referents together. The `multiply` actor operates similarly in representing a mathematical product. (One might argue that the names are not entirely appropriate; if a relationship is being specified, then it is the **sum** *relationship*, not the *operation* of **plus** that is being specified.) Another actor shown is the `lookup` actor, which performs the operations necessary to ensure that a particular concept referent will have a particular value obtained from a database. Section 14.3.1 will describe this actor in greater detail.

Figure 14.2 also shows an additional feature of conceptual graphs; namely, the notion of a *context*. The concepts representing the atoms in the molecule are grouped together into a context labeled `Molecule`. This construct not only allows relationships and features between entire groups of related concepts, it also enables future work in "possible worlds" (in the sense of Kripke [Kripke, 2003]) and in microtheories [Guha, 1991].

Figure 14.3 shows the general framework of active knowledge systems as

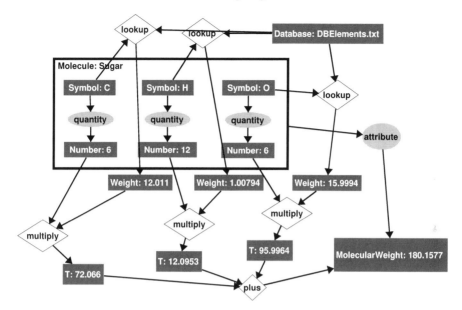

FIGURE 14.2: Graph representing a sugar molecule.

they relate to conceptual graphs. A graph appears on what is called a *sheet of assertion*, which comprises the formal model in Figure 14.3 indicates several kinds of actors that may be used to provide meaning to the elements of a model. Actors can provide temporal information (e.g., what is the current time?), they can provide results from external sensors, and they can provide database lookups (see section 14.3.1 below). They can also form the basis for communication between agents and support interchange of knowledge with other systems. Some of these functions will be illustrated later in section 14.3.

The next section gives more detail about the notion of grounding with respect to a knowledge model using conceptual graphs.

14.2   Grounding a Knowledge Model

This section explains what we mean by a *grounded* model. In essence, grounding is the process of establishing, for every symbol in the model, some individual (or relationship) in the "real world" that it corresponds to. In modeling terms, we do not say "real world"; we use the more formal term *universe of discourse* (or sometimes just *universe*). In this chapter, we will sometimes call it the system's *environment*. We will also borrow some terms and ideas from the Common Logic Standard [ISO/IEC, 2007]. That standard recognizes

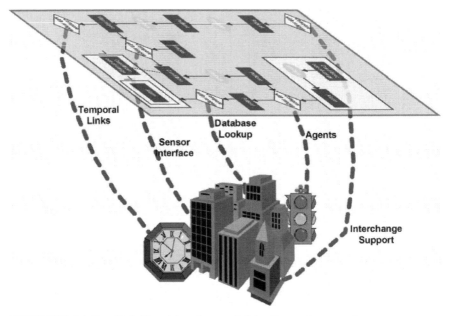

FIGURE 14.3:   Relationship of a model to its environment.

FIGURE 14.4:   Transformations in a model.

that relations and functions, while not strictly mapped to individuals, will still map to relations and functions in the environent that are useful; these make up a potentially larger realm called the *universe of reference*. A logic system thereby provides predicates and functions that allow the representation in a model of relationships between elements in a system. The rules of logic ensure that the truth of these predicates is preserved; e.g., if I assert that dog(Spot) is true and then perform any number of logical transformations, a (consistent) system should never allow me to derive not dog(Spot). This is a necessary property of any system that is meant to preserve truth values. Most knowledge-based systems therefore operate as in Figure 14.4. Elements in the model represent individuals, predicates, functions, etc.

Note that in Figure 14.4, the symbols are merely organized within the model. There is a larger question that often goes unanswered (at least in a formal way): what do the predicates mean? Logicians remind us that a predicate is assumed to be a primitive relation between symbols; a logical system's only obligation is to preserve those relationships throughout its logical oper-

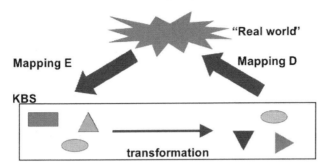

FIGURE 14.5: Grounding a model.

ations. The symbols themselves, however, are arbitrary, at least insofar as the logic operations are concerned. This presents a difficulty for a system builder, whose symbols are not arbitrary at all! Symbols used in a knowledge base stand for something, usually in the environment. Figure 14.5 is meant to remind us that symbols in a model were put there for a particular reason, which must be known when symbols are "pulled out" after transformations.

Figure 14.5 therefore suggests how a model relates to its environment. Things in the environment are represented in a knowledge-based system (shown as "KBS") through a mapping E (for "encoding") that allows their representations to be transformed within the system. Once transformed, new representations can arise within the system. These new representations' symbols can then be mapped back to the environment through a mapping D (for "decoding") that captures their interpretation. This model is intentionally simplistic, ignoring some important philosophical issues, in order to provide a basis for the discussion in this chapter.

It is crucial for a knowledge system to preserve truth internally, and of course the system builder requires the same of any system they build. Furthermore, the system builder needs for the system to preserve the symbols' meanings as they exist in the environment. The mapping D in Figure 14.5 represents this requirement: namely, the ability to "decode" the (transformed) symbols into the real things they represent.

An interpretation must first provide a mapping from symbols in the model to its domain (universe) of individuals. This is shown in Figure 14.6, where the large arrows indicate the mapping of an interpretation. Conceptual graphs provide for an *individual marker* in a concept box; e.g., Person: #3867 where #3867 is an individual marker denoting a particular individual (in this case, of type Person. The formal definition for individual markers in conceptual graphs states that there must be a way to associate each distinct marker with a specific individual. Incidentally, there is no requirement that the mapping from markers to individuals be either one-to-one or onto: more than one marker may denote the same individual, and there may be individuals that are never denoted in a particular model. The notion of an individual marker

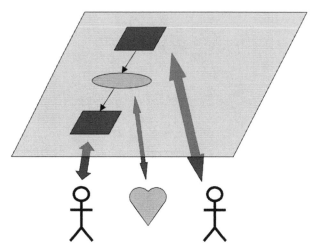

FIGURE 14.6:   Meaning in a conceptual graph.

can be extended to include any kind of identifier; a useful one is the Universal Resource Identifier (URI) [Mealling and Denenberg, 2002].

Another more subtle issue is also suggested in Figure 14.6 by the mapping/arrow/interpretation shown between a relation in the graph and the relationship (e.g., "love") between individuals in the environment. Again, URI's can provide a definitive association to a predefined relationship, but in this case there is no specific "thing" to which the mapping can refer. For the purposes of this chapter, we do not insist that interpretations map to entities or any physically identifiable things; we require only that there be a way to establish whether the relationship is a legitimate one, i.e., one that exists in the environment between two or more actual individuals in that environment.

A more precise way of defining the relationship between a concept and its interpretation (i.e., its "counterpart" in the environment) is to use two actors for the interpretation procedure: one actor to provide the mapping from an individual in the environment to its concept in the graph; another actor to provide the reverse mapping from a concept in the graph to the individual(s) required by its interpretation. Seen as a graph, the formal structure looks like Figure 14.7, where the actor `int-decode` provides the formal interpretation function, and the actor `int-encode` provides the inverse of the interpretation function.

Ignoring the actors, the concepts and relations form a typical conceptual graph representing *A cat Albert is sitting on a hat.* The identity of the referents is not yet known, since there could be many cats named "Albert." Which one do we mean? And what hat is being referred to by "#6824"? These are the questions that interpretation actors are meant to answer. Formally answering these questions is what makes a graph *grounded* in the sense that we are using in this chapter.

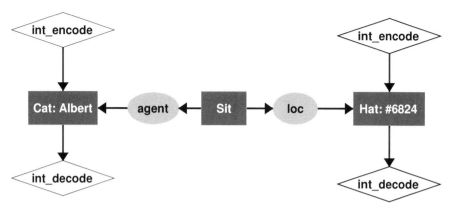

FIGURE 14.7: Interpretation actors in a conceptual graph.

The actors may need some explanation because neither of them appears to be the typical input-output functional actor that we saw before. Let us consider the `int_encode` actor first. It has no input links in the graph; it therefore appears to be an actor that fires "spontaneously" and produces a value out of nowhere. This is exactly what the encoding procedure should do. From the environment (which amounts to "nowhere" from the logical point of view) a connection arises to some symbol in the graph model. Likewise, the `int_decode` actor takes as input some symbol in the model and apparently "consumes" it. (Of course, actors do not remove or add concepts to a graph.) From a functional point of view, any actor with no output concepts does not change the identity of any concepts in the graph, since that is only possible if a concept is an output from an actor. In a moment, we will take up the question of what the `int_decode` actor really does.

Figure 14.8 illustrates the relationship between the interpretation actors and the environment. In this picture, each actor does in fact have an input and an output, and therefore operates like any function that provides a mapping: each explcitly enforces the heretofore-implicit association between a symbol in the graph and its interpretation (in this case, the individuals referred to by the symbols).

An interesting analogy can be made here with the short novel *Flatland: A Romance of Many Dimensions* [Abbott, 1998], which tells the story of a two-dimensional world, whose inhabitants are all two-dimensional geometric shapes; they cannot comprehend anything that might lie outside the (literal) plane of their existence. In a similar fashion, a conceptual graph exists on a "plane" of assertion that is necessarily unaware of anything outside of itself. Actors provide that "third dimension" of connection with an outside environment.

What does the `int_decode` actor do? It appears to map from a symbol to an individual in the environment. We could claim that we already "know"

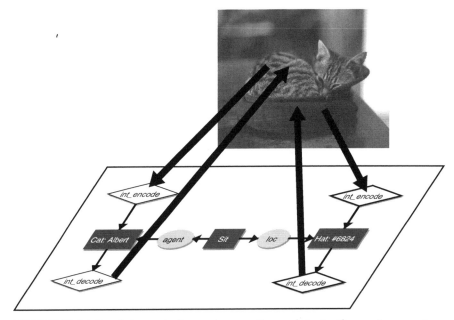

FIGURE 14.8: Interpretation actors as they relate to the environment.

which individual is denoted (e.g., which cat), because we can map from that individual to our system's symbol (e.g., `Cat: Albert`). But remember that an (internal) actor can change a concept's referent. If that happens, then it may be the case that a different individual is now being referred to (i.e., a different cat). We need to allow for the situation where the interpretation function now points to a different individual.

We will use the term *fully grounded concept* to mean a concept that has a fully specified `int_decode` actor and a fully specified `int_encode` actor linked to it.

An interpretation must also provide mappings to relationships. We have already suggested some ways this is done, although we did not describe it in terms of interpretations *per se*. Starting with the simplest actor in Figure 14.1, every actor introduced in this chapter provides an interpretation of one kind or another. The `plus` actor provides an interpretation of what an arithmetic sum means (assuming that the function of the actor faithfully executes the addition operation). But what about standard CG relations (i.e., not actors)?

A reasonable way to imagine a relation's interpretation is to consider it similarly to the interpretation of a concept; i.e., there would be an "actor" linking a CG relation with some outside function or program. But how do we incorporate this idea into the existing syntax and structure of conceptual graphs? Such a structure would violate one of the syntactic constraints of a CG; namely, that relations can only be linked to concepts or contexts and

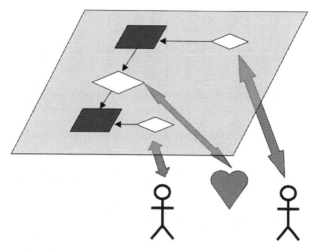

FIGURE 14.9: Actors as interpretations in a conceptual graph.

not other relations or actors. One way to deal with that problem would be to allow a relation to be attached to an actor, although that would involve major changes in the way that current conceptual graph theory and tools work. A simpler change, although it appears quite radical, is to do away with relations altogether and only have actors. That is, replace every relation in the graph with a corresponding actor that embodies the relation in the environment that is being formalized in the model. This idea is suggested informally in Figure 14.9. We continue to explore which of the two strategies would be most effective.

Now that we have established a way to ground a conceptual graph for the purposes of active knowledge systems, let's look at how they might be used. The next section contains several examples of the ways active knowledge systems might work.

14.3   Applications of Active Knowledge Systems

One of the best ways to understand why active knowledge systems are useful is to see some example applications. This section shows several different examples: section 14.3.1 shows how database semantics can be represented for use in reasoning; section 14.3.2 shows how sensors can enable a reasoning system to use values from its environment to reason about real time characteristics; section 14.3.3 shows an application supporting human communication in a collaborative environment, and section 14.3.4 shows a knowledge base that can dynamically restructure itself in response to its changing domain.

Name	Position	Yrs Experience	Degree	Major	Pct. Stock
Karen Jones	VP Marketing	18	MBA	Marketing	3
Kevin Smith	VP Technology	12	MSE	Engineering	4
Keith Williams	VP Finance	15	BS	Accounting	3
...	...	...	...	...	...

TABLE 14.1:   An Employee Relation in a Conventional Database.

14.3.1   Representing Knowledge Using a Database

This section describes how to represent database information in a form that can be used in a knowledge base for reasoning. There is an active effort in the database community to capture *metadata*; i.e., data about the data in a knowledge base so that organizations may reliably share their data. [ISO/IEC, 2006] is an international effort to standardize the representation of such metadata for sharing purposes. These efforts primarily focus on the possible values contained in a data base, with attention also being paid to the concepts represented by those values. Conceptual graphs are a good example of how the data in a database may be attached to concepts.

Here is a simple illustration. Suppose a database contains company records and has an employee relation as in Table 14.1. Because we humans can read and understand the column (field) labels, and because the first few entries' values seem to match our understanding, we tend to think that the relation table has meaning, but it really has very little inherent semantics. For example, each row appears to capture information about a different person, each of whom is employed by the organization that is being modeled by the database. As another example, values associated with **Yrs Experience** might mean "experience in business environments," or "experience in that particular position," or "experience at that particular organization." Since the strings associated with **Name** seem to be Western English person names. And of course, replacing these (apparently meaningful) names with arbitrary identifiers like "bq598" would not alter the formal characteristics of database accesses at all.

The conceptual graph to represent these is shown in Figure 14.10. It denotes the existence of a single employee's characteristics shown as concepts. Note there are no relationships between any of the concepts; in fact, the employee herself is not even shown. While all these concepts are in fact associated with a single employee, their mere co-existence is the only thing that can be shown by such a graph. For example, there is no explicit employee or person anywhere! It therefore has almost no semantics that we could use for reasoning.

To further make the point that Figure 14.10 does not show any semantics, let us consider a conceptual graph that does. Figure 14.11 shows some typical concepts we might use to represent an employee, with relationships between them. In the example, Karen Jones is the employee, whose work for some company has lasted for 18 years, in the position of vice-president for

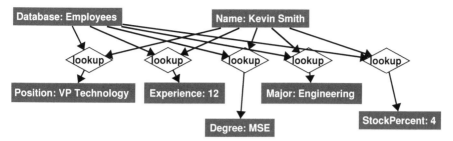

FIGURE 14.10: Graph representing a database lookup actor.

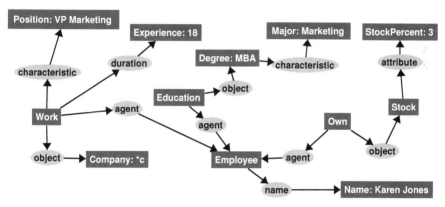

FIGURE 14.11: Graph representing knowledge about an employee.

marketing, etc.

We could easily imagine a knowledge base of employees consisting of a set of graphs, each one with the same structure as Figure 14.11, and each one with the actual name and characteristics of an employee. While in theory this could be achieved manually (by entering one identically-structured graph for each employee), there is a much easier way using actors in an active knowledge base. Combining the semantics of an employee from Figure 14.11 with the ability to retrieve actual values from a database as shown in Figure 14.10, we obtain the graph in Figure 14.12 containing both the values from the database and their accompanying semantics. Using active knowledge systems, it is thereby possible to acquire both the semantics and values in one graph that can subsequently be reasoned about.

14.3.2   Reasoning in Real Time Using Sensors

Another useful application of active knowledge systems is the use of sensors to acquire specific (possibly real time) values from the environment. In a sense, all of the applications in this chapter could be considered "sensors" in that they "sense" some aspect of the world outside of the original model

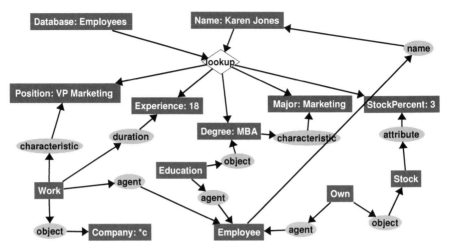

FIGURE 14.12:   Grounded knowledge about an employee.

(e.g., values in an external database). For this subsection, we consider *sensor* to mean some monitoring process's acquired value. There are many uses of sensor-based systems: rocket launch monitoring, automatic stock trading, municipal drainage systems, etc.

This section shows two simple examples; one involving a periodic clock, the other involving a periodic stock quote display. The intent is to show how an active knowledge system may interact with its environment in real time. The actors shown are simple ones, but they suggest the power of a whole set of actors designed to provide interactive capabilities.

Figure 14.13 shows several different actors. The `pulse` actor takes two inputs, an interval and a "trigger" (shown as T), which enables the pulse. (Strictly speaking, the trigger is not necessary; systems may be designed such that all actors are always activated. The `pulse` actor fires if it's enabled and repeats thereafter at the interval specified (in seconds). The `clock` actor has only one function: it grounds its output referent to the current clock time. The subgraph on the left of Figure 14.13 "refreshes" its time value every five seconds; the subgraph in the middle refreshes every ten seconds, and the one on the right grounds its output only once, when the graph is first loaded.

The graphs in Figure 14.13 illustrate a very interesting set of issues about systems that need to deal with real time values. The granularity of time must be chosen in advance by the knowledge modeler. Any given system's time value will depend on some periodic updating of the time by some means. It is possible that an active knowledge system might alter its time interval by making it the output of yet another actor that will determine (by procedures not explained here) a new interval to be applied.

The second example is in Figure 14.14. This graph has three actors: `pulse`, `stockquote`, and `displaybar`. The `pulse` actor is described above. The

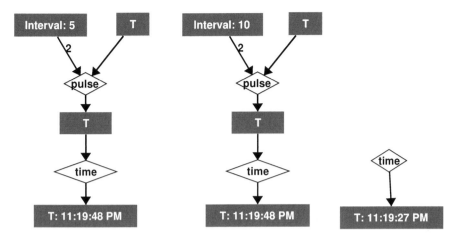

FIGURE 14.13: Graph containing a periodic clock.

`stockquote` actor accesses a network and retrieves the current stock market price for the given stock symbol (e.g., `$INDU`). Its output is the current price of the stock. Thus, we have another actor that is capable of interacting with the environment and causing the knowledge base to have symbols that reflect that environment.

The `displaybar` actor also interacts with the environment, but strictly as an output from the knowledge base. It causes a window to be displayed with the current stock price displayed on a scale, as shown on the right side of Figure 14.14. This is one way that a knowledge system could communicate with its human users (of course, there are many others). As the internal knowledge representation changes by reacting to the environment (i.e., obtaining changing stock prices), the external display reflects those changes. This in itself is not especially novel, except insofar as it does not require any additional structure or semantics to be added to the knowledge base.

14.3.3 Supporting Human Communication

Active knowledge systems can support human task-oriented activity, since a pro-active model can seek out clarification, attempt to identify ambiguity and misunderstanding. Work to support pragmatic aspects of communication is not yet completely understood, but it is clear that static models do not capture the dynamic aspects of humans trying to communicate in changing environments. Some details of this effort can be found in [Delugach, 2003].

One key activity in human communication is when one person makes a request of another. This can be modeled by an actor called `send_request` that knows (a) the sender, (b) the receiver, and (c) the message being sent. Figure 14.15 shows a request for a report from a student to a professor. Notice that there is no input to this actor; it merely "reflects" into the knowledge base

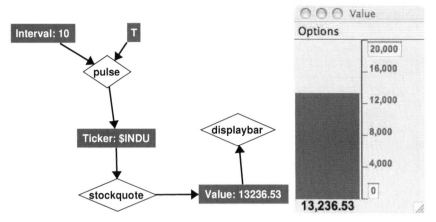

FIGURE 14.14:　Graph displaying a periodic stock quote.

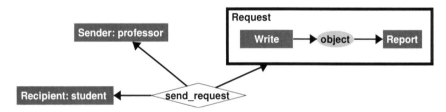

FIGURE 14.15:　Acquiring knowledge about a request.

what has happened in the environment.

As in all actively grounded graphs, the graph in Figure 14.15 does not explicitly show the specification of the **send_request** actor; e.g., how does it "know" when to "fire" and what are its outputs? The specification of the actor could be supported in a computer-supported medium where all request are done through a common mechanism (webpage or collaborative software) such that the sender, receiver, and request are all well-specified. This is not strictly required of course; we could always imagine some (very busy!) humans monitoring all requests and entering them into our system. Either way, the result would be that the request in the environment becomes formally known in the knowledge system.

Note that the requested report in Figure 14.15 has no characteristics specified. Here we assume that the knowledge system has existing models that characterize some of the normal relationships between a professor and a student. In this case, we want to show that the request must be acted upon by the student. Figure 14.16 shows this knowledge; in brief, it says that if a professor requests that a student write a report, and that student is in a class taught by the same professor, then the student must (at least start to) write the report.

Once the graph in Figure 14.15 is joined with the graph in Figure 14.16,

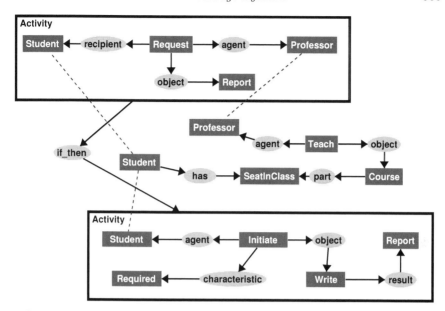

FIGURE 14.16: Knowledge about student and professor roles and obligations.

the knowledge system (if it is grounded) now contains a requirement that the student must initiate the writing of a report. Furthermore, since the model is grounded, an actual student in the environment is subject to that requirement. Since every concept has two implicit actors: one for assuring that the model reflects a changed environment, the other for assuring that the environment is aware of changes in the model (see Section 14.2), the actual student can be made aware of the new requirements; namely, that he/she must initiate the writing of a report.

The full details of this interaction appear in [Delugach, 2006]. As a final illustration of the potential to be found in active knowledge systems, Figure 14.17 shows the actor `survey` that operates as follows:

Input—concepts representing the following:

- A network that stores university papers

- A parameter stating that the median length is being sought

- A fully grounded report concept (one that is linked to `int_encode` and `int_decode` actors) denoting the student's particular report.

Output—a concept representing the following:

- A length that represents the median length of the reports found on the network

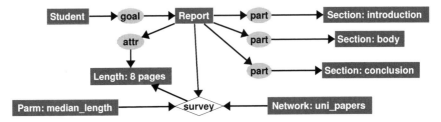

FIGURE 14.17:   An actor that surveys the Web for a typical paper length.

Obviously this actor is quite powerful! This is what active knowledge systems can do for their human users: act on their given knowledge and develop new knowledge that can be associated with existing concepts. Since the existing concept ┃ `Student` ┃ is assumed to be linked to `int_encode` and `int_decode` actors (i.e., it is fully grounded), the active knowledge system is thereby able to convey the new knowledge to that individual student.

This example shows the power of incorporating reasoning techniques and inference with the power of active knowledge systems that not only are grounded in the environment, but are capable of informing the environment of that knowledge (whatever that might mean) when inferences are made. Because of this capability, the system can initiate actions flowing from the new knowledge.

14.3.4   Self-Modifying Knowledge Bases

All of the previous examples showed cases where actors would "look up" information from outside the knowledge base and incorporate specific referents (values) into existing concepts in a graph. In these cases, the overall structure of the graph did not change; i.e., the graphs would contain the same number and types of concepts, and their associated relations would also stay the same, with only the referents changed. We have already shown the effectiveness of using these graphs in active knowledge systems; however, keeping the same structure constrains their use. This part of the chapter will show an additional capability with a special kind of actor, called a *demon* [Delugach, 1991] that allows entire sections of graphs to be asserted or retracted as the structure of the knowledge itself changes.

For this example, we will use two simple demons. The `assert` and `retract` demons, respectively, add and remove knowledge from a knowledge base. This is a basic capability for an active knowledge system. In fact, these support knowledge interchange by permitting external knowledge to be added to a knowledge base. Figure 14.18 shows an illustration of how this would operate.

The `assert` demon in Figure 14.18 indicates that the situation of *Cat Albert is sitting on a hat* is to be added to the knowledge base. Once again, we have an actor with no apparent input. As before, the actual input of the demon is

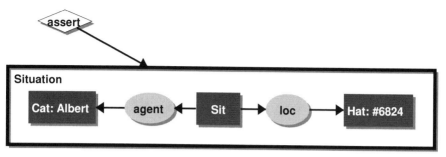

FIGURE 14.18:   Asserting knowledge into a knowledge base.

outside the sheet of assertion; it is grounded in some environment procedure that is capable of detecting a situation (in this case, that a particular cat is sitting on a particular hat) and forming the graph of the situation to be asserted.

Note that the demon is responsible for only asserting things that it can detect. It must only "fire" when the knowledge it is asserting has been determined to exist in the environment. Nothing in the description of active knowledge systems will obviate the need to ensure that asserted knowledge is accurate and desired. Put another way, it is still required that the *assert* demon provide grounding for the graphs that it asserts. An additional issue, not discussed here, is how to deal with potential conflicts or inconsistencies that may result from the new knowledge.

The **retract** demon appears in a similar fashion to the **assert** demon in Figure 14.18, except that the graph to be retracted serves as the input, not the output. There is an additional input from the environment (which cannot be shown on the sheet of assertion) that provides the "trigger" for the retraction. Graphs are only retracted by the demon when some external process in the environment determines that the graph to which it points is no longer valid and enables the appropriate input to the **retract** demon. The process of retraction may also lead to inconsistencies, which remains an important research issue.

14.3.5   Future Research Issues

This section of the chapter served to illustrate the concepts of active knowledge systems and suggest some of the ways they can be used. There are many varied research issues stemming from this approach that deserve further study because of the potential benefit to knowledge system users. The exercises in 14.5 provide a hint of what some of them are.

14.4   Conclusion

This chapter has introduced a notion called *active knowledge systems*, which is an approach to knowledge-based systems that inter*act* (not just inter-*face*) with their environment. While just about every system does have some effect in the environment, that effect is not formalized. The active knowledge system approach begins the process of establishing a formal grounding between a system's internal model and its external environment. This grounding, based on model theory, has the potential to support reliable reasoning in a system, even in the situation of a changing environment.

A deeper question arises in a system that is capable of changing its content. In principle, every inference that was previously derived must be rechecked against the new information. Newly acquired knowledge might be inconsistent with previous knowledge. The active knowledge system approach does not immediately offer any solutions to this question above and beyond other systems. It can be argued that managing inconsistency is a vital capability that every usable system needs, however, and therefore poses another interesting set of questions to be studied.

14.5   Student Projects

14.5.1   Project 1

Show that you understand the relationship between a database (or data source) and a conceptual graph.

1. Choose a database or other source (e.g., a spreadsheet) whose data is familiar to you. Specify the source.

2. Draw graphs that represent the semantics of at least two tables or relations from the database using actors.

3. Show at least two instances of each graph populated with actual values from the database.

4. Explain how the graphs are related to each other.

5. Discuss how you might use the graphs (with instantiated values) to reason about the system in question.

14.5.2 Project 2

Not every system needs to have active knowledge system features in order to be useful. Even some non-trivial systems do not require sensors, etc. of the kind described in this chapter.

- Identify three reasonable knowledge-based systems, describe each one and its scope, and tell whether it requires the features of active knowledge systems.

- For each system, you should specify what kinds of problems it is intended to solve, what kind of representations it uses, where those representations come from, and what kind of reasoning can be performed using those representation. Justify your answers, giving examples where appropriate.

14.5.3 Project 3

For some knowledge-based system with which you are familiar, design and implement one or more actors that will incorporate some of the principles outlined in this chapter. Keep track of significant problems that you encounter during the process and identify whether the problems are primarily implementation (i.e., programming) issues or whether the structure and purpose of the knowledge base does not correspond well to the notion of using actors. Comment on the usefulness of the actors you have implemented, and discuss whether their capabilities could be obtained in another way beside the one you chose.

14.5.4 Project 4

One difficulty in acquiring knowledge from external sources is that it can result in some of the knowledge being inconsistent with other knowledge in the system. Describe at least two ways that knowledge, even if from trusted sources, can nevertheless be inconsistent. Look up in the research literature several strategies for dealing with inconsistency in a knowledge base. Characterize each strategy as to whether it is compatible with the active knowledge systems approach. One way you can do this is to sketch out how actors would be designed and built to handle or otherwise accommodate each strategy. Evaluate each strategy for its appropriateness and choose one that you think could best deal with the problems of inconsistency. Justify your decision.

References

E. A. Abbott. *Flatland: A Romance of Many Dimensions (1884).* Penguin Classics, 1998.

S. Abney. Partial parsing via finite-state cascades. In *Proceedings of the ESSLLI '96 Robust Parsing Workshop*, pages 8–15, 1996.

A. Almuhareb and M. Poesio. Attribute-based and value-based clustering: An evaluation. In *Proceedings of the Conference on Empirical Methods in Natural Language Processing (EMNLP)*, pages 158–165, 2004.

U. Andelfinger. *Diskursive Anforderungsanalyse. Ein Beitrag zum Reduktionsproblem bei Systementwicklungen in der Informatik.* Peter Lang, Frankfurt, 1997.

G. Angelova and S. Mihov. Finite state automata and simple conceptual graphs with binary conceptual relations. In *Contributions to ICCS-2008, the 16th International Conference on Conceptual Structures, Toulouse, France, July 2008*, CEUR Workshop Proceedings, 2008.

A. Arnauld and P. Nicole. *La Logique ou l'Art de Penser.* G. Desprez, Paris, 1662/1683. Reprinted by Fromann, Stuttgart 1965.

J.-P. Aubert, J.-F. Baget, and M. Chein. Simple conceptual graphs and simple concept graphs. In [Schärfe et al., 2006], pages 87–101.

O. Aubert, P.-A. Champin, and Y. Prie. Integration of Semantic Web Technology in an Annotation-based Hypervideo System. In *Proc. of First International Workshop on Semantic Web Annotations for Multimedia, part of the 15th World Wide Web Conference*, 2006b.

F. Baader, R. Molitor, and S. Tobies. The guarded fragment of conceptual graphs. RWTH LTCS-Report. See also [Baader et al., 1999]. Available at: http://lat.inf.tu-dresden.de, 1998.

F. Baader, R. Molitor, and S. Tobies. Tractable and decidable fragments of conceptual graphs. In [Tepfenhart and Cyre, 1999], pages 480–493. Excerpt of [Baader et al., 1998].

F. Baader, B. Ganter, B. Sertkaya, and U. Sattler. Completing description logic knowledge bases using formal concept analysis. In M. M. Veloso, editor, *IJCAI*, pages 230–235, 2007.

R. Baeza-Yates and B. Ribeiro-Neto. *Modern Information Retrieval.* Addison-

Wesley, 1999.

J.-F. Baget. Rdf entailment as a graph homomorphism. In *International Semantic Web Conference (ISWC 2005)*, volume 3729 of *Lecture Notes in Computer Science*, pages 82–96. Springer, 2005a.

J.-F. Baget. Hypergraphs and conjunctive types for efficient projection algorithms. In [de Moor et al., 2003], pages 229–241.

J.-F. Baget. A simulation of co-identity with rules in simple and nested graphs. In [Tepfenhart and Cyre, 1999], pages 442–455.

J.-F. Baget. Homomorphismes d'hypergraphes pour la subsumption en RDF/RDFS. *Langages et modèles à objets 2004 (actes 10e conférence), RSTI - L'objet (numéro spécial)*, 10(2-3):203–216, 2004.

J.-F. Baget. RDF entailment as a graph homomorphism. In Y. Gil, E. Motta, V. R. Benjamins, and M. A. Musen, editors, *International Semantic Web Conference*, volume 3729 of *Lecture Notes in Computer Science*, pages 82–96. Springer, Berlin – Heidelberg – New York, 2005b. ISBN 3-540-29754-5.

J.-F. Baget and M.-L. Mugnier. The Complexity of Rules and Constraints. *Journal of Artificial Intelligence Research (JAIR)*, 16:425–465, 2002a.

J.-F. Baget and M.-L. Mugnier. Extensions of Simple Conceptual Graphs: The Complexity of Rules and Constraints. *JAIR*, 16:425–465, 2002b.

J.-F. Baget and E. Salvat. Rule dependencies in backward chaining of conceptual graph rules. In [Schärfe et al., 2006], pages 102–116.

M. Barbut and B. Monjardet. Ordre et classification. In *Algébre et Combinatoire*, volume II, Paris, 1970. Collection Hachette Université. Librairie Hachette.

R. Basili, M. Pazienza, and M. Vindigni. Corpus-driven unsupervised learning of verb subcategorization frames. *AI*IA-97*, 1997.

S. Bechhofer, F. van Harmelen, J. Hendler, I. Horrocks, D. L. McGuinees, P. F. Patel-Schneider, and L. A. Stein. OWL Web Ontology Language Reference. http://www.w3.org/TR/owl-ref, 2004.

K. Becker, G. Stumme, R. Wille, U. Wille, and M. Zickwolff. Conceptual information systems discussed through an IT-security tool. In R. Dieng and O. Corby, editors, *Knowledge engineering and knowledge management: methods, models, and tools*, volume 1937, pages 352–365, Berlin – Heidelberg – New York, 2000. Springer-Verlag.

P. Becker. Multi-dimensional representations of conceptual hierarchies. In G. Stumme and G. Mineau, editors, *Supplementary Proceedings of the 9th Intl. Conf. on Conceptual Structures*, pages 33–46. Department of Computer Science, University Laval, 2001.

P. Becker. Numerical analysis in conceptual information systems with ToscanaJ. In [Eklund, 2004], pages 96–103.

P. Becker. Docco: Document retrieval with Formal Concept Analysis. In [Ganter et al., 2005a].

P. Becker. Tupleware: Formal Concept Analysis for relational structures. In [Ganter et al., 2005a].

P. Becker. Formal Concept Analysis, Third international conference, ICFCA 2005, Lens, France, February 14-18, 2005, Proceedings. In [Ganter and Godin, 2005].

P. Becker and R. J. Cole. Querying and analysing document collections with Formal Concept Analysis. In *Proceedings of the 8th Australasian Document Computing Symposium*, Canberra, December 2003.

P. Becker and J. Hereth Correia. The ToscanaJ suite for implementing Conceptual Information Systems. In [Ganter et al., 2005b], pages 324–348.

P. Becker, J. Hereth, and G. Stumme. ToscanaJ – an open source tool for qualitative data analysis. In *Advances in Formal Concept Analysis for Knowledge Discovery in Databases. Proc. Workshop FCAKDD of the 15th European Conference on Artificial Intelligence (ECAI 2002). Lyon, France.*, 2002.

G. Birkhoff. *Lattice theory.* American Mathematical Society, Providende, first edition, 1940.

H. H. Bock and P. Ihm, editors. *Classification, data analysis, and knowledge organisation*, Heidelberg, 1991. Springer.

G. Boole. *An investigation of the laws of thought, on which are founded the mathematical theories of logic and probabilities.* Macmillan, 1854. Reprinted by Dover Publ., New York 1958.

C. Brewter and K. O'Hara. Knowledge representation with ontologies: The present and future. *IEEE Intelligent Systems*, 13:72–81, January-February 2004.

P. Buitelaar and P. Cimiano, editors. *Ontology Learning and Population: Bridging the Gap between Text and Knowledge.* Number 166 in Frontiers in Artificial Intelligence. IOS Press, 2008. to appear.

P. Burmeister. ConImp – Ein Programm zur Formalen Begriffsanalyse. In [Stumme and Wille, 2000], pages 25–56.

P. Burmeister. Formal Concept Analysis with ConImp: Introduction to the basic features, 2003. Translated and largely extended version of [Burmeister, 2000]. Available at http://www.mathematik.tu-darmstadt.de/~burmeister/ConImpIntro.pdf.

S. A. Caraballo. Automatic construction of a hypernym-labeled noun hierarchy from text. In *Proceedings of the 37th Annual Meeting of the Association for Computational Linguistics (ACL)*, pages 120–126, 1999.

O. Carloni, M. Leclère, and M.-L. Mugnier. Introducing Graph-based Reasoning into a Knowledge Management Tool: an Industrial Case. In *Proceedings of IEA / AIE'06*, volume 4031 of *LNCS*, pages 590–599. Springer, 2006.

C. Carpineto and G. Romano. A lattice conceptual clustering system and its application to browsing retrieval. *Machine Learning*, 24:95–122, 1996.

R. Chaffin and D. J. Herrmann. The nature of semantic relations: a comparison of two approaches. In M. W. Evens, editor, *Relational Models of the Lexicon*. Cambridge University Press, 1988.

M. Chein. Numéro spécial graphes conceptuels. *Revue d'intelligence artificielle*, 10(1), 1996.

M. Chein and D. Genest. CGs Applications: Where are we 7 years after the first ICCS? In *Proceedings of the 8th International Conference on Conceptual Structures (ICCS'2000)*, volume 1867 of *Lecture Notes in Artificial Intelligence*, pages 127–139. Springer, 2000.

M. Chein and M.-L. Mugnier. *Graph-based Knowledge Representation and Reasoning. Computational Foundations of Conceptual Graphs*. Springer, 2008.

M. Chein and M.-L. Mugnier. Conceptual Graphs: Fundamental Notions. *Revue d'Intelligence Artificielle*, 6(4):365–406, 1992.

M. Chein and M.-L. Mugnier. Positive Nested Conceptual Graphs. In D. L. et al, editor, *Proc. ICCS'97*, volume 1257 of *LNAI*, pages 95–109. Springer, 1997.

M. Chein and M.-L. Mugnier. Conceptual graphs are also graphs. Technical report, LIRMM, Université Montpellier II, 1995. Rapport de Recherche 95003.

M. Chein and M.-L. Mugnier. Representer des connaissances et raisonner avec des graphes. *Revue d'intelligence Artificielle 10*, pages 7–56, 1996.

M. Chein and M.-L. Mugnier. Concept types and coreference in simple conceptual graphs. In [Wolff et al., 2004], pages 303–318.

M. Chein, M.-L. Mugnier, and G. Simonet. Nested Graphs: A Graph-based Knowledge Representation Model with FOL Semantics. In *Proc. of KR'98*, pages 524–534. Morgan Kaufmann, 1998. Revised version available at http://www.lirmm.fr/~mugnier/.

W. Chen, M. Kifer, and D. S. Warren. Hilog: A foundation for higher-order logic programming. *Journal of Logic Programming*, 15(3):187–230, 1993.

S. Christiaens and A. de Moor. Tool interoperability from the trenches: the case of DOGMA-MESS. In A. de Moor, S. Polovina, and H. Delugach, editors, *Proc. of the First Conceptual Structures Tool Interoperability Workshop (CS-TIW 2006), Aalborg, Denmark, July 2006*, pages 103–118. Aalborg University Press, 2006.

P. Cimiano and S. Staab. Learning concept hierarchies from text with a guided agglomerative clustering algorithm. In *Proceedings of the ICML Workshop on Learning and Extending Ontologies with Machine Learning Methods*, 2005.

P. Cimiano and J. Völker. Text2onto - a framework for ontology learning and data-driven change discovery. In A. Montoyo, R. Munoz, and E. Metais, editors, *Proceedings of the 10th International Conference on Applications of Natural Language to Information Systems (NLDB)*, volume 3513 of *LNCS*, pages 227–238, 2005.

P. Cimiano and J. Wenderoth. Learning qualia structures from the web. In T. Baldwin, A. Korhonen, and A. Villavicencio, editors, *Proceedings of the ACL Workshop on Deep Lexical Acquisition*, pages 28–37, 2005.

P. Cimiano and J. Wenderoth. Automatic acquisition of ranked qualia structures from the web. In *Proceedings of the Annual Meeting of the Association for Computational Linguistics (ACL)*, pages 888–895, 2007.

P. Cimiano, S. Staab, and J. Tane. Automatic acquisition of taxonomies from text: FCA meets NLP. In *Proceedings of the ECML/PKDD Workshop on Adaptive Text Extraction and Mining*, pages 10–17, 2003.

P. Cimiano, A. Hotho, and S. Staab. Clustering ontologies from text. In *Proceedings of the 4th International Conference on Language Resources and Evaluation (LREC)*, pages 1721–1724, 2004.

P. Cimiano, A. Hotho, and S. Staab. Learning concept hierarchies from text corpora using formal concept analysis. *Journal of Artificial Intelligence Research*, 24:305–339, 2005.

P. Cimiano, A. Mädche, S. Staab, and J. Völker. Ontology learning. In S. Staab and R. Studer, editors, *Handbook of Ontologies*, International Handbooks on Information Systems. Springer, second edition, 2007.

R. Cole and P. Eklund. Browsing semi-structured web texts using formal concept analysis. In *Proceedings 9th International Conference on Conceptual Structures*, LNAI 2120, pages 319–332, Berlin, 2001. Springer.

R. Cole and G. Stumme. CEM - a conceptual email manager. In *7th International Conference on Conceptual Structures*, pages 438–452. Springer Verlag, 2000.

R. Cole, P. Eklund, and G. Stumme. CEM - a program for visualization and discovery in email. In *Principles of Data Mining and Knowledge Discovery: 4th European Conference, PKDD 2000, Lyon, France, September 2000. Proceedings*, pages 367 – 374, 2000.

R. J. Cole and P. Becker. Navigation spaces for the conceptual analysis of software structure. In [Ganter and Godin, 2005].

R. J. Cole and P. W. Eklund. Analyzing an email collection using formal concept analysis. In *European Conf. on Knowledge and Data Discovery, PKDD'99*, number 1704 in LNAI, pages 309–315. Springer Verlag, 1999.

R. J. Cole, P. W. Eklund, and G. Stumme. Document retrieval for email search and discovery using formal concept analysis. *Applied Artificial Intelligence*, 17(3):257–280, 2003.

R. J. Cole, J. Ducrou, and P. W. Eklund. Automated layout of small lattices using layer diagrams. In *Proc. of the 4th Int. Conference on Formal Concept Analysis*, volume 3874/2006 of *LNAI*, pages 291–305. Springer, 2006.

Computer Science and Telecommunications Board. National collaboratories: Applying information technology for scientific research. Technical report, National Academy Press, Washington, D.C., 1993.

O. Corby, R. Dieng, and C. Hebert. A Conceptual Graph Model for W3C RDF. In *Proceedings of ICCS'00*, volume 1867 of *LNAI*. Springer, 2000.

H. Cunningham, K. Humphreys, R. J. Gaizauskas, and Y. Wilks. Gate - a general architecture for text engineering. In *Proceedings of Applied Natural Language Processing (ANLP)*, pages 29–30, 1997.

J. Daciuk, S. Mihov, B. Watson, and R. Watson. Incremental construction of minimal acyclic finite state automata. *Journal of Computational Linguistics*, 26(1):3–16, 2000.

F. Dau. Negations in simple concept graphs. In [Ganter and Mineau, 2000], pages 263–276.

F. Dau. Concept graphs and predicate logic. In [Delugach and Stumme, 2001], pages 72–86.

F. Dau. *The Logic System of Concept Graphs with Negations and its Relationship to Predicate Logic*, volume 2892 of *Lecture Notes in Artificial Intelligence*. Springer, Berlin – Heidelberg – New York, November 2003a.

F. Dau. Concept graphs without negations: Standardmodels and standardgraphs. In [de Moor et al., 2003], pages 243–256. This paper is a part of [Dau, 2003a] as well.

F. Dau. Types and tokens for logic with diagrams: A mathematical approach. In [Wolff et al., 2004], pages 62–93.

F. Dau. RDF as graph-based, diagrammatic logic: Syntax, semantics, calculus, normalforms. In F. Esposito, Z. W. Ras, D. Malerba, and G. Semeraro, editors, *Foundations of Intelligent Systems*, volume 4203 of *Lecture Notes in Computer Science*. Springer, Berlin – Heidelberg – New York, 2006. ISBN 3-540-45764-X.

F. Dau, M.-L. Mugnier, and G. Stumme, editors. *Conceptual Structures: Common Semantics for Sharing Knowledge: 13th International Conference on Conceptual Structures, ICCS 2005*, volume 3596 of *LNCS*, 2005. Springer Verlag.

B. A. Davey and H. A. Priestley. *Introduction to lattices and order.* Cambridge University Press, Cambridge, second edition, 2002.

D. Davidson. The logical form of action sentences. Reprinted in [Davidson, 1980], pp. 105–148., 1967.

D. Davidson, editor. *Essays on Actions and Events.* Clarendon Press, Oxford, 1980.

A. de Moor. Improving the testbed development process in collaboratories. In *Proc. of the 12th International Conference on Conceptual Structures (ICCS 2004), Huntsville, Alabama, USA*, pages 261–274, 2004.

A. de Moor. The pragmatic evaluation of tool system interoperability. In H. Pfeiffer, A. Kabbaj, and D. Benn, editors, *Proc. of the Second Conceptual Structures Tool Interoperability Workshop (CS-TIW 2007), Sheffield, UK, July 2007*, pages 1–19, Bristol, UK, 2007. Research Press International.

A. de Moor and A. Anjewierden. A socio-technical approach for topic community member selection. In C. Steinfield, B. Pentland, M. Ackerman, and N. Contractor, editors, *Proc. of the Third Communities and Technologies Conference, Michigan State University, 2007*, pages 225–244. Springer, Berlin, 2007.

A. de Moor and W. van den Heuvel. Making virtual communities work: Matching their functionalities. In *Proc. of the 9th International Conference on Conceptual Structures, Stanford (ICCS 2001), July 30-August 3, 2001*, Lecture Notes in Artificial Intelligence. Springer-Verlag, 2001.

A. de Moor and H. Weigand. Formalizing the evolution of virtual communities. *Information Systems*, 32:223–247, 2007.

A. de Moor, W. Lex, and B. Ganter, editors. *Conceptual Structures for Knowledge Creation and Communication*, volume 2746 of *LNAI*, Berlin – Heidelberg – New York, 2003. Springer-Verlag.

H. Delugach. Conceptual graphs homepage. http://www.conceptualgraphs.org/, 2007.

H. S. Delugach. Dynamic assertion and retraction of conceptual graphs. In E. Way, editor, *Proc. Sixth Annual Wkshop on Conceptual Graphs*, pages 15–26, Binghamton, New York, 1991. SUNY Binghamton, New York.

H. S. Delugach. Specifying multiple-viewed software requirements with conceptual graphs. *Jour. Systems and Software*, 19:207–224, 1992.

H. S. Delugach. Towards building active knowledge systems with conceptual graphs. In A. de Moor, W. Lex, and B. Ganter, editors, *Conceptual Structures for Knowledge Creation and Communication: 11th Intl. Conf. on Conceptual Structures (ICCS 2003)*, volume LNAI 2746 of *Lecture Notes in Artificial Intelligence*, pages 296–308. Springer-Verlag, Berlin, 2003.

H. S. Delugach. Active knowledge systems for the pragmatic web. In M. Schoop, A. de Moor, and J. Dietz, editors, *Pragmatic Web: Proc. of the First Intl. Conf. on the Pragmatic Web*, volume P-39, pages 67–80. Gesellschaft fuer Informatik, Stuttgart, Germany, 2006.

H. S. Delugach and G. Stumme, editors. *Conceptual Structures: Broadening the Base*, volume 2120 of *LNAI*, Stanford, USA, July,, 2001. Springer, Berlin – Heidelberg – New York.

Deutsches Institut für Normung. *Begriffe und Benennungen. Allgemeine Grundsätze*. Beuth, Köln, 1979a.

Deutsches Institut für Normung. *Begriffssysteme und ihre Darstellung*. Beuth, Köln, 1979b.

J. Dölling. Commonsense ontology and semantics of natural language. *Zeitschrift für Sprachtypologie und Universalienforschung*, 46(2):133–141, 1993. http://www.uni-leipzig.de/~doelling/publikationen.html.

S. Domingo and P. Eklund. Evaluation of concept lattices in a web-based mail browser. In F. Dau, M.-L. Mugnier, and G. Stumme, editors, *Proceedings of the International Conference on Conceptual Structures*, LNCS3596/2005, pages 281–294. Springer, 2005.

J. Ducrou and P. Eklund. An intelligent user interface for browsing and search MPEG-7 images using concept lattices. *Int. Journal of Foundations of Computer Science*, 2007.

J. Ducrou, B. Vormbrock, and P. Eklund. FCA-based Browsing and Searching of a Collection of Images. In *Proceedings of 14th International Conference on Conceptual Structures*, LNAI4068, pages 203–214. Springer, 2006.

V. Duquenne. Contextual implications and representation properties. In [Ganter et al., 1987], pages 213–239.

H. Dyvik. Translations as semantic mirrors: from parallel corpus to wordnet. *Language and Computers*, 49:311–326, 2004.

J. Edmund M. Clarke, O. Grumberg, and D. A. Peled. *Model Checking*. MIT Press, Cambridge, MA U.S.A., 1999.

P. Edwards, S. Jackson, G. Bowker, and C. Knobel. Understanding infrastructure: Dynamics, tensions, and design - report of a workshop on "History & theory of infrastructure: Lessons for new scientific cyberinfrastructures". Technical report, University of Michigan, 2007.

M. Ehrig and Y. Sure. FOAM - framework for ontology alignment and mapping. results of the ontology alignment initiative. In B. Ashpole, M. Ehrig, J. Euzenat, and H. Stuckenschmidt, editors, *Proc. of the Workshop on Integrating Ontologies*, volume 156, pages 72–76, 2005.

P. Eklund, editor. *Concept Lattices: Second International Conference on Formal Concept Analysis, ICFCA 2004 Sydney, Australia, February 23-26, 2004 Proceedings*, volume 2961 of *Lecture Notes in Artificial Intelligence*, Berlin, 2004. Springer-Verlag.

P. Eklund. Structured ontology and information retrieval for email search and discovery. In *International Symposium on Methodologies for Intelligent Systems(ISMIS02)*, volume 2393 of *Lecture Notes in Artificial Intelligence*, pages 11–27. Springer Verlag, 2002.

P. Eklund and R. Wille. Semantology as a basis for conceptual knowledge processing. In S. O. Kuznetsov and S. Schmidt, editors, *5th International Conference on Formal Concept Analysis*, volume 4390 of *LNCS*, pages 18–38. Springer Verlag, 2007.

P. Eklund, B. Groh, G. Stumme, and R. Wille. A contextual-logic extension of toscana. In [Ganter and Mineau, 2000], pages 453–467.

P. Eklund, J. Ducrou, and P. Brawn. Concept lattices for information visualization: Can novices read line diagrams. In P. Eklund, editor, *Proceedings of the 2nd International Conference on Formal Concept Analysis - ICFCA'04*. Springer, February 2004.

G. Ellis. *Managing Complex Objects*. PhD thesis, University of Queensland, Australia, 1995.

G. Ellis, R. Levinson, W. Rich, and J. F. Sowa, editors. *Conceptual Structures: Applications, Implementation, and Theory*, volume 954 of *LNAI*, 1995. Springer, Berlin – Heidelberg – New York.

D. Eschenfelder, W. Kollewe, M. Skorky, and R. Wille. Ein Erkundungssystem zum Baurecht: Methoden der Entwicklung eines TOSCANA-Systems. In [Stumme and Wille, 2000], pages 254–272.

B. Falkenhainer, K. D. Forbus, and D. Gentner. The structure-mapping engine: Algorithm and examples. *Artificial Intelligence*, 41:1–63, 1989.

N. Fanizzi, L. Iannone, I. Palmisano, and G. Semeraro. Concept formation in expressive description logics. In *Proc. of the 15th European Conference on Machine Learning (ECML'04)*. Springer Verlag, 2004.

S. Farrar and J. Bateman. General ontology baseline. SFB/TR8 internal report I1-[OntoSpace] D1, Collaborative Research Center for Spatial Cognition, University of Bremen, Germany, 2004.

D. Faure and C. Nedellec. A corpus-based conceptual clustering method for verb frames and ontology. In P. Velardi, editor, *Proceedings of the LREC Workshop on Adapting lexical and corpus resources to sublanguages and applications*, pages 5–12, 1998.

C. Fellbaum, editor. *WordNet: An Electronic Lexical Database and Some of its Applications*. MIT press, 1998.

C. J. Fillmore. The case for case. In E. Bach and R. T. Harms, editors, *Universals in Linguistic Theory*, pages 1–88. Holt, Rinehart and Winston, New York, 1968.

J. Firth. *A synopsis of linguistic theory 1930-1955*. Studies in Linguistic Analysis, Philological Society, Oxford. Longman, 1957.

J. A. Fodor and E. Lepore. What can't be valued, can't be valued, and it can't be supervalued either,. *Journal of Philosophy*, 93:516–536, 1996.

K. D. Forbus, D. Gentner, and K. Law. Mac/fac: A model of similarity-based retrieval. *Cognitive Science*, 19(2):141–205, 1994.

G. Frege. *Begriffsschrift, eine der arithmetischen nachgebildete Formelsprache des reinen Denkens*. Louis Nebert, Halle a. S., 1879.

G. Frege. Über Sinn und Bedeutung. *Zeitschr. f. Phil. und phil. Kritik*, 100: 25–50, 1892.

N. Fuhr. Models in Information Retrieval, 2000.

L. Gaillard, J. Nanard, B. Bachimont, and L. Chamming. Intentions based authoring process from audiovisual resources. In H. B. et al, editor, *Proc. of International Workshop on Semantically Aware Document Processing and Indexing-SADPI'07*, volume ISBN:78-1-15159-668-4 of *DL*, pages 21–30. ACM, 2007.

B. Gaines, L. Lee, and M. Shaw. Modeling the human factors of scholarly communities supported through the Internet and the World Wide Web. *Journal of the American Society for Information Science*, 48(11):987–1003, 1997.

P. Gamallo, A. Agustini, and G. P. Lopes. Clustering syntactic positions with similar semantic requirements. *Computational Linguistics*, 21(1):107 – 145, 2005.

B. Ganter and R. Godin, editors. *Formal Concept Analysis, Third International Conference, ICFCA 2005, Lens, France, February 14-18, 2005, Proceedings*, volume 3403 of *LNCS*, 2005. Springer Verlag.

B. Ganter and G. W. Mineau, editors. *Conceptual Structures: Logical, Linguistic and Computational Issues*, volume 1867 of *LNAI*, Berlin – Heidelberg – New York, 2000. Springer-Verlag.

B. Ganter and R. Wille. *Formale Begriffsanalyse: Mathematische Grundlagen.* Springer-Verlag, Heidelberg, 1996.

B. Ganter and R. Wille. *Formal Concept Analysis: Mathematical Foundations.* Springer-Verlag, Berlin – Heidelberg – New York, 1999a. English translation of [Ganter and Wille, 1996].

B. Ganter and R. Wille. Contextual attribute logic. In [Tepfenhart and Cyre, 1999], pages 377–388.

B. Ganter, R. Wille, and K. E. Wolff, editors. *Beiträge zur Begriffsanalyse*, Mannheim, 1987. B.I.-Wissenschaftsverlag.

B. Ganter, R. Godin, and E. Mephu Nguifo, editors. *Supplementary Proc. of the 3rd Intl. Conf. on Formal Concept Analysis*, 2005a. University of Artois.

B. Ganter, G. Stumme, and R. Wille, editors. *Formal Concept Analysis: Foundations and Applications*, volume 3626 of *LNCS*. Springer Verlag, Berlin – Heidelberg – New York, 2005b.

C. Gasperin, P. Gamallo, A. Agustini, G. Lopes, and V. de Lima. Using syntactic contexts for measuring word similarity. In *Proceedings of the ESSLLI Workshop on Semantic Knowledge Acquisition and Categorization*, 2001.

P. Gehring and R. Wille. Semantology: basic methods for knowledge representations. In [Schärfe et al., 2006], pages 215–228.

M. R. Genesereth and R. Fikes. Knowledge interchange format, version 3.0 reference manual. TR Logic-92-1, Computer Science Department, Stanford University, 1992.

D. Genest and M. Chein. A content-search information retrieval process based on conceptual graphs. *Knowledge and Information Systems (KAIS)*, 8:292–309, 2005.

G. Gentzen. Untersuchungen über das logische Schließen, 1935. Translated into English in [Gentzen, 1969].

G. Gentzen. Investigations into logical deduction. In M. E. Szabo, editor, *The Collected Papers of Gerhard Gentzen*, pages 68–131. North-Holland Publishing Co., Amsterdam, 1969. Translated by the editor.

R. Göckel. *Lexicon philosophicum, quo tanquam clave philosophicae fores*

382 References

aperiuntur. Georg Olms Verlag, Hildesheim, 1613.

R. Godin, C. Pichet, and J. Gecsei. Design of a browsing interface for information retrieval. In *Proceedings of the 12th annual international ACM SIGIR conference on Research and development in information retrieval.* ACM Press, 1989.

A. C. Graesser, S. E. Gordon, and L. E. Brainerd. QUEST: A model of question answering. *Computers and Mathematics with Applications,* 23(6–9):733–745, 1992.

G. Grefenstette. *Explorations in Automatic Thesaurus Construction.* Kluwer, 1994.

G. Grefenstette. SEXTANT: Exploring unexplored contexts for semantic extraction from syntactic analysis. In *Proceedings of the Annual Meeting of the Association for Computational Linguistics (ACL),* pages 324–326, 1992.

P. Grenon, B. Smith, and L. Goldberg. Biodynamic ontology: Applying bfo in the biomedical domain. In D. M. Pisanelli, editor, *Ontologies in Medicine,* number 102 in Studies in Health Technology and Informatics, pages 20–38. IOS Press, 2004.

A. Großkopf and G. Harras. Begriffliche Erkundung semantischer Strukturen von Sprechaktverben. In [Stumme and Wille, 2000].

T. R. Gruber. Toward principles for the design of ontologies used for knowledge sharing. In *Formal Analysis in Conceptual Analysis and Knowledge Representation.* Kluwer, 1993.

N. Guarino. Formal ontology, conceptual analysis and knowledge representation. *International Journal of Human-Computer Studies,* 43:625–640, November/December 1995.

R. Guha. *Contexts: A Formalization and Some Applications.* PhD thesis, Stanford University, 1991.

J.-L. Guigues and V. Duquenne. Familles minimales d'implications informatives resultant d'un tableau de données binaires. *Math. Sci. Humaines,* pages 5–18, 1986.

M. Gurstein. Effective use: A community informatics strategy beyond the digital divide. *First Monday,* 8(12), 2003.

P. Haase and J. Völker. Ontology learning and reasoning - dealing with uncertainty and inconsistency. In P. C. G. da Costa, K. B. Laskey, K. J. Laskey, and M. Pool, editors, *Proc. of the Workshop on Uncertainty Reasoning for the Semantic Web (URSW),* pages 45–55, 2005.

H.-M. Haav. An application of inductive concept analysis to construction of domain-specific ontologies. In *Proceedings of the VLDB Pre-conference*

Workshop on Emerging Database Research in East Europe, 2003.

S. Handschuh and S. Staab. CREAM - CREAting Metadata for the Semantic Web, 2003.

Z. S. Harris. Word. *Distributional Structure*, 10(23):146–162, 1954.

N. Hartmann. Aristoteles und das Problem des Begriffs. *Abhandlungen der Preussischen Akademie der Wissenschaften: Philosophisch-historische Klasse*, Jg. 1939(5), 1939.

P. Hayes and B. McBride. RDF semantics. Technical Report http://www.w3.org/TR/rdf-mt/, W3C, 2003.

P. Hayes and C. Menzel. A semantics for the knowledge interchange format. In *Proc. IJCAI 2001 Workshop on the IEEE Standard Upper Ontology, Seattle*, 2001.

P. Hayes and C. Menzel. IKL specification document. http://www.ihmc.us/users/phayes/IKL/SPEC/SPEC.html, 2006.

D. G. Hays. Dependency theory: a formalism and some observations. *Language*, 40(4):511–525, 1964.

M. A. Hearst. Automatic acquisition of hyponyms from large text corpora. In *Proceedings of the 14th International Conference on Computational Linguistics*, pages 539–545, 1992.

J. Hereth. Formale Begriffsanalyse im Data Warehousing. Diplomarbeit. FB Mathematik, TU Darmstadt, 2000.

J. Hereth, G. Stumme, R. Wille, and U. Wille. Conceptual knowledge discovery and data analysis. In [Ganter and Mineau, 2000], pages 421–437.

D. Hindle. Noun classification from predicate-argument structures. In *Proceedings of the Annual Meeting of the Association for Computational Linguistics (ACL)*, pages 268–275, 1990.

J. Hintikka. Surface semantics: definition and its motivation,. In H. Leblanc, editor, *Truth, Syntax and Modality*, pages 128–147. North-Holland, Amsterdam, 1973.

W. Hodges. *A Shorter Model Theory*. Cambridge University Press, Cambridge, UK, 1997. ISBN 0521587131.

L. Hollink, G. Schreiber, J. Wielemaker, and B. Wielinga. Semantic annotation of image collections, 2003.

J. Hopcroft. An n log n algorithm for minimizing the states in a finite automaton. In Z. Kohavi, editor, *The Theory of Machines and computations*, pages 189–196. Academic Press, 1971.

J. Howse, F. Molina, S.-J. Shin, and J. Taylor. On diagram tokens and types. In M. Hegarty, B. Meyer, and N. H. Narayanan, editors, *Diagrams*, volume 2317 of *Lecture Notes in Computer Science*, pages 146–160. Springer, Berlin – Heidelberg – New York, 2002.

D. J. D. Hughes. Proofs without syntax. *Annals of Mathematics*, 2006.

ISO/IEC. ISO/IEC 11179 - Information technology - Metadata Registries (parts 1-6), 2006.

ISO/IEC. ISO/IEC 24707 – Information technology – Common Logic (CL) - a framework for a family of logic-based languages, 2007.

M. Johnston and F. Busa. Qualia structure and the compositional interpretation of compounds. In *Proceedings of the ACL SIGLEX workshop on breadth and depth of semantic lexicons*, 1996.

F. L. J.Y. Nie. Toward a broader logical model for information retrieval, 1998.

A. Kabbaj. Interoperability: The next steps for Amine platform. In *Proc. of the Second Conceptual Structures Tool Interoperability Workshop (CS-TIW 2007), Sheffield, UK, July 2007*, pages 65–69, 2007.

A. Kabbaj. SMGC: un système de manipulation des graphes conceptuels. M.sc. thesis, Univ. Laval, Canada, 1987.

A. Kabbaj. Toward a conceptual actor language for conceptual graph theory. In [Mineau et al., 1993], pages 147–160.

A. Kabbaj. Self-organizing knowledge bases: The integration based approach. In *Proc. Of the Int. KRUSE Symposium: Knowledge Retrieval, Use, and Storage for Efficiency, Santa Cruz, CA, USA*, pages 64–68, 1995.

A. Kabbaj. *Un systeme multi-paradigme pour la manipulation des connaissances utilisant la theorie des graphes conceptuels*. PhD thesis, Univ. De Montreal, Canada, 1996.

A. Kabbaj. Synergy: a conceptual graph activation-based language. In [Tepfenhart and Cyre, 1999], pages 198–213.

A. Kabbaj. Synergy as an hybrid object-oriented conceptual graph language. In [Tepfenhart and Cyre, 1999], pages 247–261.

A. Kabbaj. Development of intelligent systems and multi-agents systems with amine platform. In [Schärfe et al., 2006], pages 286–299.

A. Kabbaj and C. Frasson. An incremental model of memory formation for a multi-strategy learning environment. In [Mineau et al., 1993], pages 30–44.

A. Kabbaj and M. Janta-Polczynski. From PROLOG++ to PROLOG+CG: A CG object-oriented logic programming language. In [Ganter and Mineau, 2000], pages 540–550.

A. Kabbaj, C. Frasson, M. Kaltenbach, and J.-Y. Djamen. A conceptual and contextual object-oriented logic programming: The PROLOG++ language. In W. M. Tepfenhart, J. P. Dick, and J. F. Sowa, editors, *Conceptual Structures: Current Practices, Second International Conference on Conceptual Structures, ICCS '94, College Park, Maryland, USA, August 16-20, 1994, Proceedings*, volume 835 of *Lecture Notes in Computer Science*, pages 251–274. Springer, 1994. ISBN 3-540-58328-9.

A. Kabbaj, B. Moulin, J. Gancet, D. Nadeau, and O. Rouleau. Uses, improvements, and extensions of Prolog+CG. In H. S. Delugach and G. Stumme, editors, *Conceptual Structures: Broadening the Base, 9th International Conference on Conceptual Structures, ICCS 2001, Stanford, CA, USA, July 30-August 3, 2001, Proceedings*, volume 2120 of *LNAI*, pages 346–359, Berlin – Heidelberg – New York, 2001. Springer-Verlag. ISBN 3-540-42344-3.

A. Kabbaj, K. Bouzouba, K. E. Hachimi, and N. Ourdani. Ontologies in amine platform: Structures and processes. In [Schärfe et al., 2006], pages 300–313.

H. Kamp. A theory of truth and semantic representation. In J. A. G. Groenendijk, T. M. V. Janssen, and M. B. J. Stokhof, editors, *Formal Methods in the Study of Language*, Mathematical Centre Tracts, pages 277–322. Mathematisch Centrum Amsterdam, Amsterdam, 1981.

H. Kamp and U. Reyle. *From Discourse to Logic*. Kluwer, Dordrecht, 1993.

I. Kant. *Logic*. Dover, Mineola, 1988.

U. Kaufmann. Begriffliche Analyse von Daten über Flugereignisse. Implementierung eines Erkundungs- und Analysesystems mit TOSCANA. Diplomarbeit. FB Mathematik, TU Darmstadt, 1996.

M. Keeler and H. Pfeiffer. Building a pragmatic methodology for KR tool research and development. In *Proc. of the 14th International Conference on Conceptual Structures (ICCS 2006), Aalborg, Denmark, July 16-21, 2006*, pages 314–330, 2006.

G. N. Kerdiles. *Saying it with Pictures: A Logical Landscape of Conceptual Graphs*, volume DS 2001-09. ILLC Dissertation Series, 2001.

G. N. Kerdiles. Dynamic semantics for conceptual graphs. In [Tepfenhart and Cyre, 1999], pages 494–507.

M. Kim and P. Compton. Formal concept analysis for domain-specific document retrieval systems. In *Lecture Notes in Computer Science*, volume 2256, page 237, 2001a.

M. Kim and P. Compton. Incremental development of domain-specific document retrieval systems. In *1st International Conference on Knowledge*

Capture, 2001b.

U. Kipke and R. Wille. Begriffsverbände als Ablaufschema zur Gegenstands-bestimmung. In O. O. P. O. Degens, H.-H. Hermes, editor, *Die Klassifikation und ihr Umfeld*, pages 164–170. Indeks-Verlag, Frankfurt, 1986.

S. Klein and R. F. Simmons. Syntactic dependence and the computer generation of coherent discourse. *Mechanical Translation*, 7, 1963.

J. Klinger. Simple semiconcept graphs: A boolean logic approach. In [Delugach and Stumme, 2001].

J. Klinger. Semiconcept graphs with variables. In U. Priss, D. Corbett, and G. Angelova, editors, *Conceptual Structures: Integration and Interfaces*, volume 2393 of *LNAI*, Borovets, Bulgaria, July, 15–19, 2002. Springer, Berlin – Heidelberg – New York.

J. Klinger. *The Logic System of Protoconcept Graphs*. Shaker Verlag, Aachen, 2005. Dissertation, Darmstadt University of Technology.

B. Koester. FooCA - Web information retrieval with Formal Concept Analysis. Diploma Thesis, Darmstadt University of Technology, 2006.

B. Kohler-Koch. Zur Empirie und Theorie internationaler Regime. In B. Kohler-Koch, editor, *Regime in den internationalen Beziehungen*, pages 15–85. Nomos, Baden-Baden, 1989.

W. Kollewe, M. Skorsky, F. Vogt, and R. Wille. TOSCANA — ein Werkzeug zur begrifflichen Analyse und Erkundung von Daten. In [Wille and Zickwolff, 1994], pages 267–288.

J. L. Kolodner and C. K. Riesbeck, editors. *Experience, Memory, and Reasoning*. Lawrence Erlbaum Associates, Hillsdale, NJ, 1986.

S. Kripke. *Naming and Necessity*. Blackwell Publishing, 2003.

F. Kronlid. Modes of explanation - aristotelian philosophy and pustejovskyan linguistics. Ms. University of Göteborg, 2003.

M. Leclère and M. Mugnier. Some algorithmic improvements for the containment problem of conjunctive queries with negation. In *Proceedings of ICDT'07 (International Conference on Database Theory)*, volume 4353 of *Lecture Notes in Computer Science*, pages 401–418. Springer, 2007.

M. Leclère and M.-L. Mugnier. Simple conceptual graphs with atomic negation and difference. In [Schärfe et al., 2006], pages 331–345.

K. Lengnink. Ähnlichkeit als Distanz in Begriffsverbänden. In R. W. G Stumme, editor, *Begriffliche Wissensverarbeitung: Methoden und Anwendungen*, pages 57–71. Springer, 2001.

H. Levesque, F. Pirri, and R. Reiter. Foundations for the situation calculus.

Electronic Transactions on Artificial Intelligence, 2(3-4):159–178, 1998.

B. Levin. *English Verb Classes and Alternations: A Preliminary Investigation.* University of Chicago Press, 1993.

R. A. Levinson and G. Ellis. Multilevel hierarchical retrieval. *Knowledge Based Systems*, 5(3):233–244, 1992.

C. Lindig. Fast concept analysis. In [Stumme, 2000], pages 152–161.

C. Lindig and G. Snelting. Formale begriffsanalyse im software engineering. In [Stumme and Wille, 2000], pages 151–175.

LOGOS. Knowledge-on-Demand for Ubiquitous Learning. http://www.logosproject.com/, 2008.

J. Lorhard. *Ogdoas Scholastica.* Sangalli, 1606.

D. Lukose, editor. *Conceptual Structures: Fulfilling Peirce's Dream*, volume 1257 of *LNAI*, 1997. Springer, Berlin – Heidelberg – New York.

P. Luksch and R. Wille. A mathematical model for conceptual knowledge systems. In H. Bock and P. Ihm, editors, *Classification, data analysis, and knowledge organisation*, pages 156–162. Springer, 1991.

M. Luxenburger. Implications partielles dans un contexte. *Mathématiques, informatique et sciences humaines*, 113(29e année):35–55, 1991.

M. Luxenburger. *Implikationen, Abhängigkeiten und Galois–Abbildungen.* PhD thesis, TH Darmstadt, Aachen, 1993.

K. Mackensen. *Simplizität. Genese und Wandel einer musikästhetischen Kategorie des 18. Jahrhunderts.* Bärenreiter, Kassel, 2000.

P. Martin. CGKAT: a knowledge acquisition tool and an information retrieval tool which exploits structured documents, conceptual graphs and ontologies. In *Proc. CGTOOLS'97.* Univ. of Washington, 1997.

P. Martin and P. Eklund. Embedding Knowledge in Web Documents. In *Proceedings of the 8th Int. World Wide Web Conference (WWW-8)*, pages 1403–1419, 1999.

P. Martin et al. CG Tools Wikipedia Article. http://en.wikipedia.org/wiki/CG_tools, 2007.

G. Mazzola. *Symmetrie in Kunst, Natur und Wissenschaft. Band 3 - Spiel, Natur und Wissenschaft.* Roether, Darmstadt, 1986.

J. McCarthy. Notes on formalizing context. In *IJCAI-93: 13th International Joint Conference on Artificial Intelligence*, pages 555–560. Morgan Kaufmann, 1993.

D. L. McGuinness and F. van Harmelen. OWL Web Ontology Language

Overview. `http://www.w3.org/TR/2004/REC-owl-features-20040210/`, 2004.

R. McKinley. *Categorical Models of First-Order Classical Proofs*. PhD thesis, University of Bath, 2006.

M. Mealling and R. Denenberg. RFC 3305 - Report from the Joint W3C/IETF URI Planning Interest Group: Uniform Resource Identifiers (URIs), URLs, and Uniform Resource Names (URNs): Clarifications and recommendations, 2002.

S. Mihov. *Minimal Acyclic Automata: Constructions, Algorithms, Applications*. PhD thesis, Bulgarian Academy of Sciences, Sofia, 2000. (Abstract in English at http://www.lml.bas.bg/ stoyan).

G. Miller, R. Beckwith, C. Fellbaum, D. Gross, and K. J. Miller. Introduction to wordnet: an on-line lexical database. *International Journal of Lexicography*, 3(4):235–244, 1990.

G. Mineau, R. Missaoui, and R. Godinx. Conceptual modeling for data and knowledge management. *Data & Knowledge Engineering*, 33:137–168, 2000.

G. W. Mineau, B. Moulin, and J. F. Sowa, editors. *Conceptual Structures: Theory and Applications. ICCS'93, Quebec, Canada, Proceedings*, volume 699 of *LNAI*, Berlin – Heidelberg – New York, 1993. Springer-Verlag.

N. Moreau, M. Leclère, M. Chein, and A. Gutierrez. Formal and Graphical Annotations for Digital Objects. In H. B. et al, editor, *Proc. of International Workshop on Semantically Aware Document Processing and Indexing-SADPI'07*, volume ISBN:78-1-15159-668-4 of *DL*, pages 69–78. ACM, 2007.

B. Moulin and A. Kabbaj. SMGC: a tool for conceptual graphs processing. *The Journal for the integrated study of artificial intelligence, cognitive science and applied epistemology*, 7:23–47, 1990.

M.-L. Mugnier. Knowledge representation and reasonings based on graph homomophism. In [Ganter and Mineau, 2000], pages 172–192.

M.-L. Mugnier. On generalization / specialization for conceptual graphs. *Journal of Experimental and Theoretical Computer Science*, 7:325–344, 1995. (Also Available as Research Report LIRMM 93-003, January 1993).

M.-L. Mugnier and M. Chein, editors. *Conceptual Structures: Theory, Tools and Applications, 6th International Conference on Conceptual Structures, ICCS '98, Montpellier, France, August 10-12, 1998, Proceedings*, volume 1453 of *LNAI*, 1998. Springer, Berlin – Heidelberg – New York. ISBN 3-540-64791-0.

M.-L. Mugnier and M. Chein. Polynomial algorithms for projection and matching. In H. Pfeiffer and T. Nagle, editors, *Conceptual Structures: Theory and Implementation, Proceedings of the 7th Annual Workshop on Conceptual Graphs (AWCG'92), New Mexico State University, USA, July 1992*, volume 754 of *LNAI*, pages 239–251. Springer-Verlag, 1992.

M.-L. Mugnier and M. Leclère. On querying simple conceptual graphs with negation. *Data and Knowledge Engineering*, 60:468–493, 2007.

M.-L. Mugnier and E. Salvat. Sound and complete forward and backward chaining of conceptual graph rules. In P. W. Eklund, G. Ellis, and G. Mann, editors, *Conceptual Structures: Knowledge Representation as Interlingua*, volume 1115 of *LNAI*, pages 248–262. Springer, Berlin – Heidelberg – New York, 1996.

T. E. Nagle, J. A. Nagle, L. L. Gerholz, and P. W. Eklund, editors. *Conceptual Structures: Current Research and Practice*, New York, 1992. Ellis Horwood. ISBN: 0-13-175878-0.

M. Nanard, J. Nanard, J. Chauche, and P. King. A structural computing approach to the production of multimedia document series. *New Review on Hypermedia and Multimedia*, 12(2):165–190, 2006.

C. Nedellec. Corpus-based learning of semantic relations by the ILP system Asium. In J. Cussens and S. Dzeroski, editors, *Learning Language in Logic*, volume 1925 of *Lecture Notes in Computer Science*, pages 259–278. Springer, 1999. ISBN 3-540-41145-3.

A. Newell and H. A. Simon. *Human Problem Solving*. Prentice-Hall, Englewood Cliffs, NJ, 1972.

P. Øhrstrøm, J. Andersen, and H. Schärfe. What has happened to ontology. In [Dau et al., 2005], pages 425–438.

L. J. Old. Unlocking the semantics of roget's thesaurus using formal concept analysis. In [Eklund, 2004], pages 236–243.

C. S. Peirce. Description of a notation for the logic of relatives. Reprinted in *Writings of Charles S. Peirce*, Indiana University Press, Bloomington, vol. 2, pp. 359-429., 1870.

C. S. Peirce. On the algebra of logic. *American Journal of Mathematics*, 3: 15–57, 1880.

C. S. Peirce. On the algebra of logic. *American Journal of Mathematics*, 7: 180–202, 1885.

C. S. Peirce. Reasoning and the logic of things. In K. L. Ketner, editor, *The Cambridge Conferences Lectures of 1898*. Harvard University Press, Cambridge, MA, 1898/1992.

C. S. Peirce. Vague. In J. M. Baldwin, editor, *Dictionary of Philosophy and*

Psychology, page 748. MacMillan, New York, 1902.

C. S. Peirce. Manuscripts on existential graphs. In *Collected Papers of Charles Sanders Peirce*, volume 4, pages 320–410. Harvard University Press, Cambridge, MA, 1906.

C. S. Peirce. Manuscript 514, with commentary by J. F. Sowa . http://www.jfsowa.com/peirce/ms514.htm, 1909.

C. S. Peirce. *Reasoning and the logic of things*. Harvard University Press, Cambridge, Mass, 1992.

F. Pereira, N. Tishby, and L. Lee. Distributional clustering of english words. In *Proceedings of the 31st Annual Meeting of the Association for Computational Linguistics (ACL)*, pages 183–190, 1993.

U. Petersen, H. Schärfe, and P. Øhrstrøm. Online course in knowledge representation using conceptual graphs. http://www.huminf.aau.dk/cg/, 2001-2008.

W. Petersen. How formal concept lattices solve a problem of ancient linguistics. In [Dau et al., 2005], pages 337–352.

W. Petersen. A set-theoretical approach for the induction of inheritance hierarchies. *Electronic Notes in Theoretical Computer Science*, 51, 2002.

J. Peterson. *Petri Net Theory and the Modeling of Systems*. Prentice Hall PTR Upper Saddle River, NJ, USA, 1981.

C. A. Petri. Grundsätzliches zur beschreibung diskreter prozesse. Presented at the 3. Colloquium über Automatentheorie, Hannover, 1965, 1965.

K. Petridis, S. Bloehdorn, C. Saathoff, N. Simou, V. Tzouvaras, S. Handschuh, Y. Avrithis, Y. Kompatsiaris, S. Staab, and M. G. Strintzis. Knowledge representation and semantic annotation for multimedia analysis. *IEEE Proceedings on Vision, Image and Signal Processing - Special issue on the Integration of Knowledge, Semantics and Digital Media Technology*, 153(3): 253–394, 2006.

M. F. Porter. An algorithm for suffix stripping. *Program*, 14(3):130–137, 1980.

P. H. Portner. *What is Meaning? - Fundamentals of Formal Semantics*. Blackwell Publishing, 2006.

S. Prediger. Nested concept graphs and triadic power context families: A situation–based contextual approach. In [Ganter and Mineau, 2000], pages 263–276.

S. Prediger. *Kontextuelle Urteilslogik mit Begriffsgraphen – Ein Beitrag zur Restrukturierung der Mathematischen Logik*. Shaker Verlag, Aachen, 1998a. Dissertation, Darmstadt University of Technology.

S. Prediger. Simple concept graphs: A logic approach. In [Mugnier and Chein, 1998], pages 225–239. ISBN 3-540-64791-0.

J. Preece. *Online Communities: Designing Usability, Supporting Sociability.* John Wiley & Sons, New York, 2000.

A. Preller, M.-L. Mugnier, and M. Chein. Logic for Nested Graphs. *Computational Intelligence: An International Journal,* 14-3:335–357, 1998.

U. Priss. Classification of meronymy by methods of relational concept analysis. In *Proceedings of the 1996 Midwest Artificial Intelligence Conference, Bloomington, Indiana,* 1996.

U. Priss. *Relational Concept Analysis: Semantic Structures in Dictionaries and Lexical Databases (PhD Thesis).* Verlag Shaker, Aachen, 1998a.

U. Priss. The formalization of wordnet by methods of relational concept analysis. In [Fellbaum, 1998], pages 179–196.

U. Priss. Efficient implementation of semantic relations in lexical databases. *Computational Intelligence,* 15(1):79–87, 1999.

U. Priss. Linguistic applications of formal concept analysis. In [Ganter et al., 2005b], pages 149–160.

U. Priss. Formal concept analysis in information science. *Annual Review of Information Science and Technology,* 40:521–543, 2006.

U. Priss and L. J. Old. Metaphor and information flow. In *Proceedings of the 12th Midwest Artificial Intelligence and Cognitive Science Conference,* pages 99–104, 2001.

U. Priss and L. J. Old. Modelling lexical databases with formal concept analysis. *Journal of Universal Computer Science,* 10(8):967–984, 2004.

U. Priss and L. J. Old. Conceptual exploration of semantic mirrors. In [Ganter and Godin, 2005], pages 21–32.

U. Priss and L. J. Old. An application of relation algebra to lexical databases. In [Schärfe et al., 2006], pages 388–400.

G. J. Probst, S. Raub, and K. Romhardt. *Managing knowledge: building blocks for success.* Wiley, New York, 1999.

J. Pustejovsky. The generative lexicon. *Computational Linguistics,* 17(4): 209–441, 1991.

M. R. Quillian. Semantic memory. In M. Minsky, editor, *Semantic Information Processing,* pages 227–270. MIT Press, Cambridge, MA, 1968.

W. V. O. Quine. Reduction to a dyadic predicate. *J. Symbolic Logic,* 19(3): 180–182, 1954. Reprinted in [Quine, 1995], pp. 224–226.

W. V. O. Quine. *Selected Logic Papers*. Harvard University Press, Cambridge, MA, enlarged edition, 1995.

R. Raban and H. S. Delugach. Animating conceptual graphs. In D. Lukose, H. S. Delugach, M. Keeler, L. Searle, and J. F. Sowa, editors, *Conceptual Structures: Fulfilling Peirce's Dream*, volume 1257 of *Lecture Notes in Artificial Intelligence*, pages 431–445. Springer-Verlag, 1997.

L. Reeve and H. Han. Survey of semantic annotation platforms. In *SAC '05: Proceedings of the 2005 ACM symposium on Applied computing*, pages 1634–1638. ACM Press, 2005.

P. Resnik. Selectional preference and sense disambiguation. In *Proceedings of the ACL SIGLEX Workshop on Tagging Text with Lexical Semantics: Why, What, and How?*, 1997.

D. D. Roberts. *The Existential Graphs of Charles S. Peirce*. Mouton, The Hague, Paris, 1973.

T. Rock and R. Wille. Ein TOSCANA-Erkundungssystem zur Literatursuche. In [Stumme and Wille, 2000], pages 239–253.

P. M. Roget. *Roget's International Thesaurus*. Thomas Crowell, New York, 3rd edition, 1962.

S. Rudolph. Exploring relational structures via FLE. In K. E. Wolff, H. D. Pfeiffer, and H. S. Delugach, editors, *Conceptual Structures at Work: 12th International Conference on Conceptual Structures*, volume 3127 of *LNCS*, pages 196 – 212, Huntsville, AL, USA, 2004. Springer.

S. Rudolph. Acquiring generalized domain-range restrictions. In R. Medina and S. Obiedkov, editors, *Proceedings of the 6th International Conference on Formal Concept Analysis (ICFCA'08)*, volume 4933 of *Lecture Notes in Artificial Intelligence*, pages 32–45. Springer, 2008.

S. Rudolph. *Relational Exploration - Combining Description Logics and Formal Concept Analysis for Knowledge Specification*. Universitätsverlag Karlsruhe, 2006. Dissertation.

SAPHIR. Système d'Assistance à la Publication Hypermédia par spécification d'Intention et modélisation Rhétorique. http://semioweb.msh-paris.fr/saphirdoc/, 2008.

J. Saquer and J. S. Deogun. Concept aproximations based on rough sets and similarity measures. In *Int. J. Appl. Math. Comput. Sci.*, volume 11, pages 655 – 674, 2001.

Q. Sarraf and G. Ellis. Business rules in retail: The Tesco.com story. *Business Rules Journal*, 7(6), 2006. http://www.brcommunity.com/a2006/n014.html.

S. Sawyer. A market-based perspective on information systems development. *Communications of the ACM*, 44(11):97–102, 2001.

R. C. Schank, editor. *Conceptual Information Processing*. North-Holland Publishing Co., Amsterdam, 1975.

R. C. Schank. *Dynamic Memory: A Theory of Learning in Computers and People*. Cambridge University Press, 1982.

R. C. Schank. *Tell Me a Story: A new look at real and artificial memory*. MacMillan, 1991.

H. Schärfe, P. Hitzler, and P. Øhrstrøm, editors. *Conceptual Structures: Inspiration and Application, Proceedings of the 14th International Conference on Conceptual Structures, ICCS'06*, volume 4068 of *LNAI*, 2006. Springer Verlag.

H. Schmid. LoPar: Design and implementation. Arbeitspapiere des Sonderforschungsbereiches 340 149, IMS Stuttgart, July 2000.

H. Schmid. Probabilistic part-of-speech tagging using decision trees. In *Proceedings of the International Conference on New Methods in Language Processing*, 1994.

A. T. Schreiber, B. Dubbeldam, J. Wielemaker, and B. Wielinga. Ontology-based photo annotation. *IEEE Intelligent Systems*, May/June, 2001.

E. Schröder. *Vorlesungen über die Algebra der Logik, Band 1*. Chelsea, New York, 2nd edition, 1966.

D. Schuler. A pattern language for living communication. In *Participatory Design Conference (PDC'02), Malmo, Sweden, June*, 2002.

T. B. Seiler. *Begreifen und Verstehen. Ein Buch über Begriffe und Bedeutungen*. Verlag Allgemeine Wissenschaft, Mühltal, 2001.

O. Selz. *Über die Gesetze des geordneten Denkverlaufs*. Spemann, Stuttgart, 1913.

O. Selz. *Zur Psychologie des produktiven Denkens und des Irrtums*. Friedrich Cohen, Bonn, 1922.

S.-J. Shin. *The Iconic Logic of Peirce's Graphs*. Bradford Book, Massachusetts, 2002.

G. Simonet. Two fol-semantics for simple and nested conceptual graphs. In [Mugnier and Chein, 1998], pages 240–254. ISBN 3-540-64791-0.

B. Smith. Formal ontology, common sense and cognitive science. *International Journal of Human-Computer Studies*, 43:641–667, 1995.

B. Smith. Ontology. In L. Floridi, editor, *lackwell Guide to the Philosophy of*

Computing and Information, pages 155–166, Oxford, 2003. Blackwell.

B. Smith. Beyond concepts, or: Ontology as reality representation. In A. Varzi and L. Vieu, editors, *Formal Ontology and Information Systems. Proceedings of the Third International Conference (FOIS 2004)*, page 73–84, Amsterdam, 2004. IOS Press.

J. F. Sowa. *Conceptual structures: information processing in mind and machine*. Addison-Wesley, Reading, Mass., 1984. ISBN 0-201-14472-7.

J. F. Sowa. *Knowledge Representation: Logical, Philosophical, and Computational Foundations*. Brooks Cole, Pacific Grove, CA, 2000a.

J. F. Sowa. Conceptual graphs summary. In [Nagle et al., 1992], pages 3–51. ISBN: 0-13-175878-0.

J. F. Sowa. Conceptual graphs: Draft proposed american national standard, 2000b. Old version: http://www.jfsowa.com/cg/cgdpansw.htm, New version: http://www.jfsowa.com/cg/cgstandw.htm. See also [Sowa, 1999].

J. F. Sowa. Syntax, semantics, and pragmatics of contexts. In [Ellis et al., 1995], pages 1–15. See also [Sowa, 2001].

J. F. Sowa. Logic: Graphical and algebraic. manuscript, Croton-on-Hudson, 1997a.

J. F. Sowa. Peircean foundation for a theory of context. In [Lukose, 1997], pages 41–64. See also [Sowa, 2001].

J. F. Sowa. Conceptual graphs: Draft proposed american national standard. In [Tepfenhart and Cyre, 1999], pages 1–65. See also [Sowa, 2000b].

J. F. Sowa. Semantic foundations of contexts. This paper is a revised merger of [Sowa, 1995] and [Sowa, 1997b]. Available at: http://www.jfsowa.com/ontology/contexts.htm, 2001.

J. F. Sowa. Conceptual graphs for a database interface. *IBM Journal of Research and Development*, 20(4):336–357, 1976.

J. F. Sowa. Laws, facts, and contexts: Foundations for multimodal reasoning. In V. F. Hendricks, K. F. Jørgensen, and S. A. Pedersen, editors, *Knowledge Contributors*, pages 145–184. Kluwer Academic Publishers, Dordrecht, 2003.

J. F. Sowa. Worlds, models, and descriptions. *Studia Logica*, 84(2):323–360, 2006. Special Issue *Ways of Worlds II*.

J. F. Sowa. *Conceptual Structures: Information Processing in Mind and Machine*. Addison-Wesley, 1984.

J. F. Sowa and A. K. Majumdar. Analogical reasoning. In [de Moor et al., 2003], pages 16–36.

N. Spangenberg. *Familienkonflikte eßgestörter Patientinnen. Eine empirische Untersuchung mit der Repertory Grid Technik.* Habilitationsschrift, Universität Gießen, 1990.

N. Spangenberg and K. E. Wolff. Datenreduktion durch die Formale Begriffsanalyse von Repertory Grids. In J. W. Scheer and A. Catina, editors, *Einführung in die Repertory Grid-Technik, Band 2, Klinische Forschung und Praxis*, volume 2, pages 38–54. Verlag Hans Huber, 1993.

C. Sporleder. A galois lattice based approach to lexical inheritance hierarchy learning. In *Proceedings of the ECAI Workshop on Machine Learning and Natural Language Processing for Ontology Engineering (OLT)*, 2002.

J. Stahl and R. Wille. Preconcepts and set representations of concepts. In W. Gaul and M. Schader, editors, *Classification as a tool of research*, pages 431–438. North-Holland, Amsterdam, 1986.

J. Stewart. Theorem proving using existential graphs. Ms thesis, Computer and Information Science, University of California at Santa Cruz, 1996.

S. Strahringer and R. Wille. Towards a structure theory for ordinal data. In M. Schader, editor, *Analyzing and modeling data and knowledge*, pages 129–139. Springer-Verlag, Heidelberg, 1992.

R. Studer, V. R. Benjamins, and D. Fensel. Knowledge engineering: Principles and methods. *Data Knowledge Engineering*, 25(1-2):161–197, 1998.

G. Stumme, editor. *Working with Conceptual Structures – Contributions to ICCS 2000*, Mannheim, 2000. Shaker Verlag.

G. Stumme and A. Maedche. FCA-merge: Bottom-up merging of ontologies. In *Proc. 17th International Conference on Artificial Intelligence (IJCAI '01)*, pages 225–230, 2001.

G. Stumme and R. Wille, editors. *Begriffliche Wissensverarbeitung: Methoden und Anwendungen.* Springer-Verlag, Berlin – Heidelberg – New York, 2000.

G. Stumme, R. Wille, and U. Wille. Conceptual knowledge discovery in databases using formal concept analysis methods. In J. M. Zytkow and M. Quafofou, editors, *Principles of Data Mining and Knowledge Discovery*, volume 1510 of *LNAI*, pages 450–458, Heidelberg, 1998. Springer-Verlag.

Y. Sure, S. Bloehdorn, P. Haase, J. Hartmann, and D. Oberle. The SWRC ontology - semantic web for research communities. In C. Bento and G. D. Amilcar Cardoso, editors, *Proceedings of the 12th Portuguese Conference on Artificial Intelligence - Progress in Artificial Intelligence (EPIA 2005)*, number 3803 in LNCS, pages 218–231, 2005.

V. Takàcs. Two applications of galois graphs in pedagogical research. Manuscript of a lecture at the TH Darmstadt 1984 (Bericht in [Wille,

1984a]), 1984.

J. Tane, P. Cimiano, and P. Hitzler. Query-based multicontexts for knowledge base browsing: an evaluation. In [Schärfe et al., 2006], pages 413–426.

J. Tappe. Simple concept graphs with universal quantifiers. Master's thesis, Darmstadt University of Technology, 2000a.

J. Tappe. Simple concept graphs with universal quantifiers. In [Stumme, 2000], pages 95–104.

A. Tarski. The concept of truth in formalized languages. In A. Tarski, editor, *Logic, Semantics, Metamathematics*, pages 152–278. Hackett Publishing Co., Indianapolis, second edition, 1933.

W. Tepfenhart and W. Cyre, editors. *Conceptual Structures: Standards and Practices, 7th International Conference on Conceptual Structures, ICCS'99, Blacksburg, VA, USA, July 1999, Proceedings*, volume 1640 of *LNAI*, Berlin – Heidelberg – New York, 1999. Springer-Verlag.

L. Tesnière. *Éléments de Syntaxe structurale*. Librairie C. Klincksieck, Paris, 1959. Second edition, 1965.

The Gene Ontology Consortium. Gene Ontology: tool for the unification of biology. *Nature Genetics*, 25:25–29, 2000.

M. Uschold and M. Grüninger. Ontologies: principles, methods, and applications. *Knowledge Engineering Review*, 11(2):93–155, 1996.

B. C. van Fraassen. Singular terms, truth-value gaps, and free logic. *Journal of Philosophy*, 64:481–495, 1966.

C. van Rijsbergen. *Information retrieval*. Butterworths, 1979.

N. Vogel. Ein begriffliches Erkundungssystem für Rohrleitungen. Diplomarbeit. fb mathematik, TU Darmstadt, 1995.

F. Vogt. *Formale Begriffsanalyse mit C++: Datenstrukturen und Algorithmen*. Springer-Verlag, Berlin – Heidelberg – New York, 1996.

F. Vogt and R. Wille. TOSCANA — a graphical tool for analyzing and exploring data. In R. Tamassia and I. G. Tollis, editors, *Graph Drawing*, pages 226–233, Berlin – Heidelberg – New York, 1995. Springer-Verlag.

F. Vogt, C. Wachter, and R. Wille. Data analysis based on a conceptual file. In [Bock and Ihm, 1991], pages 131–142.

J. Völker, P. Hitzler, and P. Cimiano. Acquisition of OWL DL axioms from lexical resources. In *Proc. of the 4th European Semantic Web Conference (ESWC'07)*. Springer, 2007a.

J. Völker, D. Vrandecic, Y. Sure, and A. Hotho. Learning disjointness. In

Proc. of the 4th European Semantic Web Conference (ESWC'07). Springer, 2007b.

B. Vormbrock and R. Wille. Semiconcept and protoconcept algebras: the basic theorems. In [Ganter et al., 2005b], pages 34–48.

S. Walter. Das Generative Lexikon: Pustejovsky Qualia Structures und die aitiai des Aristoteles (Lexikalische Semantik im Rückgriff auf antike Gedanken). University of the Saarland, 2001.

E. C. Way. Special issue on conceptual graphs. *Journal of Experimental and Theoretical Artificial Intelligence (JETAI)*, 4(2), 1992.

E. W. Weisstein. Bell number. From *MathWorld*—A Wolfram Web Resource. http://mathworld.wolfram.com/BellNumber.html, 2007. See also Math Forum web page: http://mathforum.org/advanced/robertd/bell.html.

M. Wermelinger. Conceptual graphs and first-order logic. In [Ellis et al., 1995], pages 323–337.

R. Wille. Conceptual graphs and formal concept analysis. In [Lukose, 1997], pages 290–303.

R. Wille. Begriffliche Wissensverarbeitung: Theorie und Praxis. *Informatik Spektrum*, 23:357–369, 2000a.

R. Wille. Boolean concept logic. In [Ganter and Mineau, 2000], pages 317–331.

R. Wille. Begriffliche Wissensverarbeitung in der Wirtschaft. Information - Wissenschaft und Praxis. *Organ der Deutschen Gesellschaft für Informationswissenschaft und Informationspraxis e.V.*, 53:149–160, 2002.

R. Wille. Conceptual content as information - basics for conceptual judgment logic. In [de Moor et al., 2003], pages 1–15.

R. Wille. Preconcept algebras and generalized double boolean algebras. In [Eklund, 2004], pages 1–13.

R. Wille. Implicational concept graphs. In [Wolff et al., 2004], pages 52–61.

R. Wille. Formal concept analysis as mathematical theory of concepts and concept hierarchies. In [Ganter et al., 2005b], pages 1–33.

R. Wille. Conceptual knowledge processing in the field of economics. In [Ganter et al., 2005b], pages 226–249.

R. Wille. Methods of conceptual knowledge processing. In R. Missaoui and J. Schmid, editors, *Proceedings of ICFCA 2006*, volume 3874 of *Lecture Notes in Computer Science*, pages 1–29. Springer, 2006.

R. Wille. Restructuring lattice theory: an approach based on hierarchies of concepts. In I. Rival, editor, *Ordered Sets*, pages 445–470. Reidel,

Dordrecht-Boston, 1982.

R. Wille. Liniendiagramme hierarchischer Begriffssysteme. In H. H. Bock, editor, *Anwendungen der Klassifikation: Datenanalyse und numerische Klassifikation*, pages 32–51. Indeks-Verlag, Frankfurt, 1984a. Translated into English in [Wille, 1984b].

R. Wille. Line diagrams of hierachical concept systems. *International Classification*, 11:77–86, 1984b.

R. Wille. Bedeutungen von Begriffsverbänden. In [Ganter et al., 1987], pages 161–211.

R. Wille. Knowledge acquisition by methods of formal concept analysis. In E. Diday, editor, *Data Analysis and Learning Symbolic and Numeric Knowledge*, pages 365–380. Nova Science Publisher, New York–Budapest, 1989.

R. Wille. Concept lattices and conceptual knowledge systems. *Computers & Mathematics with Applications*, 23:493–515, 1992.

R. Wille. Begriffsdenken: Von der griechischen Philosophie bis zur künstlichen Intelligenz heute. In *Dilthey-Kastanie*, pages 77–109. Ludwig-Georgs-Gymnasium, Darmstadt, 1995.

R. Wille. Conceptual landscapes of knowledge: A pragmatic paradigm for knowledge processing. In G. Mineau and A. Fall, editors, *Proceedings of the International Symposium on Knowledge Representation, Use, and Storage Efficiency, Simon Fraser University, Vancouver 1997*, pages 2–13, 1997. also in [Wille, 1999].

R. Wille. Conceptual landscapes of knowledge: A pragmatic paradigm for knowledge processing. In W. Gaul and H. Locarek-Junge, editors, *Classification in the Information Age*, pages 344–356. Springer, Berlin-Heidelberg, 1999.

R. Wille and M. Zickwolff, editors. *Begriffliche Wissensverarbeitung — Grundfragen und Aufgaben*, Mannheim, 1994. B.I.–Wissenschaftsverlag.

M. E. Winston, R. Chaffin, and D. Herrmann. A taxonomy of part-whole relations. *Cogn. Science*, 11:417–444, 1987.

L. Wittgenstein. *Philosophical Investigations*. Blackwell Publishing, 1953. German Text, with a Revised English Translation.

K. E. Wolff and M. Stellwagen. Conceptual optimization in the productions of chips. In J. Janssen and C. H. Skiadas, editors, *Applied stochastic models and data analysis*, volume II, pages 1054–1064. World Scientific Publ. Comp., 1993.

K. E. Wolff, H. D. Pfeiffer, and H. S. Delugach, editors. *Conceptual Structures at Work: 12th International Conference on Conceptual Structures, ICCS*

2004, Huntsville, AL, USA, July 19-23, 2004. Proceedings, volume 3127 of *Lecture Notes in Computer Science*, Berlin – Heidelberg – New York, 2004. Springer-Verlag.

D. Wunderlich. *Arbeitsbuch Semantik*. Athenaeum, Königstein im Taunus, 1980.

S. Yevtushenko. *Computing and Visualizing Concept Lattices*. PhD thesis, Darmstadt University of Technology, 2004. URL `http://elib.tu-darmstadt.de/diss/000488/`.

S. A. Yevtushenko. System of data analysis "concept explorer". In *Proceedings of the 7th national conf, on Artificial Intelligence KII-2000*, pages 127–134, Russia, 2000. In Russian.

J. J. Zeman. *The Graphical Logic of C. S. Peirce*. PhD thesis, University of Chicago, 1964. Available at: `http://www.clas.ufl.edu/users/jzeman/`.

Index

Printed and bound by CPI Group (UK) Ltd, Croydon, CR0 4YY

25/10/2024

01779329-0001